HUMORAL CONTROL
OF GROWTH AND DIFFERENTIATION

VOLUME I
Vertebrate Regulatory Factors

CONTRIBUTORS

Ruth Hogue Angeletti

Pietro U. Angeletti

Frederick F. Becker

William S. Bullough

Natalie S. Cohen

Philip Ferris

T. N. Frederickson

Anthony S. Gidari

P. F. Goetinck

Albert S. Gordon

Klaus Havemann

Robert A. Kuna

David M. Lapin

Rita Levi-Montalcini

Joseph LoBue

Ronald A. Malt

Donald Metcalf

Norman Molomut

T. T. Odell, Jr.

Arnold D. Rubin

Manfred Schmidt

Edward F. Schultz

Edgar A. Tonna

Esmail D. Zanjani

HUMORAL CONTROL
OF GROWTH
AND DIFFERENTIATION

VOLUME I

Vertebrate Regulatory Factors

Edited by

JOSEPH LOBUE

Department of Biology
Graduate School of
Arts and Science
New York University
Washington Square Campus
New York, New York

ALBERT S. GORDON

Department of Biology
Graduate School of
Arts and Science
New York University
Washington Square Campus
New York, New York

1973

ACADEMIC PRESS   *New York and London*

A Subsidiary of Harcourt Brace Jovanovich, Publishers

ACADEMIC PRESS, INC.
111 Fifth Avenue, New York, New York 10003

United Kingdom Edition published by
ACADEMIC PRESS, INC. (LONDON) LTD.
24/28 Oval Road, London NW1

Library of Congress Cataloging in Publication Data

LoBue, Joseph.
    Humoral control of growth and differentiation.

    Includes bibliographies.
    CONTENTS: 1.   Vertebrate regulatory factors.
    1.   Hormones.   2.   Cell proliferation.   3.   Cell
differentiation.   4.   Cellular control mechanisms.
I.   Gordon, Albert, joint author.   II.   Title.   [DNLM:
1.   Cell differentiation.   2.   Cells—Growth & development.   3.   Growth substances.   4.   Hormones—Physiology.
5.   Homeostasis.   6.   Vertebrates.   WK515 L799h]
QH604.L6   574.3   73-2067
ISBN 0–12–453801–0

CONTENTS

I   CHALONES

1.   Chalone Control Systems

William S. Bullough

II   BLOOD CELL FORMATION AND RELEASE

2.   Erythropoietin: The Humoral Regulator of Erythropoiesis

Albert S. Gordon, Esmail D. Zanjani, Anthony S. Gidari, and Robert A. Kuna

3. Humoral Regulation of Neutrophil Production and Release
Edward F. Schultz, David M. Lapin, and Joseph LoBue

4. Humoral Regulation of Eosinophil Production and Release
Natalie S. Cohen, Joseph LoBue, and Albert S. Gordon

5. The Colony Stimulating Factor (CSF)
Donald Metcalf

6. Humoral Regulation of Thrombocytopoiesis
T. T. Odell, Jr.

10. Possible Feedback Inhibition of Leukemic Cell Growth:
    Kinetics of Shay Chloroleukemia Grown in Diffusion
    Chambers and Intraperitoneally in Rodents

   Philip Ferris, Joseph LoBue, and Albert S. Gordon

III HUMORAL CONTROL OF ORGANS
AND TISSUE GROWTH

11. The Nerve Growth Factor

   *Ruth Hogue Angeletti, Pietro U. Angeletti, and
   Rita Levi-Montalcini*

12. Humoral Aspects of Liver Regeneration

   Frederick F. Becker

LIST OF CONTRIBUTORS

Numbers in parentheses indicate the pages on which the authors' contributions begin.

RUTH HOGUE ANGELETTI (229), Laboratorio di Biologie Cellulare (CNR), Rome, Italy

PIETRO U. ANGELETTI (229), Istituto Superiore di Sanita, Rome, Italy

FREDERICK F. BECKER (250), Department of Pathology, New York University School of Medicine, New York, New York

WILLIAM S. BULLOUGH (1), Mitosis Research Laboratory, Birkbeck College, University of London, London, England

NATALIE S. COHEN (69), Department of Biochemistry, University of Southern California School of Medicine, Los Angeles, California

PHILIP FERRIS (213, 361), Waldemar Medical Research Foundation, Inc., Woodbury, New York

T. N. FREDERICKSON (139), Department of Pathobiology, University of Connecticut, Storrs, Connecticut

ANTHONY S. GIDARI (25, 165), Department of Medicine, Downstate Medical Center, Brooklyn, New York

P. F. GOETINCK (139), Department of Animal Genetics, University of Connecticut, Storrs, Connecticut

ALBERT S. GORDON (25, 62, 165, 213), Department of Biology, Graduate School of Arts and Science, New York University, Washington Square Campus, New York, New York

KLAUS HAVEMANN (183), Medizinische Klinik der Universität Marburg, Marburg, Germany

ROBERT A. KUNA (25, 165), Department of Biology, Graduate School of Arts and Science, New York University, Washington Square Campus, New York, New York

DAVID M. LAPIN (52), Department of Biological Sciences, Fairleigh Dickinson University, Teaneck, New Jersey

RITA LEVI-MONTALCINI (229), Department of Biology, Washington University, St. Louis, Missouri

JOSEPH LoBUE (52, 69, 213, 361), Department of Biology, Graduate School of Arts and Science, New York University, Washington Square Campus, New York, New York

RONALD A. MALT (257), Massachusetts General Hospital Surgical Services, Massachusetts General Hospital, and Shriners Burns Institute, and Department of Surgery, Harvard Medical School, Boston, Massachusetts

DONALD METCALF (91), Cancer Research Unit, Walter and Eliza Hall Institute, Melbourne, Australia

NORMAN MOLOMUT (361), Waldemar Medical Research Foundation, Inc., Woodbury, New York

T. T. ODELL, JR. (119), Biology Division, Oak Ridge National Laboratory, Oak Ridge, Tennessee

ARNOLD D. RUBIN (183), Department of Medicine (Hematology), Mount Sinai School of Medicine of the City University of New York, New York, New York

MANFRED SCHMIDT (183), Department of Medicine (Hematology), Mount Sinai School of Medicine of the City University of New York, New York, New York

EDWARD F. SCHULTZ (52), Department of Medicine (Hematology), Mount Sinai School of Medicine of the City University of New York, New York, New York

EDGAR A. TONNA (275), Institute for Dental Research, Brookdale Dental Center of New York University, College of Dentistry, New York, New York

ESMAIL D. ZANJANI (25, 165), Department of Physiology, Mount Sinai School of Medicine of the City University of New York, New York, New York

PREFACE

The purpose of Volume I of this two-volume work is to introduce the reader to the fascinating subject of humoral control of growth and differentiation in the vertebrate organism. The emphasis is placed on those chemical messengers which have for a number of reasons received relatively little attention in standard treatises on endocrine physiology. Humoral regulation of hematopoiesis is examined extensively not only because the editors have experience in this field but because so much information is available on humoral control of blood cells.

Clearly, not all facets of this broad field could possibly be covered, so that considerable editorial selection was necessary. The ultimate contents of this work were thus predicated by a number of factors. Among these were that topics for inclusion had to be associated with a sufficient body of literature to allow a substantive, critical, up-to-date review to be conducted. Moreover, contributors had to be recruited who were highly competent and enthusiastic in their interest in the development of such a treatise. Finally, the interests of the editors had to be satisfied, we hope with objectivity.

The observant reader will become increasingly aware of an interesting reality as he progresses through this book. That is, it will be seen that each topic is presented in its own unique state and degree of completeness, and this varies considerably from chapter to chapter. This phenomenon, as the experienced investigator so well knows, results from the uneven development and differences in quality and quantity of data which exist for a given subject. Considered from a different aspect it means that the treatise may be taken to represent a record—a comparative chronometer—against which the present scientific status of the subject may be measured. For example, consider the nerve growth factor (NGF) and erythropoietin (Ep). These are well-established humoral differentiating principles for which a tremendous body of literature exists. Therefore, we expect much will be learned about these agents from a diligent reading of the appropriate chapters. Conversely, contrast what is known for NGF and Ep with the information available on those humors suspected of regulating compensatory renal and liver growth and these areas are seen to be in their scientific infancy.

Finally, we believe the reader will be impressed by the fact that for all the diversity of material presented, a common and recurrent theme will emerge. This theme will be a stimulator–inhibitor, humorally based, feedback regulation of growth and differentiation.

We would like to express our gratitude to Mrs. Blanche Ciotti, a devoted secretary, who gave so much of her time and effort to the expert typing of manuscripts and the handling of the mountains of correspondence which developed during the planning phases of the work. Thanks are also due to a wife, Catherine LoBue, and a young son, Philip, who were invaluable aids in the mechanical aspects of subject indexing. We are also sincerely grateful to the staff of Academic Press for their friendly advice and cooperation during all phases of the production of this treatise.

Joseph LoBue
Albert S. Gordon

CONTENTS OF VOLUME II

NONVERTEBRATE NEUROENDOCRINOLOGY AND AGING

I

CHALONES

1

CHALONE CONTROL SYSTEMS

William S. Bullough

I. INTRODUCTION

It is fundamental to the very survival of any metazoan animal that the levels of function of all the constituent tissues shall be properly balanced and coordinated. At all times each tissue must have a metabolic potentiality that is adequate to meet the demands that are put upon it. Such a metabolic potentiality depends both on the tissue mass and on the ability of the tissue cells to modify their functional effort to changing circumstances (Bullough, 1967).

In a mitotic tissue the mass is primarily determined by the number of cells present. Such a tissue is potentially capable of changing its mass by increasing or decreasing its cell number, and this occurs readily in hormone-dependent tissues. However, in most of the tissues of a full-grown mammal the rate of new cell production merely offsets the rate of old cell loss in such a way that the tissue mass remains remarkably constant. Furthermore when tissue damage leads to abnormal cell loss, there follows an abnormal burst of mitosis which quickly restores the normal tissue mass. It is therefore clear that within each mitotic tissue there must be some mechanism which constantly operates to maintain the tissue mass in proper relation to body mass. This is the mechanism of cellular homeostasis which is discussed here.

The few non-mitotic tissues that exist in the body (for instance, the skeletal and cardiac muscles and the nerve cells) are clearly derived from mitotic tissues. In them the appropriate numbers of cells are established by mitotic activity early in life. The mitotic potential is then lost and all subsequent growth in tissue mass is dependent on growth in cell size. Such cells often have a continuing ability to adjust their size to the metabolic demand, as is well known in the case of the skeletal muscles. This is the mechanism of metabolic homeostasis, which is known to be controlled at the DNA–RNA–enzyme levels, and which is not discussed here.

II. MITOTIC CONTROL: POSITIVE FEEDBACK

In mammals the earliest observations on the mechanisms controlling tissue replacement by mitosis were made on skin wounds and on liver and kidney regeneration (see Bullough, 1965, 1967). The most dramatic results were those obtained after partial hepatectomy or after unilateral nephrectomy. When part of the liver is removed the remnant grows rapidly until its mass is again normal, when growth ceases (Bucher, 1963); when one kidney is removed the other grows to almost twice its size, when again growth ceases. These responses are strictly organ-specific: partial hepatectomy stimulates only liver growth, while unilateral nephrectomy stimulates only kidney growth.

It has often been believed that these responses depend on a positive feedback mechanism. In the case of a skin wound the dead or damaged cells have been thought to produce a mitosis-stimulating "wound hormone," which naturally disappears when the wound has healed. Fifty years of search for such a hormone has however failed. In the case of a liver remnant or of a single remaining kidney it has been postulated that, because the cells must work harder, growth by mitosis will continue until the work load per cell is once more normal. This idea of increased metabolic load leading to increased tissue mass is evidently derived from the common observation that increased muscular effort leads to the development of increased muscular mass. However, this is a false analogy since increased muscular work leads only to increased muscle cell size; the non-mitotic muscles cannot increase their cell number. In spite of much research there is still no positive evidence that an increased workload can lead to increased mitotic activity; attempts to demonstrate this, for instance by inducing an increased rate of liver detoxication (see Argyris, 1969), can be criticized on the grounds that the treatments used may have caused liver damage, which was the real cause of the observed mitotic stimulus.

The question of the existence of positive feedback mechanisms of one kind or another still remains open.

III. MITOTIC CONTROL: NEGATIVE FEEDBACK

A more recent idea has been that tissue mass is normally determined by a negative feedback system. Thus Mercer (1962, Fig. 1.), on theoretical grounds, suggested that in the epidermis the distal keratinizing cells may produce a messenger molecule which diffuses into the basal epidermal layer to inhibit mitosis; that when the epidermis reaches an appropriate thickness, enough of this mitotic inhibitor is produced to prevent the formation of more epidermal cells; and that epidermal thickness (or mass) must therefore oscillate, like a thermostat, above and below the normal figure.

Such a negative feedback system could also explain the consequences of partial hepatectomy and of unilateral nephrectomy. Thus if the kidneys produce a kidney-specific mitotic inhibitor, the concentration of which is in balance between the kidneys and the blood, the removal

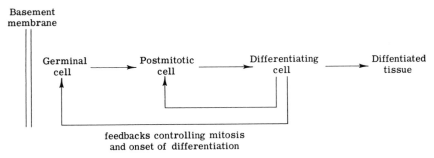

Fig. 1. Diagram illustrating a hypothetical negative feedback mechanism for the control of epidermal mitotic activity (after Mercer, 1962).

of one kidney would result in the blood concentration being halved. This would increase the diffusion gradient from the remaining kidney, reduce the chalone concentration in that kidney, and thus permit rapid growth by mitosis.

The earliest indication that this theory is indeed basically correct came from some observations by Saetren (1956), who showed that both the liver and the kidney produce organ-specific mitosis inhibitors, an observation which has since been confirmed (Simnett and Chopra, 1969; Scaife, 1970).

IV. MITOTIC CONTROL IN EPIDERMIS

Bullough and Laurence (1960a, 1964a) then began a critical series of experiments on mitotic control in mouse epidermis. They showed first that a positive feedback, based on a mitosis stimulating "wound hormone," does not exist, and that the high mitotic activity developing alongside an epidermal wound is due to the loss of a previously present mitotic inhibitor, which is strictly tissue-specific in its antimitotic action. This inhibitor has in fact all the characteristics postulated for the messenger molecule in the negative feedback system shown in Fig. 1, and it has been named the epidermal chalone (pronounced Kalōn).

The epidermal chalone is extractable in water solution from macerated epidermis, and when injected *in vivo* or added to an *in vitro* medium it selectively inhibits epidermal mitotic activity. It can be purified by ethanol fractionation (Bullough *et al.*, 1964) followed by electrophoresis at pH 3 and by cold dialysis (Hondius Boldingh and Laurence, 1968), and it is evidently a protein or a glycoprotein with a molecular weight of about 30,000.

V. EPIDERMAL CHALONE AND STRESS HORMONES

In the *in vitro* experiments with mouse epidermis, Bullough and Laurence (1964a, b) also found that the antimitotic action of added epidermal chalone was only evident in the presence of traces of adrenalin, and this has since proved to be the basis of an invaluable diagnostic test.

This observation has also provided a solution to the old problem of diurnal mitotic rhythms. It has long been known that the mitotic rate in epidermis, and in at least 20 other tissues (Bullough, 1965), fluctuates in a regular manner throughout the 24 hours, and that the highest epidermal mitotic activity is typical of rest or sleep (Bullough, 1948). The lowest epidermal mitotic activity is found in active animals, in which the concentration of adrenalin in the blood reaches a high level (Bullough and Laurence, 1964a, b). As would be expected, adrenalectomy destroys the diurnal mitotic cycle and results in a constantly high epidermal mitotic rate (Bullough and Laurence, 1966).

It is also well known that the other stress hormone, the glucocorticoid hormone from the adrenal cortex, has an antimitotic action. In the case of the epidermis this has been partly analyzed by Bullough and Laurence (1968a), who have suggested that it may act by reducing the permeability of the cell membrane and the rate of chalone loss and so cause an increase in the intracellular chalone concentration. They have also shown that a glucocorticoid hormone in some way prolongs the activity of the intracellular adrenalin. This type of action, through the chalone mechanism, is evidently the rationale for the use of glucocorticoid hormones as antimitotic agents in such hyperplastic skin conditions as psoriasis.

It must be emphasised that adrenalin and the glucocorticoid hormones are not themselves antimitotic; they function only by strengthening the antimitotic action of the endogenous chalone.

VI. OTHER CHALONE SYSTEMS

Many tissue-specific chalone systems are now known, and the epidermal chalone has been found to inhibit mitosis not only in the surface epidermis but also in the lens epithelium (Voaden, 1968), the lining epithelia of the mouth (Randers Hansen, 1967), and probably also in the esophagus (Bullough and Laurence, 1964a). However, the various epidermal derivatives have their own individual chalone systems. The hair follicle has a complex system which has so far defied detailed analysis (see Bullough, 1965, 1967), and there is a sebaceous gland chalone (Bullough and Laurence, 1970a) and an eccrine gland chalone (Bullough and Deol, 1972). Also in the skin the melanocytes, which occur among the epidermal cells, are controlled by a melanocyte chalone (Bullough and Laurence, 1968c), and there is evidence that the cells of the dermal connective tissues are controlled by one or more chalone systems (Bullough and Laurence, 1960b). Recently Houck (1971) has extracted a fibroblast chalone from connective tissue cells *in vitro*.

In the blood, extensive work by Rytömaa and his associates has established the existence of chalone systems in both the granulocytic (Rytömaa and Kiviniemi, 1968a) and erythrocytic cell populations

(Kivilaakso and Rytömaa, 1971). Although the control systems regulating the numbers of circulating granulocytes and erythrocytes are certainly complex, in each case they include a negative feedback mechanism in which chalone released from the mature cells acts back through the blood to inhibit the mitotic activity of the granulocytic and erythrocytic precursor cells in the bone marrow. Recent work has shown that the same is true of the circulating lymphocytes, which release a lymphocytic chalone to inhibit the production of new lymphocytes (Bullough and Laurence, 1970b; Lasalvia *et al.*, 1970; Houck *et al.*, 1971).

Similar mechanisms have been described in the liver (Saetren, 1956; Scaife, 1970) and the kidney (Saetren, 1956; Simnett and Chopra, 1969), and there is preliminary evidence of the existence of several other chalone systems.

It now seems probable that all those body tissues that are capable of mitosis, whether normally or after damage or after hormone stimulation, will be found to be under the control of specific chalone mechanisms.

VII. CHARACTERISTICS OF CHALONES

A chalone is defined as a substance that is produced within a tissue to control by inhibition the mitotic activity of that same tissue. It is strictly tissue-specific and is unable to inhibit mitosis in any other tissue.

In addition it is commonly characterized by the fact that its action is strengthened in the presence of the two stress hormones, although exceptions to this are provided by the granulocytic and erythrocytic chalones.

Chemically the epidermal chalone has a protein nature and a molecular weight of about 30,000 (Hondius Boldingh and Laurence, 1968) and this is probably also true of the sebaceous gland and melanocytic chalones. The lymphocytic chalone is also basically a protein and it has a molecular weight of about 50,000 (Houck *et al.*, 1971). Again the granulocytic and erythrocytic chalones are exceptional; they are polypeptides, or glycopolypeptides, with a molecular weight of less than 4000 (Rytömaa and Kiviniemi, 1968a; Paukovits, 1971).

Chalones are evidently constantly produced within the tissue cells and constantly lost from them into the surrounding fluids; this has been demonstrated in the case of mature granulocytes suspended in a saline medium (Rytömaa and Kiviniemi, 1968a). The chalone concentration within a tissue is clearly determined by the rate of synthesis minus the rate of loss. It has been suggested, though so far without any direct evidence, that the rate of synthesis may be higher in mature postmitotic cells. It has also been suggested, again without any direct evidence, that the rate of loss may be influenced by specific enzymatic degradation. Certainly it may be expected that messenger molecules whose function is constantly to monitor the state of the tissue will have a short half-life. However, this could be simply achieved by a relatively high rate of chalone loss from the tissue cells, and it has been established that chalones, still in their active form, are present not only in the blood but also in the urine (Bullough and Laurence, 1971) and, in the case of the eccrine gland chalone, in the sweat (Bullough and Deol, 1972).

VIII. THE CHALONE MECHANISM

A chalone mechanism has at least three component parts: the process and rate of chalone synthesis, the transport of chalone from cell to cell, and the cellular response to the chalone concentration (Bullough, 1967). In any tissue the rate of chalone synthesis is probably constant, at least in the postmitotic mature cells. Regarding chalone transport, it is clear that chalone escapes in large amounts from the cells and that other cells of the same tissue can then absorb it. Thus chalone produced by the epidermis on one side of the mouse ear crosses the central connective tissues to influence the mitotic activity in the epidermis on the other side of the ear (Bullough and Laurence, 1960a), and in pathological conditions the chalone escaping from a growing epidermal carcinoma depresses the mitotic activity of the whole of the normal epidermis (Bullough and Deol, 1971a). In the granulocytic, erythrocytic, and lymphocytic systems, chalones from the mature cells are carried in the blood and lymph to be absorbed by the mitotic cells in distant parts of the body. Also in all the chalone systems so far studied, the subcutaneous or intraperitoneal injection of a chalone-containing extract leads quickly to a mitotic depression in the appropriate tissue however distant it may be.

The manner in which cells take up circulating chalone molecules may be the basis of the tissue specificity of chalone action (Bullough, 1967). Clearly, since the process of mitosis is essentially the same in all types of cells, any antimitotic substance should inhibit mitosis equally in all tissues. However, if to a generalized antimitotic molecule there is linked a specialized molecule which is able to pass the cell membrane of only one type of tissue cell, the situation becomes explicable. One difficulty with this theory is that in tumor cells which have evidently suffered considerable membrane modification the tissue specificity of chalone uptake is not lost. Thus a tumor responds by mitotic inhibition only to the chalone of its tissue of origin (Bullough and Deol, 1971b).

The most complex part of a chalone mechanism is probably the response mechanism within the tissue cells. It is evident that the activity of any type of tissue cell must be determined in relation to some critical chalone concentration: a fall in chalone concentration below this level allows the cell to enter the mitotic cycle, while a rise in chalone concentration above this level directs the cell into the postmitotic state. In other words, with a subthreshold chalone concentration, genes are activated to direct the syntheses on which the mitotic process depends, while with a suprathreshold chalone concentration, the mitotic genes are silenced and the genes directing postmitotic aging and tissue function are activated instead.

IX. THE EPIDERMAL RESPONSE TO THE EPIDERMAL CHALONE

It is next necessary to analyze the way in which the chalone mechanism influences tissue mass. In the past it has commonly been assumed that a chalone system functions in the manner of the relatively simple negative feedback mechanism shown in Fig. 1. However, in the case of mouse epidermis there are two important points of counter-evidence. First, while the thickness of the epidermis normally remains remarkably constant (which indicates that the control mechanism must be very precise), the dose-response curve for the epidermal chalone indicates that such a degree of precision would not be expected (Bullough and Laurence, 1964a).

Second, and more important, it has been clearly shown that extra chalone, or extra stress, which slows the mitotic rate also slows the rate of aging of the postmitotic cells (Bullough and Ebling, 1952), while conversely an increase in the mitotic rate is accompanied by an increase in the rate of postmitotic aging (Ebling 1957; see Fig. 2). Indeed so precisely matched are these coincident changes that, within wide limits, any chronic change in the mitotic rate causes no change at all in the epidermal thickness (Bullough, 1973). Obviously the mass of mouse epidermis is not determined solely by a simple chalone negative feedback mechanism, and the same also applies to rat epidermis (see Bullough, 1972).

However, the epidermis of other species, including pig and man, does show changes in epidermal thickness after changes in the mitotic rate, and this is also true in mouse and rat after the epidermal mitotic rate has increased beyond a certain point. Indeed the normal epidermis of

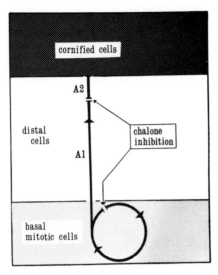

Fig. 2. Diagram showing how the epidermal chalone inhibits both the mitotic cycle in the basal epidermal cells and the aging of the postmitotic cells.

pig and man shows close similarities to the mildly hyperplastic epidermis of mouse and rat: the mitotic rate is relatively high and the epidermis is relatively thick. Furthermore, when human epidermis becomes hypoplastic, as in senility, it comes to resemble normal mouse epidermis and any further fall in the mitotic rate is not accompanied by any further epidermal thinning.

These various situations and reactions are illustrated in Fig. 3, and recently Bullough (1973) has provided an explanation. In the thinner epidermis of the mouse, with its constant thickness, the chalone concentration is relatively high, the mitotic rate relatively low, the rate of postmitotic aging relatively slow, and the basal cell layer has a flat junction with the dermis (Fig. 4). In this basal layer there are two types of cell which can be differentially stained (Christophers, 1971a): one type is mitotic and the other is postmitotic. Mitosis takes place in such a way that both the newly formed cells remain in the basal layer (Bullough and Laurence, 1964b) in which the pressure therefore rises. Some adjacent postmitotic cell, which evidently has a weaker grip on the dermis, is then pushed out. With reduced mitotic activity there is less pressure so that fewer postmitotic cells are pushed out, and simultaneously the nonbasal postmitotic cells take longer to rise to the surface and to become keratinized; the converse is true with increased mitotic activity. Thus the epidermal thickness remains unchanged.

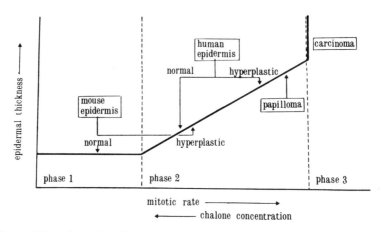

Fig. 3. The relationship between epidermal mitotic activity and epidermal mass (thickness).

However, with increasing mitotic activity not only do the mitotic cells divide more often but a greater proportion of the basal cells become involved in the mitotic cycle. Thus a decreasing proportion of the basal cells are able to become postmitotic and to move out, until finally their numbers become inadequate to relieve the pressure generated by mitosis. At this point the postmitotic cells disappear from the basal layer (Christophers, 1971b), which then begins to fold (as in man) or to double (as in hyperplasia in mouse). There is then an increased *number* of basal mitotic cells per unit area of the skin, and although postmitotic aging also proceeds faster, there is a consequent increase in the *number* of overlying postmitotic cells (Fig. 4). The epidermis therefore thickens. From this point onward, the lower the chalone concentration, the higher the mitotic rate, the higher the pressure in the basal layer, the greater the degree of folding (or doubling), the greater the number of basal cells, the greater the number of overlying postmitotic cells, and the thicker the epidermis (see Tosti *et al.,* 1959, 1969). Thus, in this type of epidermis, the thickness is directly related to the mitotic rate.

Toward its limit, as for instance in psoriasis in man, the mitotic rate is so high and the degree of folding is so great, that the ratio of the inner epidermal surface to the outer epidermal surface reaches 4 or 5:1 (Tosti *et al.,* 1959).

Although most of the available data come from studies of epidermis it is already clear that similar reactions occur in other tissues. Thus sebaceous glands with a relatively low mitotic rate, as in mouse, do not shrink any further if the mitotic rate is chronically reduced (Bullough and Ebling, 1952). Conversely, beyond a certain point, increasing

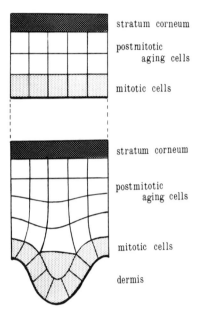

Fig. 4. The upper diagram represents the phase 1 type of epidermis of a normal mouse; the lower diagram represents the phase 2 type epidermis seen in man.

mitotic activity leads to such reduced numbers of basal postmitotic cells that the glands begin to increase in mass. In this case the basal layer does not fold or double; it merely increases in area like the surface of an inflating balloon. This may be the common reaction pattern in most if not all mitotic tissues.

X. WOUND HEALING AND TISSUE REGENERATION

In terms of the chalone mechanism the only difference between the processes of wound healing, as in skin, and of tissue regeneration, as in liver, is that the former is a local reaction to small-scale damage while the latter is a general reaction to large scale damage.

Wound healing has been most closely studied in the skin (see Dunphy and Van Winkle, 1969), and the simplest situation is that created by a cut through the epidermis into the dermis. The healing process has two components: first, the continuity of the epidermis is reestablished by the migration of a thin sheet of adjacent epidermal cells across the wound cavity (Winter, 1964); second, the normal epidermal thickness is reestablished by a burst of high mitotic activity. Remarkably little is known about the migratory movement. It is usually explained in terms of "contact inhibition" (Abercrombie, 1964) but this is merely a form of words used to cover a lack of knowledge. It is, however, clear that a migrating cell does not also enter mitosis (Bullough, 1966, 1969), and this may be simply because mitosis is a process with high energy demands (Bullough, 1962) which cannot be met while so much energy is being used for movement.

The burst of mitosis that develops round an epidermal wound may be primarily the outcome of damage to the cell membranes, which allows undue quantities of the chalone to escape, together perhaps with a reduced capacity for chalone synthesis within the damaged cells (Bullough, 1965, 1967). Evidently the chalone content of the adjacent cells falls by about 50% (Bullough and Laurence, 1968b) and a mass of newly produced cells is quickly formed. This mitotic reaction can be inhibited by chalone injections (Frankfurt, 1971) or by excessive stress (Bullough, 1969).

Although it has not been directly demonstrated, it is clear from what has been said that the high mitotic activity adjacent to a wound must also be accompanied by an increased rate of postmitotic cell aging. However, even if the newly formed postmitotic epidermal cells should begin to die and so to form keratin in as little as 4 or 5 days (Bullough, 1972), this period would normally be adequate for the healing of the wound and for the reestablishment of the usual epidermal thickness. Then with the normal chalone concentration reestablished, as evidenced by the fall in the mitotic rate, the life span of the new postmitotic cells is extended to its normal length (see Bullough, 1973).

This is a highly efficient process as indeed would be expected since there must always have been a high selective advantage in favor of those individuals that were able to recover rapidly from accidental skin damage.

If tissue damage is extensive, as for instance in liver after the ingestion of toxic substances (or experimentally after partial hepatectomy), the mitotic reaction is no longer local but involves the whole tissue. In liver the mitotic reaction to damage remains local until about 10% of the liver mass has been destroyed; at this point the reaction becomes general and its intensity is in direct proportion to the amount of liver tissue destroyed (or excised). This is evidently the result of a general fall of the liver chalone concentration in the body as a whole, and the general rise in mitotic activity is rapid. A mass of new cells is quickly formed, and again, although these cells must show an increased rate of postmitotic aging, the time available for full liver restoration is more than adequate. Thus in the rat the normal postmitotic life span of a liver cell is about 400 days while with a high mitotic rate it is reduced to about 26 days (MacDonald, 1961). In liver regeneration the normal mass is restored well before this shorter time. The mitotic rate then falls and the newly formed cells acquire their normal long life span.

In pathological conditions, however, when local skin damage or general liver damage is continuous, the mitotic rate remains high. The tissue mass may then become larger than normal but the shortened life expectancy of the newly formed cells prevents excessive growth. This is an important safety mechanism, which also operates in constantly stimulated hormone-dependent tissues and in the various benign and chronic tumors. Other types of tumors, however, continue to grow and so to endanger life.

XI. CARCINOMATA

The causes of carcinogenesis may be numerous and diverse, but all growing forms of cancer share only one important characteristic which distinguishes them from their tissues of origin: their rate of cell production by mitosis exceeds their rate of cell loss (Bullough and Deol, 1971b).

On the assumption, made here, that a certain critical concentration of chalone must develop in a cell before it can cease its mitotic activity and switch to the postmitotic aging state, it is clear that, in theory, situations could arise in which this critical chalone level could not be attained. There could, for instance, be a failure to synthesize enough chalone, a failure to retain within the cell the chalone that is synthesized, or a failure to respond adequately to the chalone that is there. In optimum conditions the situation then developing would be an explosive form of growth similar to that seen in a fast-growing carcinoma.

However, conditions are not usually optimum and tumor growth is often remarkably slow. There are several reasons for this: an immune response to the tumor may result in continuing cell injury and death (see Weiss, 1971); the blood supply commonly deteriorates in the tumor center; and with increasing tumor mass many cells do manage to enter the postmitotic aging phase and die (Bullough and Deol, 1971b).

XII. THE PATTERN OF TUMOR GROWTH

The 11 tumors in six species (including granulocytic leukemia in man) so far studied have all proved similar in containing the chalone of their tissue of origin and in reacting by mitotic inhibition when treated with

extra amounts of this chalone (see Bullough and Deol, 1971b). Other
similar results have been obtained by Bichel (1971a, b), and the available evidence suggests that the high mitotic activity in these tumors
was due primarily to a low chalone concentration. The only estimates
of tumor chalone concentration so far attempted have indicated, both
in a rabbit epidermal carcinoma (Bullough and Laurence, 1968b) and
in a rat granulocytic leukemia (Rytömaa and Kiviniemi, 1968b), that
the intracellular chalone content was 10% or less of normal and that
this was due to an abnormally high rate of chalone loss through defective
cell membranes.

As a general rule while the tumors studied remained very small, only
mitotic cells were present (Bullough and Deol, 1971b). However, when
the tumors became larger typical postmitotic cells appeared centrally
and these as they died formed a central necrotic mass (Fig. 5). Often
these postmitotic cells showed typical tissue syntheses: thus in growing
melanomata the postmitotic cells may synthesize melanin (Bullough and
Laurence, 1968c) while in growing epidermal carcinomata they may
synthesize keratin (Bullough and Deol, 1971a). If the capacity for such
syntheses has been lost, the postmitotic cells nevertheless age and die
(Bullough and Deol, 1971b).

Such aging and death is evidently caused by a progressive rise in
the chalone concentration in the central tumor region, which may be
related both to the increasing tumor mass and to a weakening blood
supply. As a general rule, the peripheral tumor region, which remains
mitotic, can be seen to have an excellent capillary supply which could
efficiently drain away the chalone; the central tumor region can often

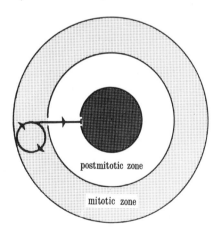

Fig. 5. The typical structure of a partly grown carcinoma with a central mass
of dead cells (After Bullough and Deol, 1971b; compare Fig. 2).

be seen to have a poor blood supply, due apparently to vessel deterioration, so that less chalone may be drained away. Sometimes the central blood vessel deterioration may progress so far that the surrounding tumor cells die of malnutrition.

An important consequence of continuing tumor growth combined with excessive chalone loss is a rising chalone concentration in the body as a whole, as can be demonstrated first by the falling mitotic rate within the tumor's tissue of origin (Bullough and Laurence, 1968b; Bullough and Deol, 1971a) and later by the falling growth rate of the tumor itself. In theory, given a limited body space and a large enough tumor, the chalone concentration should rise so high that all further tumor growth is prevented. In any chronic tumor this point is actually reached, but "in any acute tumour, death occurs before the inhibition becomes strong enough to stop growth completely" (Bullough, 1971). However, it seems clear that the growth of most, if not all, tumors follows a sigmoid curve (Laird, 1964, 1969; Burns, 1969; Bichel, 1970, 1971a) so that according to Laird (1964) "tumours are still responsive to a greater or lesser extent . . . to some feedback control" and according to Burns (1969) "tumors may regulate their growth as normal . . . tissues do by the production of a homologous specific mitotic inhibitor." The latest work of Bichel (1971a, b) has clearly shown the existence of tumors of "specific mitotic inhibitors," which have all the characteristics of chalones.

It follows that it should be possible by injections of the appropriate chalones to stop, and even to reverse, tumor growth. In recent years this has been done in the case of three experimental tumors: two melanomata of mouse and hamster (Mohr *et al.*, 1968) and one granulocytic leukemia of rat (Rytömaa and Kiviniemi, 1969, 1970). In the most successful cases a permanent cure was obtained. The fact that this may also be achieved in man is shown by the inhibition of ten cases of human granulocytic leukemia *in vitro* by means of the granulocytic chalone. Indeed there is good reason to believe that chalones will prove to have a considerable medical value, although all attempts at clinical testing must await the development of techniques for the mass production of chalones in a pure enough state.

XIII. GENERAL CONCLUSIONS

The first "chemical messenger" to be extracted and tested experimentally was the hormone secretin (Bayliss and Starling, 1902), which is produced by the duodenal cells and carried by the blood to the pancreas, where it stimulates the secretion of the pancreatic digestive enzymes. Since then many hormone systems have been discovered, many hormones have been chemically defined, and the science of endocrinology has been established. It is however clear that the hormones are of secondary importance in animal organization: they have been added to the basic organization to enable certain target tissues to respond appropriately to events originating outside the animal (see Bullough, 1967). These events vary in scale from simple food intake, which leads, for instance, to secretin secretion, to complex environmental variations, which lead to the secretion of a sequence of reproductive hormones and so to the onset of a breeding season.

The first chalone to be extracted and to be fully tested experimentally was that of the epidermis (Bullough and Laurence, 1964a), and since then chalones have been found in all those mitotic tissues in which they have been sought. The chalones are also "chemical messengers" and as such they too fall within the ambit of endocrinology. Their role is primary in that no differentiated tissue in which cell replacement is possible could continue to exist in their absence. Chalone systems are fundamental to vertebrate organization and probably to the organization of other types of metazoan animals as well.

All the known "chemical messengers" act as links in homeostatic mechanisms and they form an obvious hierarchy. The chalones, together with their associated mechanisms, maintain the mitotic tissues at an appropriate mass and therefore at an appropriate level of function; the hormones act by modifying the rate of function, and often also the mass, of their specific target tissues; and the pheromones act by modifying tissue development and function on other animals of the same social group.

The endocrinology of the future must consider the nature and manner of action of this whole hierarchy of "chemical messengers" and not just those of the hormones alone. Such a study is clearly important for its own sake, but it is also urgent because of its practical value to medicine in the treatment of a wide variety of syndromes which result from the breakdown of homeostatic mechanisms.

REFERENCES

Abercrombie, M. (1964). Behaviour of cells toward one another. *Advan. Biol. Skin* **5**, 95–112.
Argyris, T. S. (1969). Enzyme induction and control of growth. *In* "Repair and Regeneration" (J. E. Dunphy and W. Van Winkle, eds.), pp. 201–216. McGraw-Hill, New York.

Bayliss, W. M., and Starling, E. H. (1902). The mechanism of pancreatic secretion. *J. Physiol. (London)* **28**, 325–353.

Bichel, P. (1970). Tumor growth inhibiting effect of JB-I ascitic fluid. *Eur. J. Cancer* **6**, 291–296.

Bichel, P. (1971a). Autoregulation of ascites tumour growth by inhibition of the G-1 and G-2 phase. *Eur. J. Cancer* **7**, 349–355.

Bichel, P. (1971b). Feedback regulation of growth of ascites tumours in parabiotic rats. *Nature (London)* **231**, 449–450.

Bucher, N. L. R. (1963). Regeneration of mammalian liver. *Int. Rev. Cytol.* **15**, 245–300.

Bullough, W. S. (1948). Mitotic activity in the adult male mouse. *Proc. Roy. Soc. B* **135**, 212–233.

Bullough, W. S. (1962). The control of mitotic activity in adult mammalian tissues. *Biol. Rev.* **37**, 307–342.

Bullough, W. S. (1965). Mitotic and functional homeostasis. *Cancer Res.* **25**, 1683–1727.

Bullough, W. S. (1966). Cell replacement after tissue damage. *In* "Wound Healing" (C. Illingworth, ed.), pp. 43–59. Churchill, London.

Bullough, W. S. (1967). "The Evolution of Differentiation." Academic Press, New York.

Bullough, W. S. (1969). Epithelial repair. *In* "Repair and Regeneration" (J. E. Dunphy and W. Van Winkle, eds.), pp. 35–46. McGraw-Hill, New York.

Bullough, W. S. (1971). The actions of the chalones. *Agents Actions* **2**, 1–7.

Bullough, W. S. (1973). The epidermal chalone mechanism. *Natl. Cancer Inst. Mon.* **38**, in press.

Bullough, W. S., and Deol, J. U. R. (1971a). Chalone-induced mitotic inhibition in the Hewitt keratinising epidermal carcinoma of the mouse. *Eur. J. Cancer* **7**, 425–431.

Bullough, W. S., and Deol, J. U. R. (1971b). The pattern of tumour growth. *Symp. Soc. Exp. Biol.* **25**, 255–275.

Bullough, W. S., and Deol, J. U. R. (1972). Chalone control of mitotic activity in eccrine sweat glands. *Brit. J. Derm.* **86**, 586–592.

Bullough, W. S., and Ebling, F. J. (1952). Cell replacement in the epidermis and sebaceous glands of the mouse. *J. Anat.* **86**, 29–34.

Bullough, W. S., and Laurence, E. B. (1960a). The control of epidermal mitotic activity in the mouse. *Proc. Roy. Soc. B* **151**, 517–536.

Bullough, W. S., and Laurence, E. B. (1960b). The control of mitotic activity in the mouse skin. Dermis and hypodermis. *Exp. Cell Res.* **21**, 394–405.

Bullough, W. S., and Laurence, E. B. (1964a). Mitotic control by internal secretion: the role of the chalone-adrenalin complex. *Exp. Cell. Res.* **33**, 176–194.

Bullough, W. S., and Laurence, E. B. (1964b). The production of epidermal cells. *Symp. Zool. Soc. London* **12**, 1–23.

Bullough, W. S., and Laurence, E. B. (1966). The diurnal cycle in epidermal mitotic duration and its relation to chalone and adrenalin. *Exp. Cell. Res.* **43**, 343–350.

Bullough, W. S., and Laurence, E. B. (1968a). The role of glucocorticoid hormones in the control of epidermal mitosis. *Cell. Tiss. Kinet.* **1**, 5–10.

Bullough, W. S., and Laurence, E. B. (1968b). Control of mitosis in rabbit V × 2 epidermal tumours by means of the epidermal chalone. *Eur. J. Cancer* **4**, 587–594.

Bullough, W. S., and Laurence, E. B. (1968c). Control of mitosis in mouse and hamster melanomata by means of the melanocyte chalone. *Eur. J. Cancer* **4**, 607–615.

Bullough, W. S., and Laurence, E. B. (1970a). Chalone control of mitotic activity in sebaceous glands. *Cell. Tiss. Kinet.* **3**, 291–300.

Bullough, W. S., and Laurence, E. B. (1970b). The lymphocytic chalone and its antimitotic action on a mouse lymphoma *in vitro. Eur. J. Cancer* **6**, 525–531.

Bullough, W. S., and Laurence, E. B. (1971). Unpublished.

Bullough, W. S., Hewett, C. L., and Laurence, E. B. (1964). The epidermal chalone: a preliminary attempt at isolation. *Exp. Cell Res.* **36**, 192–200.

Burns, E. R. (1969). On the failure of self-inhibition of growth of tumors. *Growth* **33**, 24–45.

Christophers, E. (1971a). Cellular architecture of the stratum corneum. *J. Invest. Derm.* **56**, 165–169.

Christophers, E. (1971b). Personal communication.

Dunphy, J. E., and Van Winkle, W. (eds.) (1969). "Repair and Regeneration." McGraw-Hill, New York.

Ebling, F. J. (1957). The action of testosterone on the sebaceous glands and epidermis in castrated and hypophysectomized male rats. *J. Endocrinol.* **15**, 297–306.

Frankfurt, O. S. (1971). Epidermal chalone. Effect on cell cycle and on development of hyperplasia. *Exp. Cell Res.* **64**, 140–144.

Hondius-Boldingh, W., and Laurence, E. B. (1968). Extraction, purification and preliminary characterisation of the epidermal chalone. *Eur. J. Biochem.* **5**, 191–198.

Houck, J. C. (1971). Personal communication.

Houck, J. C., Irausquin, H., and Leikin, S. (1971). Lymphocyte DNA synthesis inhibition. *Science* **173**, 1139–1141.

Kivilaakso, E., and Rytömaa, T. (1971). Erythrocytic chalone, a tissue-specific inhibitor of cell proliferation in the erythron. *Cell. Tiss. Kinet.* **4**, 1–9.

Laird, A. K. (1964). Dynamics of tumour growth. *Brit. J. Cancer* **18**, 490–502.

Laird, A. K. (1969). Dynamics of growth in tumors and in normal organisms. *Nat. Cancer Inst. Mon.* **30**, 15–28.

Lasalvia, E., Garcia-Giralt, E., and Macieira-Coelho, A. (1970). Extraction of an inhibitor of DNA synthesis from human peripheral blood lymphocytes and bovine spleen. *Eur. J. Clin. Biol. Res.* **15**, 789–792.

MacDonald, R. A. (1961). Lifespan of liver cells. *Arch. Int. Med.* **107**, 335–343.

Mercer, E. H. (1962). The cancer cell. *Brit. Med. Bull.* **18**, 187–192.

Mohr, U., Althoff, J., Kinzel, V., Süss, R., and Volm, M. (1968). Melanoma regression induced by chalone: a new tumour inhibiting principle acting *in vivo. Nature* (*London*) **220**, 138–139.

Paukovits, W. R. (1971). Control of granulocyte production: separation and chemical identification of a specific inhibitor (chalone). *Cell Tiss. Kinet.* **4**, 539–547.

Randers Hansen, E. (1967). Mitotic activity and mitotic duration in tongue and gingival epithelium of mice. *Odontol. Tidsk.* **75**, 480–487.

Rytömaa, T., and Kiviniemi, K. (1968a). Control of granulocyte production. *Cell Tiss. Kinet.* **1**, 329–340, 341–350.

Rytömaa, T., and Kiviniemi, K. (1968b). Control of DNA duplication in rat chloroleukaemia by means of the granulocytic chalone. *Eur. J. Cancer* **4**, 595–606.

Rytömaa, T., and Kiviniemi, K. (1969). Chloroma regression induced by the granulocytic chalone. *Nature* (*London*) **222**, 995–996.

Rytömaa, T., and Kiviniemi, K. (1970). Regression of generalised leukaemia in rat induced by the granulocytic chalone. *Eur. J. Cancer* **6**, 401–410.

Saetren, H. (1956). A principle of auto-regulation of growth. *Exp. Cell Res.* **11**, 229–232.

Scaife, J. F. (1970). Liver homeostasis: an *in vitro* evaluation of a possible specific chalone. *Experientia* **26**, 1071–1072.

Simnett, J. D., and Chopra, D. P. (1969). Organ specific inhibitor of mitosis in the amphibian kidney. *Nature (London)* **222**, 1189–1190.

Tosti, A., Scerrato, R., and Fazzini, M. L. (1959). Saggio di esplorazione biometrica dell' epidermide umana. *Ann. Ital. Derm. Sifil.* **14**, 185–247.

Tosti, A., Nazzaro, P., and Porro, M. (1969). Histodynamics of the neoplastic and pseudoneoplastic growth of epidermis. *Ital. Gen. Rev. Derm.* **9**, 1–10.

Voaden, M. J. (1968). A chalone in the rabbit lens. *Exp. Eye Res.* **7**, 313–325.

Weiss, D. W. (ed.) (1971). "Immunological parameters of host-tumor relationships." Academic Press, New York.

Winter, G. D. (1964). Movement of epidermal cells over the wound surface. *Advan. Biol. Skin* **5**, 113–127.

II

BLOOD CELL FORMATION AND RELEASE

2

ERYTHROPOIETIN: THE HUMORAL
REGULATOR OF ERYTHROPOIESIS*

*Albert S. Gordon, Esmail D. Zanjani, Anthony S. Gidari, and
Robert A. Kuna†*

I. INTRODUCTION

A. Hypoxia: The Fundamental Stimulus for Erythropoiesis

One of the most readily apparent examples of homeostasis in the
higher animal relates to the maintenance of the dimensions of the circu-

* Original work reported in this Chapter was supported by U.S. Public Health
Service Research Grants 5 R01 HE03357–15 from the National Heart and Lung
Institute and 1 R01 AM15525–01 from the National Institute of Arthritis and
Metabolic Diseases.
† Predoctoral U.S. Public Health Service Trainee (Grant 5 T01 HE05645–07).

lating erythron. When erythrocytes are removed or destroyed, compensatory changes occur which result in replacement of these elements. As the red blood cell mass increases, reactions are set into operation which retard the production of these cells. Such information suggests the existence of a sensitively attuned control of the rates of red cell production and elimination. It became evident early that this regulatory influence might have been related to the relative availability of oxygen to the organism (Grant and Root, 1952). Indeed, it is now extensively documented that oxygen deficiency (hypoxia) constitutes the fundamental stimulus for erythropoiesis in higher animals (Gordon, 1959). It also appears to stimulate hemoglobin production in at least one species of invertebrate (Fox et al., 1951; Hildemann and Keighley, 1955). Four types of hypoxia exist: hypoxic (low oxygen pressures), anemic, stagnant and histotoxic. Each of these has been demonstrated to evoke increased erythropoiesis (Gordon, 1959; Krantz and Jacobson, 1970).

Since hypoxia is the fundamental physiological trigger for erythropoiesis, it follows that the relation between oxygen supply and demand within the organism should constitute the crucial determinant for governing red cell production (Gordon, 1954; Jacobson and Goldwasser, 1957; Crafts and Meineke, 1959). Ample evidence is available to support this proposal. Thus hypophysectomy and starvation which lower oxygen need relative to demand lead to a reduction in erythropoiesis. Similarly, hyperoxia and excess numbers of red cells which increase oxygen supply also result in a decrease in red cell production. On the other hand, hormones of thyroidal origin and possibly growth hormone which augment the oxygen requirement in relation to oxygen supply cause an accelerated rate of erythropoiesis. In this category are the anemias which stimulate red cell production by lowering oxygen available relative to demand.

The mechanism by which oxygen deficiency stimulates erythropoiesis has been subjected to intensive scrutiny. Although a direct stimulatory effect of hypoxia on the erythroid-forming tissues was first inferred, Grant and Root (1947) demonstrated that dogs subjected to acute or chronic bleedings did not exhibit a decrease in the pO_2 or percent oxygen saturation values of marrow blood at times when enhanced erythropoiesis was occurring in this tissue. Similarly, no drop was detected in the percent O_2 saturation values of marrow blood from anemic subjects or patients with primary or secondary polycythemia (see review by Grant and Root, 1952). It was studies of this kind which pointed to the possibility that the increased erythropoiesis induced by hypoxia is mediated through humoral mechanisms.

B. Endocrine Influences on Erythropoiesis

A role of the endocrine system in erythropoiesis might well have been anticipated since it is clear that hormones significantly influence a wide spectrum of biological reactions. Thus it would be expected that these actions should be imparted directly or indirectly to the blood-forming tissues. Evidence for a relation of hormones to erythropoiesis was derived from early experiments indicating that ablation of the hypophysis in rats results in the development of an anemia within several weeks following surgery (Vollmer *et al.*, 1939; Crafts, 1941). The posthypophysectomy anemia is ameliorated by administration of the lacking hormonal factors. These include anterior hypophyseal, thyroid, adrenal cortical, and testicular hormones (Meyer *et al.*, 1940; Vollmer *et al.*, 1942; Evans *et al.*, 1961; Gordon, 1968). Combinations of these hormones are more effective in correcting the anemia and the bone marrow hypoplasia than single factors. Thus a mixture of thyroxine, cortisone and growth hormone can overcome the peripheral blood and marrow changes that characterize adenohypophyseal deficiency (Crafts and Meineke, 1959). A combination of growth hormone, testosterone and thyroxine also achieves this restorative effect (Gordon, 1954) thus indicating that the ameliorative actions are not restricted to one particular combination of hormones.

The evidence has thus implicated the anterior hypophysis and several endocrine target organs in the regulation of erythropoiesis. In this regard, the question as to whether the endocrine system is obligatory for erythropoiesis has been satisfactorily answered by the demonstration that hypophysectomy, adrenalectomy, and orchidectomy or combinations of these procedures modify to some extent, but do not abolish, the enhancement of red cell production that occurs as a result of hypoxia (Feigin and Gordon, 1950; Piliero, 1959). On the basis of this evidence, the conclusion was reached that the role of the hormones was facultative rather than obligatory in the regulation of erythropoiesis and that their action probably was to reinforce rather than to exert a primary influence on erythropoiesis (Gordon, 1957).

II. ERYTHROPOIETIN

A. Factors Influencing Its Production

The recognition and general acceptance of erythropoietin (Ep) as a humoral principle of cardinal importance has refocused attention on the mechanism by which hypoxia, hormonal factors and other agents influence erythropoiesis. Ep is a glycoprotein hormone that stimulates red cell production in higher animals including man. It is found in elevated amounts in the plasma, serum, lymph, and urine of animals subjected to various forms of hypoxia (Gordon, 1959; Fisher, 1969; Gordon and Zanjani, 1970a) as well as in oxygen deficiency states in man (Fisher, 1969; Gordon and Zanjani, 1970a). A finding of considerable significance is that it is also detectable in the plasma (Mirand et al., 1965) and urine (Adamson et al., 1966; Alexanian, 1966; Van Dyke et al., 1966) of normal humans, thus supporting the proposal that it is not only concerned in emergency or "panic" erythropoiesis but in the normal daily replacement of red cells as well. A sex difference (Alexanian, 1966; Van Dyke et al., 1966) and a diurnal variation (Adamson et al., 1966) in its production in humans is suggested from its daily excretion patterns. The finding that anti-Ep immune serum can practically eliminate erythropoiesis in normal mice constitutes additional evidence that Ep is importantly involved in the day-to-day formation of erythrocytes (Schooley and Garcia, 1962).

B. Methods of Assay

Both in vivo and in vitro procedures are currently employed for the detection of Ep. The in vivo assay most commonly used measures the increase induced by Ep in the percent ^{59}Fe incorporation into the red cells of mice made plethoric by transfusion of homologous red cells or polycythemic by exposure to reduced oxygen tensions followed by return of the mice to normal pressures for several days (Camiscoli and Gordon, 1970). With these methods, a suppression of erythropoiesis and probably Ep production eventually occurs and, in this state, the mice are more sensitive to Ep. Radioiron incorporation values may be converted into equivalent Ep units by reference to the standard curve for the International Reference Preparation (IRP) of Ep (Cotes, 1971). These methods permit quantitative detection of quantities of Ep as small as 0.05–0.10 IRP units per sample administered. Among the presently available in vitro methods are those measuring radioiron incorporation into heme of isolated marrow (Krantz and Jacobson, 1970) or of fetal liver cells (Stephenson and Axelrad, 1971) as well as hemagglutination inhibition (Lange et al., 1969) and radioimmune assay (Fisher and Roh, 1971) techniques. As is the case for other hormones, it seems probable that when the specificity of the radioimmune assay is improved, this will be the method of choice. Of importance are recent findings indicating that results obtained with the in vivo and in vitro assays for a number of Ep preparations do not always concur and therefore estimates with the two types of assay cannot necessarily be compared (Lange et al., 1969; Dukes et al., 1970).

C. Purification and Chemistry

A considerable degree of purification of Ep has been achieved. Using urine obtained from highly anemic human subjects, Espada and Gutnisky (1970) have obtained fractions with specific activities as high as 8000 units/mg of protein. Likewise, Goldwasser and Kung (1971a) have reported on preparations of Ep, extracted from anemic sheep plasma, with activities in the vicinity of 8000–9000 units/mg protein. These latter preparations are almost completely pure and the major contaminant present is desialated Ep. The finding that plasma Ep is inactivated by trypsin and sialidase suggests that it is a sialic acid-containing protein. Studies by Lowy (1970) have indicated that some residues of tryptophan, lysine and/or arginine, and possibly tyrosine are necessary parts of its polypeptide structure. On the other hand, free sulfhydryl and hydroxyl groups do not appear to be required for activity. The molecular weight of sheep plasma Ep has been estimated to be 45,800 (Goldwasser and Kung, 1971b). It contains approximately 30% carbohydrate of which 10.8% is sialic acid with the remainder of its structure being protein. Of interest is the observation that desialated Ep is inactive when administered to the animal but retains its ability to stimulate heme synthesis by isolated marrow cells (Goldwasser and Kung, 1968). It is conjectured that sialic acid may serve as an attachment site of Ep to a carrier substance or may aid in preventing destruction of Ep within the organism. A possibility worthy of exploration is that desialation of Ep at its target site (blood-forming tissues) is a prelude to the exertion of its physiological action.

D. Site(s) of Action

The primary target site of Ep is considered to be hematopoietic precursor elements now termed Ep responsive cells (ERC). Ep apparently induces these cells to differentiate into the earliest recognizable members of the nucleated erythron (Krantz and Jacobson, 1970). The mechanism by which Ep causes this inductive action has been studied in two types of *in vitro* systems: (1) adult rat bone marrow cells (Krantz *et al.*, 1963) and (2) mouse fetal liver cells (Cole and Paul, 1966). In the marrow system, Ep triggers the early production of a specific RNA assembly (messenger, ribosomal precursor and transfer) (Gross and Goldwasser, 1969) that precedes and is required for increased uptake of iron (Hrinda and Goldwasser, 1969), hemoglobin production (Gallien-Lartigue and Goldwasser, 1965), and stroma formation (Dukes *et al.*, 1963) by the reactive hematopoietic cells. It is of interest that an enhancing effect on ∂-amino levulinic acid (ALA) synthetase activity has been reported in rabbit bone marrow cultures to which Ep has been added (Necheles and Rai, 1969). As with the 5β-H steroid metabolites (Levere and Granick, 1967) Ep may function here in part as a physiological derepressor of ALA synthetase, a rate-limiting enzyme in the production of heme.

It has been demonstrated (Cole and Paul, 1966; Cole et al., 1968) that cells from the livers of $10\frac{3}{4}$ to $14\frac{1}{2}$-day-old mouse fetuses synthesize hemoglobin at a low rate in tissue culture. Upon addition of Ep during these critical times, there is a marked increase in the rate of hemoglobin production. Following this initial period of Ep sensitivity in the mouse, a time occurs at which the rate of hemoglobin synthesis remains high but cannot be further increased by Ep. The response to Ep is apparently dependent on a period of early and one of relatively late RNA synthesis with an intervening phase of DNA production (Paul and Hunter, 1969). The secondary phase of RNA synthesis, which appears to be essential for hemoglobin production, is dependent on the previous short period of DNA synthesis. Morphological studies (Chui et al., 1971) suggest that a primary action of Ep on fetal liver cells is the stimulation of RNA synthesis in early erythroblasts and that Ep is required for maintaining the precursor cell population, enabling it to continue to proliferate into hemoglobin-synthesizing elements. Of importance are recent findings indicating that cultures of disaggregated cells from whole embryos (8–12 days of age) and yolk-sac tissues of the mouse respond to Ep with increased heme synthesis (Bateman and Cole, 1971). This suggests that the differentiation of primary embryonic erythroid cells in situ may be under the influence of an Ep similar or identical to the adult type.

Evidence is unfolding that the action of Ep is not confined to a stimulation of differentiation in the ERC but that it induces other changes as well, including effects on already differentiated elements of the nucleated erythroid cell series. Thus direct effects in vitro have been described on erythroblast morphology (Borsook et al., 1968) and on DNA and heme synthesis in erythroblasts (Powsner and Berman, 1967), as well as on erythroblast divisions (Necheles et al., 1968). Additional actions relate to an early release of reticulocytes (Gordon et al., 1962; Fisher et al., 1965), a switching of hemoglobin A to C in nonanemic A/A sheep (Thurmon et al., 1970) and a stimulation of progenitor cell flow into the ERC compartment or an increased rate of ERC division (Reissmann and Samorapoompichit, 1970). A distinct vasoproliferative activity in transplanted splenic tissue follows treatment of donor mice with Ep (Feleppa et al., 1971). This observation, coupled with the finding that Ep induces an increased blood flow and opening of sinusoids in the splenic microcirculation (McClugage *et al.*, 1971), suggests that in addition to its action on the stem cells, Ep also alters the hematopoietic internal microenvironment (Trentin, 1970) to furnish an optimal site for erythropoiesis (McCuskey *et al.*, 1971). There is the possibility that these different actions may be exerted by different forms of Ep. This finds support in the observation that some Ep preparations appear to contain a number of biological components with differing *in vivo* and *in vitro* activities (Dukes *et al.*, 1970). However, it remains to be determined whether the various effects exerted by Ep are the result of specific actions of different members of an Ep complex.

E. Site of Formation and Biogenesis of Ep

It is now conceded that the kidney is the chief locus of production of Ep. This is supported by both clinical and experimental studies. Thus anemia is often an accompaniment to renal deficiency states in man (Brown and Roth, 1922; Loge *et al.*, 1950) and, on occasion, erythrocytosis is seen in patients with hypernephroma (Forssell, 1954) or hydronephrosis (Cooper and Tuttle, 1957).

Jacobson and Goldwasser (1957) and Jacobson *et al.* (1959) showed that bilateral nephrectomy in rats or rabbits markedly suppressed the production of Ep in response to different forms of hypoxia. That this decrease in Ep production could not be attributed to the toxicity of retained wastes was indicated by the findings that bilateral ureteral ligation, which resulted in a degree of uremia in rats approximately equivalent to that seen after nephrectomy, caused only a slight diminution in the ability to produce Ep in response to similar types and degrees of hypoxia (Jacobson and Goldwasser, 1957; Jacobson *et al.*, 1959).

Past attempts to extract Ep from renal tissue have not been uniformly successful and, at best, only relatively small amounts have been obtained (Gordon *et al.*, 1967; Gordon and Zanjani, 1970a). This inability to detect consistently significant amounts of Ep even in kidneys of hypoxic animals, might have stemmed from the possibility that it exists in the kidney in the form of an inactive precursor (Kuratowska *et al.*, 1964) or as an activator of Ep. In 1966, we reported on the extraction of a factor from the kidneys of hypoxic rats that was erythropoietically inactive when administered intraperitoneally alone to assay mice, but which when incubated with normal plasma or serum resulted in the production of Ep (reviewed by Gordon *et al.*, 1967). This principle was termed the renal erythropoietic factor (REF) or more recently erythrogenin (Eg) (Gordon and Zanjani, 1970b).

Fractionation procedures have been used to determine the subcellular location of Eg (Gordon and Zanjani, 1970b). Briefly, kidneys were obtained from adult female rats made hypoxic by exposure to 0.45 atm of air for 19 hours. Subcellular fractionation methods (La Bella et al., 1963) were now instituted for collection of the nuclear, heavy-mitochondrial, light-mitochondrial, and microsomal fractions. Each of these particulate fractions was suspended in either 0.02 M phosphate buffer (pH 6.8) or distilled water (not below pH 6.0) and centrifuged at 37,000 × g for 30 minutes. The supernatant fluids derived from each extract were then assayed for their erythropoiesis stimulating activity, both before and after incubation with normal rat serum (NRS) that had been dialyzed against ethylenediamine tetraacetate (EDTA). This dialysis procedure prevents the operation of an Ep-inactivating system present in the incubation mixture that is apparently cation-dependent (Gordon et al., 1967). The erythropoietic index employed was the percent incorporation of radioiron into the circulating red cells (Camiscoli and Gordon, 1970).

Table I indicates that erythropoiesis stimulating activity was not demonstrable in extracts of the nuclear or heavy-mitochondrial fractions

TABLE I

Erythropoiesis Stimulating Activity of Hypotonic Extracts of Subcellular Fractions of Rat Kidneys, after Incubation with Saline or Normal Serum[a]

Material assayed		Percent RBC-^{59}Fe incorporation[b] (mean ± 1 sem)[c]
Saline		0.96 ± 0.24
Serum		1.43 ± 0.30
0.05 IU Ep[d]		6.49 ± 0.92
0.20 IU Ep		14.06 ± 1.87
	Extract + saline	Extract + serum
Nuclear extract	1.47 ± 0.36	2.19 ± 0.33
Heavy-mitochondrial extract	0.87 ± 0.21	1.32 ± 0.35
Light-mitochondrial extract	1.97 ± 0.39	8.94 ± 1.63
Microsomal extract	0.66 ± 0.13	9.64 ± 1.08
Soluble fraction	2.33 ± 0.62	3.01 ± 0.49

[a] Equal volumes of the hypotonic extracts and either saline or EDTA-dialyzed normal rat serum were incubated for 30 minutes at 37°C.

[b] Values represent the mean percent RBC-radioiron incorporation induced by the injection of 2 ml of the incubation mixture into each of 5 exhypoxic polycythemic mice.

[c] Standard error of the mean.

[d] International Reference Preparation units of Ep.

when tested before or after incubation with EDTA-dialyzed NRS. Although extracts of the light-mitochondrial and microsomal fractions showed no capacity to stimulate erythropoiesis when given alone via the intraperitoneal route to the assay mice, appreciable activity appeared after incubation with the dialyzed NRS. The soluble fraction contained only slight activity which was not augmented following incubation with NRS. Dialyzed serum alone evoked little or no erythropoietic activity in the assay mice. Our present procedure for extraction of Eg is indicated in Chart 1.

Chart 1
*Current Procedure for Extraction of Erythrogenin from
Rat Kidneys*

1. Kidneys are removed immediately following exsanguination, decapsulated, weighed, and minced.
2. Tissue is homogenized in cold 0.25 M sucrose (10 gm/ml) with Potter-Elvehjem homogenizer as follows:
   a. The tissue is first homogenized with a loose-fitting pestle which has its diameter reduced $\frac{3}{8}-\frac{1}{2}$ mm by grinding against a stone. *The mixture should appear homogeneous at the end of this step.*
   b. The mixture is rehomogenized using the standard-sized pestles. *The duration of this step is dependent upon the state of the mixture after step 2a.*
3. The homogenate is centrifuged at 5500 $\times$ g for 10 minutes and the precipitate is discarded.
4. The supernatant fluid is recentrifuged at 30,000 $\times$ g for 35 minutes. This supernatant fluid is discarded.
5. The precipitate is mildly homogenized in the presence of 1–2 drops of Cutscum (detergent) and distilled water (2 ml/gm) while in the centrifuge tubes.
6. The resulting mixture is frozen for 48 hours.
7. On the day of the assay the mixture is thawed and centrifuged at 37,000 $\times$ g for 35 minutes.
8. The supernatant fluid contains erythrogenin and should possess 9–13 mg of protein/ml.

Chart 1. Procedure for extraction of erythrogenin from kidneys. (From Gordon and Zanjani, 1970b; courtesy of J. B. Lippincott.)

Studies on the kinetics of the Eg-serum system have suggested that Eg may be an enzyme that converts a serum substrate into the active circulating form of Ep (Zanjani *et al.*, 1967). It is actually Ep which is generated in the Eg-serum incubation system since the generated erythropoiesis-stimulating activity was completely abolished upon addition of anti-Ep globulin to the incubation mixture (Schooley *et al.*, 1970). Moreover, anti-Ep failed to antagonize the biological activity of Eg and did not inhibit the ability of serum to serve as a substrate for Eg (Schooley *et al.*, 1970). On the other hand, an anti-Eg serum, developed in rabbits against human Eg, which inhibited the action of Eg *in vitro* and depressed erythropoiesis *in vivo*, exerted no inhibitory action on the erythropoietic activity of Ep (McDonald *et al.*, 1971). These experiments support our concept that Eg, the serum substrate, and the product of their interaction, namely Ep, are three immunochemically dissimilar entities.

Evidence is now available for the existence of inhibitors of Ep in renal tissue (Erslev and Kazal, 1968; Fisher *et al.*, 1968; Kuratowska, 1968). One of these is lipid in nature (Erslev and Kazal, 1968). Upon addition of fresh serum to a human urinary Ep-lipid complex, some of the bound Ep is released. It has been proposed that Ep exists in an inactive lipid-bound form in kidneys and that a serum factor may control the release of this bound Ep into the circulating blood. This is a concept worthy of further exploration and its relation to the Eg–serum factor–Ep system is presently under investigation.

F. Feedback Relations in the Eg–Serum Factor–Ep System

As might have been anticipated, reciprocal regulatory mechanisms exist in the Eg–serum factor–Ep system. Three aspects of this control will be discussed, namely factors that stimulate and inhibit Ep and Eg production, factors that enhance and depress serum substrate production, and the mechanisms of removal or inactivation of Ep. On *a priori* grounds, it would be suspected that agents which influence erythropoiesis could operate by either affecting production of Eg, the serum factor or both.

1. Effects of Hypoxia on Ep and Eg Production

As an example, the mechanisms underlying the stimulatory effects of hypoxia on erythropoiesis will be considered.

Procedure (*Gordon et al.*, 1966). Adult male rats (250–300 gm) of the Long-Evans strain were employed. A group of 5 normal rats was subjected to 0.4 atm of air for 16 hours. A second group of 5 rats constituted the untreated unexposed controls. Each of an additional 10 rats was rendered plethoric by two successive daily intraperitoneal injections of 10 ml whole homologous blood. Of these 10 rats, 5 were subjected to 0.4 atmosphere of air for 16 hours while the remaining 5 comprised the plethoric unexposed controls. Directly following the period of hypoxia, the exposed and unexposed rats were exsanguinated and the sera and kidneys of each of the groups collected and separately pooled. Eg was extracted from the kidneys (Chart 1) and incubated as described above with dialyzed normal rat serum. The sera and the Eg–serum mixtures were examined for erythropoietic activity in exhypoxic polycythemic mice (Camiscoli and Gordon, 1970). Five to six mice were used to assay each sample. All mice received 1 ml of serum or 1 ml of incubation fluid as a single intraperitoneal injection.

In Fig. 1, bar 1 indicates that exposure to hypoxia resulted in a highly significant increase in the quantity of Ep in the serum as judged by the percent RBC-^{59}Fe incorporation values in the recipient assay mice. Note, however, that the rise in serum Ep in the hypertransfused hypoxia-exposed rats was considerably smaller (bar 2) than that noted in the nonplethorized groups exposed to the same degree of hypoxia (bar 1). Hypoxia also induced a highly significant increase in renal Eg activity in nonplethorized rats (bar 3) but was ineffective in this regard in the hypertransfused group (bar 4). Effects similar to those reported here for hypoxia have been described for other erythropoietic stimuli including androgens (Gordon *et al.*, 1966).

These studies thus show that the rise in serum Ep, evoked by hypoxia, is accompanied by and may be the result of an increased production

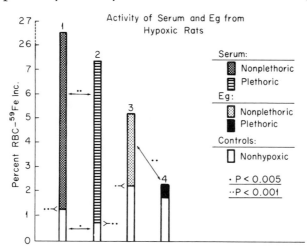

Fig. 1. Ep content of serum and Eg activity in normal and plethoric rats subjected to 0.4 atm of air and in unexposed controls. (From Gordon *et al.*, 1966; courtesy of Proc. Soc. Exp. Biol. Med.)

of Eg in the kidneys. Another point of significance is that hypertransfusion markedly inhibits the appearance of Ep in response to hypoxia (7.8% as against 26.5% for the RBC-radioiron incorporation levels evoked by the serum of the hypoxia-exposed transfused and hypoxia-subjected nontransfused groups, respectively, in the assay mice). This reduction in Ep production bears a direct relation to the marked inhibition of renal Eg activity produced by the transfusions. The mechanism by which the plethoric state operates to inhibit the production or activation of Eg is not as yet clear. Several possibilities exist (Gordon et al., 1968). One involves the possible presence of an inhibitor in the blood of hypertransfused animals (Krzymowski and Krzymowska, 1962; Whitcomb and Moore, 1965). Alternatively, it might also be a consequence of an increased oxygen supply to the kidneys due to the greater numbers of circulating red cells, at least up to certain peripheral hematocrit levels (Thorling and Erslev, 1968; Guidi and Scaro, 1970).

2. FACTORS INFLUENCING SERUM SUBSTRATE PRODUCTION AND/OR AVAILABILITY

In the course of our studies on the biogenesis of Ep, we found that although various forms of hypoxia resulted in elevated levels of Eg in the kidney, the increases noted were not of sufficient magnitude to account for the associated large elevations of circulating Ep evoked by the same stimuli. Thus subjection of rats to a bout of hypoxia induces a rise in renal Eg which, however, does not exceed 2- to 3-fold when compared to the quantity of Eg present in normal kidneys (Gordon et al., 1967). In addition, normal serum from a variety of mammalian sources contains, on the average, 0.05–0.10 IRP units/ml of substrate for Eg (Zanjani et al., 1971). (One substrate unit represents the amount of serum which yields 1.0 IRP unit of Ep when incubated with 1.0 ml of Eg for 1 hour at 37°C, a quantity extracted from 0.5 gm of hypoxic rat kidneys.) On the other hand, levels of Ep in excess of 2 IRP units/ml of plasma are known to exist in anemic animals (McDonald et al., 1970). This represents a 20- to 40-fold increase over the amount present in normal plasma. It is therefore possible that during hypoxic stimulation a large increase in plasma substrate level occurs to account for the marked rise in Ep levels.

To test this possibility, we first determined the effects of continuous hypoxia (0.45 atm of air) on substrate and Ep production in rats (Zanjani et al., 1971). Ten groups, each consisting of 4 rats, were established. Estimates of serum Ep and substrate values were made after 0, 15, 30, and 45 minutes and following 1, 2, 5, 8, 10, and 15 hours of hypoxia. To this end, for each of the 10 groups, 1 ml of serum was incubated with 1 ml of saline for 1 hour and assayed for its Ep content. In addition, 1 ml of serum, after treatment with anti-Ep to eliminate the Ep present and then with goat antirabbit γ-globulin (GARGG) to rid the serum of excess anti-Ep, was tested for its substrate content. As described above, this was determined by incubating this treated serum with 1 ml of Eg (derived from 0.5 gm hypoxic rat kidneys) and then assaying the incubation fluid for the quantity of Ep generated. Each of 5 to 6 exhypoxic polycythemic assay mice received an intraperitoneal injection of 2 ml of each of the mixtures tested.

Figure 2 indicates that Ep was not detectable in the plasma at any of the time intervals examined during the first hour of hypoxia. Although the serum substrate levels were normal at 15 minutes of hypoxia, a decrease to nondetectable levels occurred at the 30-, 45-, and 60-minute periods of exposure. Significant quantities of substrate in the serum were noted at 2 hours of hypoxia accompanied by a significant rise in the levels of Ep. Further increases in the quantity of serum substrate were seen at 5 hours and the Ep values at this time of hypoxia were higher than they had been at 2 hours. Peak values of both substrate and Ep in serum were evident at 8 hours. The levels of Ep were still relatively high at 10 hours but at this time the substrate values had decreased to those found in normal unexposed rats. At 15 hours, when a significant decrease in the amount of serum Ep was seen, the substrate values

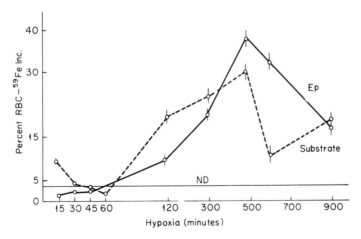

Fig. 2. Effects of continuous exposure of rats to hypoxia on the Ep and substrate content of the serum. Ordinate represents the percent RBC-radioiron incorporation values in assay exhypoxic mice. ND, nondetectable. Vertical lines through points (mean values) indicate ± 1 sem. (From Gordon and Zanjani, 1970b; courtesy of J. B. Lippincott.)

again showed a rising trend. From these experiments, the following conclusions may be reached:

a. Elevated quantities of Ep do not appear in the serum until sufficient substrate has become available (e.g., at 2, 5, and 8 hours of continuous hypoxia). Substrate levels therefore appear to constitute the major rate-limiting factor in the biogenesis of Ep.

b. A low level of substrate in the plasma may provide the mechanism for subsequent increased substrate production which in turn leads to increased formation of Ep. These decreased levels of substrate are evident at 1 and 10 hours of continuous hypoxia. We also have some preliminary evidence that Eg may also serve as a stimulus to an increase in substrate levels above baseline values.

c. The variations that occur in Ep levels in the plasma of animals exposed continuously to hypoxia (Camiscoli and Gordon, 1970) are probably due in part to fluctuations in the production and/or availability of substrate.

d. High levels of Ep in plasma may serve as a negative feedback on substrate and therefore on Ep production (e.g., seen at 10 hours after continuous hypoxia).

It seemed important to test more directly the latter possibility, i.e., that Ep, upon attaining relatively high levels in the plasma, serves to inhibit its own production. For this purpose, several groups, each composed of 6 Long-Evans rats, were established. Group 1 was subjected to continuous hypoxia corresponding to 0.45 atm of air for 5 hours and bled immediately thereafter. The rats of Group 2 were injected intravenously with 15 IRP units of human urinary Ep (obtained from a child with hypoplastic anemia) and exsanguinated 5 hours later. Group 3 was given one intravenous injection of 15 IRP units of human urinary Ep just prior to a 5-hour exposure to hypoxia and bled immediately after termination of the hypoxic period. The plasmas obtained from these 3 groups were pooled separately and assayed for their Ep and for their substrate content in exhypoxic polycythemic mice. Each mouse received 0.2 ml of plasma intraperitoneally and the erythropoietic response measured as the percent RBC-radioiron incorporation. In estimating substrate levels, plasmas were first dialyzed against 0.005 M EDTA. As before, plasma samples containing Ep were treated with anti-Ep prior to their incubation with Eg. The quantity of anti-Ep used was twice that required to neutralize the erythropoietic activity of the plasma. Again the excess anti-Ep in the samples was removed by addition of GARGG. One ml samples of Ep-free plasma were now incubated with 1 ml of Eg-containing fluid and the amount of Ep generated in the incubation mixtures was determined.

Figure 3 shows that exposure of rats to 5 hours of hypoxia caused the appearance of considerable amounts of Ep in the plasma and that this was accompanied by the presence of relatively large amounts of substrate (Group 1). The rats injected with 15 IRP units of Ep exhibited approximately 0.2 IRP units of Ep activity per 0.7 ml of plasma at 5 hours after administration of the exogenous Ep (Group 2). Substrate activity in these rats did not differ significantly from that noted in plasma samples of control untreated rats. Of importance was the observation that the Ep levels in the plasma of rats given both exogenous Ep and exposure to hypoxia (Group 3) were not different from those seen in rats given only the Ep (Group 2). It is also to be noted that injection of Ep just prior to exposure of the rats to hypoxia, induced a marked decrease in substrate level (Group 3) when compared to those in rats subjected only to hypoxia (Group 1). Figure 3 also indicates that renal Eg values were increased by hypoxia but no lowering of these levels occurred as a result of Ep injection before their exposure to reduced oxygen tensions (Zanjani *et al.*, 1968).

The data therefore establish an inhibitory feedback exerted by Ep on its own production. The mechanism appears to operate through an

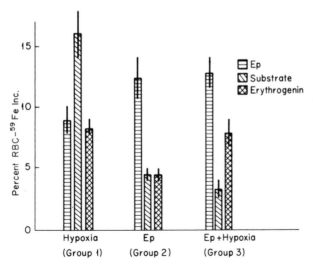

Fig. 3. Effects of exogenous Ep on endogenous Ep, plasma substrate and renal Eg levels in rats exposed to continuous hypoxia. The Ep and substrate activity of the plasma as well as the Eg levels in the kidney were determined in exhypoxic polycythemic mice and expressed as percent RBC-radioiron incorporation values. Vertical lines through tops of bars (mean values) indicate ± 1 sem. (From Gordon and Zanjani, 1970b; courtesy of J. B. Lippincott.)

ability of Ep to decrease the production, availability and/or activity of the plasma substrate for Eg.

3. Sites of Inactivation and Elimination

   a. *The Liver.* Two main mechanisms have been proposed for inactivation or metabolism of Ep. One involves a role of the liver and the other, the state of activity of the blood-forming tissues upon which Ep exerts its effects. Jacobsen *et al.* (1956) induced anemia in rabbits with phenylhydrazine and noted highest Ep activity in the plasmas of animals with severely damaged livers. The hypothesis advanced was that the hepatic injury interfered with an Ep inactivating function of the liver thus resulting in higher quantities of Ep in the circulating blood. Further support for a hepatic role was provided by Prentice and Mirand (1957) who prevented, with the hepatoxic agent phenylhydrazine, the tendency of Ep to disappear from the blood of rats exposed continuously for 48 hours to reduced oxygen tensions. Along the same lines, Mirand *et al.* (1959) demonstrated a smaller effectiveness of Ep-containing plasma to stimulate erythropoiesis when introduced directly into the hepatic portal circulation than when administered via the jugular vein. Similarly, Alpen (1962) gave several injections of anemic rabbit plasma to animals that had been pretreated with carbon tetrachloride. Their finding that the circulating red blood cell volume was 38% higher in animals receiving the carbon tetrachloride than in animals not injected with the drug would suggest that Ep was destroyed at a slower rate in the animals with liver damage. In addition, Burke and Morse (1962) demonstrated that the Ep levels in perfusates circulated through isolated livers of rats that had received carbon tetrachloride were significantly greater than in perfusates circulated through the livers of normal rats. In contradistinction, however, Fischer and Roheim (1963), could find no evidence of inactivation of Ep by perfused rat liver. This difference in result may relate in some way to the use by Burke and Morse of blood perfusates derived from rats that had been treated with cobalt to elevate the endogenous levels of Ep whereas exogenous sheep plasma Ep was used by Fischer and Roheim. It has been recently reported (Fisher and Roh, 1970) that a rapid decrease occurred in the concentration of Ep in hypoxic dog plasma perfused through isolated dog livers. When, however, the dogs supplying the livers were pretreated with SKF 525A, a drug that inhibits hepatic microsomal enzymes, no significant decrease in Ep was observed during the perfusion. These results were interpreted to mean that microsomal enzymatic destruction of Ep occurs in the perfused normal liver system. Further evidence for this contention was provided by experiments which indicated an ability of hepatic microsomal fractions from normal dogs, but not from SKF-525A treated dogs, to inhibit Ep *in vitro* (Fisher and Roh, 1970). Of interest are observations indicating that a lower rate of disappearance of exogenous Ep occurred in the plasma of germfree than in conventional mice (Mirand *et al.*, 1972). This may relate to observations that the liver is less developed, both from a morphological and a functional point of view, in germfree than in conventional mice (Thorbecke and Benacerraf, 1959; Miyakawa *et al.*, 1965).

There are also clinical cases that implicate the liver in the inactivation of Ep. Thus Mirand and Murphy (1971) have demonstrated elevated levels of plasma and urinary Ep in patients with various forms of hepatic dysfunction occurring in viral hepatitis, hepatic necrosis, toxic damage states caused by alcohol or carbon tetrachloride as well as in subjects with hepatic infiltration secondary to other systemic illnesses, conditions not consistently associated with anemia. These studies were construed as providing clinical evidence for the association of hepatic dysfunction with an inability of the liver to inactivate Ep. Of relevance are also the observations that whereas homogenates of normal human liver inactivate human Ep *in vitro*, similarly prepared homogenates of the tumorous liver tissue from a patient with hepatocellular carcinoma displaying erythrocytosis lacked this capacity to metabolize Ep (Gordon *et al.*, 1970). A study here of the relative protease activities of the normal liver and the tumorous liver tissues would appear to be indicated.

Not in accord with the concept that the liver represents a site of inactivation or degradation of Ep are experiments indicating that some phenylhydrazine-treated rabbits, which sustained little or no liver damage, showed higher levels of Ep in the plasma than some of those experiencing hepatic damage (Lowy *et al.*, 1959). Moreover, administration of hepatotoxic doses of carbon tetrachloride to rats failed to alter the rate of disappearance of injected Ep from their plasma (Keighley, 1962).

Although much of the evidence appears to implicate the liver as a site of inactivation of Ep, additional experiments are required to secure this contention. As has been pointed out (Krantz and Jacobson, 1970), the effects ascribed to the liver may relate, at least in part, to its nonspecific proteolytic activity rather than to a physiological role of this organ.

b. The Blood-Forming Tissues. A number of articles point to the possibility that the blood-forming tissues may also serve as a site of metabolism of Ep. A phenomenon reported by several investigators (Stohlman, 1959; Finne, 1965; Faura *et al.*, 1969) concerns the observation that rodents and humans exposed to oxygen deficiency show increases in plasma Ep for only a finite length of time (e.g., 18–24 hours) following which the Ep levels decline despite continuation of the hypoxic stimulus. The explanation provided by some was that this decrease was due to an augmented utilization of Ep by the increased numbers of erythropoietic elements in the blood-forming organs. However, it has been demonstrated (Fried et al., 1970) that the lowered level of plasma Ep noted after 72 hours of hypoxia is better explained in terms of a decreased production of Ep than by an enhanced utilization. After an 8-hour period at room pressure, the ability of these rats to produce Ep in response to hypoxia again became evident.

The possibility has been investigated that plasma Ep levels are higher in animals with defective marrow than in those with normal or highly erythroid marrows. In this regard, the rate of disappearance of endogenously evoked Ep from the plasma has been reported to be slower for irradiated than for control rats (Stohlman, 1959). Similar results have been obtained for exogenous Ep in both irradiated conventional and germfree mice (Mirand et al., 1972); these experiments were performed during a period before the anemia resulting from X-irradiation had occurred. Moreover, the observations of McDonald et al. (1970) indicating that transplants of bone marrow cells suppressed the elevated levels of Ep noted in the plasma of mice on the tenth day after X-irradiation may be interpreted as supporting the concept that plasma Ep values are influenced by the functional state of the marrow cells. Along these lines, Stohlman (1959) had called attention to findings that the plasma levels of Ep in anemic subjects bear a relation not only to the severity of the hypoxia but also to the functional state of the erythroid marrow. Thus higher quantities of Ep are generally detectable in the plasma of patients with refractory anemia as compared to those with hemolytic anemia. This concept was extended by Hammond and Ishikawa (1962) who demonstrated that following transfusion there was a more rapid disappearance of Ep in the plasma of patients with erythroid hyperplasia (hemolytic anemia) than in those with erythroid hypoplasia (congenital hypoplastic anemia). On the other hand, equivalent rises in plasma Ep levels were noted in patients with active and aplastic bone marrows at similar peripheral hemoglobin levels (Van Dyke and Pollycove, 1962).

Possible evidence for the marrow utilization hypothesis is derived from experiments by LoBue et al. (1968) who reported that a significant decrease in exogenous Ep levels occurred in blood perfusates recirculated for 3.5 hours through isolated hind legs of rats. However, previous experiments by Fisher et al. (1965) did not reveal a significant decrease in Ep levels in blood circulated through isolated hind legs of rabbits for a 6-hour period. This discrepancy of results in LoBue's and Fisher's experiments may relate to the fact that in LoBue's perfusions, increases in the numbers of marrow erythroblasts in the legs were evident whereas in Fisher's studies no increase in these elements was observed. On the other hand, significant evidence against the concept of marrow utilization of Ep is derived from experiments performed by Naets and Wittek (1968). They studied the effect of experimentally induced hyperplasia (Naets and Wittek, 1968) or hypoplasia and aplasia (Naets and Wittek, 1970) of the erythroid marrow on the rate of disappearance of Ep from the plasma of dogs and rats. No significant differences were noted among the different groups suggesting that utilization of Ep is not influenced by differentiation of the stem cell compartment or by the differentiated erythroid cells themselves. Somewhat similar results had been obtained previously in dogs by Bozzini (1966).

It is thus evident that the question as to whether the blood-forming tissues are significantly involved in the inactivation or metabolism of Ep is not completely resolved. Certainly, additional studies are required to control variables in the experiments cited above. These include determination of whether Ep production and excretion as well as the volumes and compartments of distribution remain constant during the course of measuring the diminishing plasma levels of Ep. On *a priori* grounds it would be expected that blood-forming cells, in responding normally to Ep, should consume at least a portion of the Ep with which it interacts. Controlled *in vitro* experiments involving exposure of Ep to bone marrow and splenic blood-forming elements, in various stages of erythroid activity, may contribute to the solution of this elusive problem although as yet no reliable methods have been devised for determining the metabolism of Ep by blood-forming cells.

c. Renal Excretion. The normal human excretion of Ep into the urine varies between 0.9 and 4.0 units/day (Adamson *et al.*, 1966; Alexanian, 1966). That the Ep appearing in urine is most likely derived from plasma is seen from experiments in which anemic patients, following transfusion, exhibited proportional decreases of Ep in plasma and urine (Medici *et al.*, 1957; Rosse and Waldmann, 1964). The renal clearance of endogenously-produced Ep was calculated to be 0.06–0.67 ml/min and the total urinary excretion amounted to only 10% of the daily loss (Rosse and Waldmann, 1964). Similar findings have been reported for the excretion of exogenous Ep. Thus, Weintraub *et al.* (1964) found, following intravenous administration of sheep Ep, a 70% reduction in the plasma levels of the hormone during the initial 3.5 hours; only 2–5% of the Ep was recovered in the urine. Calculations indicated a renal clearance of 0.1–0.6 ml/minute. Thus it would appear that renally excreted hormone constitutes only a small fraction of the amount of either endogenously evoked or exogenously administered Ep. Problems concerning the role of the kidney in the excretion of Ep that require resolution relate to the possible reabsorption of Ep by the kidney, inactivation by inhibitors in renal tissue (Erslev and Kazal, 1968; Fisher *et al.*, 1968) and direct secretion of Ep (or Eg) from the renal epithelium into the urine.

III. GENERAL CONCLUSIONS

It is evident from the foregoing discussions that feedback mechanisms operate in the production and elimination of Ep. Hypoxia appears to provide the normal fundamental mechanism for stimulation of erythropoiesis through an increase in Ep production by the kidney and possible extrarenal sites. This action seems to be mediated in part through an increased production or activation of erythrogenin (Eg). Eg interacting with a plasma substrate, erythropoietinogen, in the kidney or the plasma or both results in the generation of Ep which induces accelerated production of RBC. The increase in RBC mass induced by Ep serves to inhibit production of Eg, as seen clearly in the plethorized animal. With a decrease in the availability of Eg, Ep production and erythropoiesis are slowed. The resumption of erythropoiesis is dependent on a sufficient decline in RBC mass to permit resynthesis of Eg.

In addition, it would seem that, upon attaining certain critical levels, Ep is also capable of limiting its own production, an effect that appears to be mediated through an inhibition of the production or availability of the plasma substrate. A decline in the level of Ep as well as a depletion of substrate, in turn, triggers events that lead to a compensatory rise in the circulating levels of substrate which, upon interacting with Eg, leads to reappearance of Ep. There is some preliminary evidence to suggest that the kidney (i.e., Eg) is required for the appearance of substrate in quantities above normal baseline values. A concept depicting these facets of Ep biogenesis and its feedback control is illustrated in Fig. 4.

Inactivation or metabolism of Ep constitutes another important aspect in the evaluation of the homeodynamic regulation of erythropoiesis. Although considerable evidence points to the liver and blood-forming tissues as sites of metabolism of Ep, complete accord in this area of investigation does not exist. Excretion through the kidney appears to be a minor pathway of Ep elimination. A possible informative approach to this problem might involve the labeling of Ep with an isotope of high

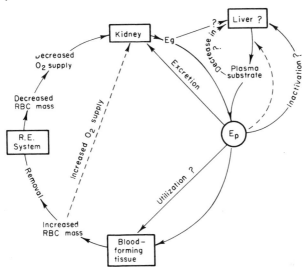

Fig. 4. Scheme of biogenesis of Ep and its feedback regulation. (Solid line indicates stimulation and dotted line, inhibition.) Removal and/or destruction of red blood cells (RBC), by decreasing blood oxygen content, stimulates production of renal Eg. This renal factor interacts with a substrate in plasma to yield the functional circulating Ep. This interaction may occur in the kidney, in the plasma, or both. Preliminary evidence suggests that Eg may stimulate substrate production. Utilization of substrate may also constitute a stimulus to production and/or release of substrate. The liver probably represents the chief source of the plasma substrate for Eg. Elevation in the quantity of circulating Ep of either exogenous or endogenous origin, serves to inhibit production of Ep; this appears to be accomplished through a reduction in the available levels of substrate. The increased RBC mass (increased oxygen supply) induced by Ep, through a negative feedback action on the kidney, reduces Eg production. Excess as well as old and effete RBC are removed by the R.E. system. Elimination through the kidney constitutes only a minor pathway for elimination of Ep. Other organs possibly concerned with elimination or inactivation of Ep include the liver and the blood-forming tissues.

specific activity and tracing chronologically the distribution of the labeled hormone in various body organs. Tissues in which the hormone appears quickly and then disappears might furnish clues on the identity and nature of the mechanism concerned with Ep metabolism.

REFERENCES

Adamson, J. W., Alexanian, R., Martinez, C. and Finch, C. A. (1966). *Blood* **28**, 354.
Alexanian, R. (1966). *Blood* **28**, 344.

Alpen, E. L. (1962). In "Erythropoiesis" (L. O. Jacobson and M. Doyle, eds.), pp. 134–135. Grune and Stratton, New York.

Bateman, A. E., and Cole, R. J. (1971). J. Embryol. Exp. Morph. 26, 475.

Borsook, H., Ratner, K., Tattrie, B., and Teigler, D. (1968). Nature (London) 217, 1024.

Bozzini, C. E. (1966). Nature (London) 209, 1140.

Brown, G. E., and Roth, G. M. (1922). Arch. Int. Med. 30, 817.

Burke, W. T., and Morse, B. S. (1962). In "Erythropoiesis" (L. O. Jacobson and M. Doyle, eds.), pp. 111–119. Grune and Stratton, New York.

Camiscoli, J. F., and Gordon, A. S. (1970). In "Regulation of Hematopoiesis" (A. S. Gordon, ed.), pp. 369–393. Appleton, New York.

Chui, D., Djaldetti, M., Marks, P., and Rifkind, R. (1971). J. Cell Biol., 21, 623.

Cole, R. J., and Paul, J. (1966). J. Embryol. Exp. Morph. 15, 245.

Cole, R. J., Hunter, J., and Paul, J. (1968). Brit. J. Haematol. 14, 477.

Cooper, W. M., and Tuttle, W. B. (1957). Ann. Int. Med. 47, 1008.

Cotes, P. M. (1971). In "Kidney Hormones" (J. W. Fisher, ed.), pp. 243–267. Academic Press, New York.

Crafts, R. C. (1941). Endocrinology 29, 596.

Crafts, R. C., and Meineke, H. A. (1959). Ann. N.Y. Acad. Sci. 77, 501.

Dukes, P. P., Takaku, F., and Goldwasser, E. (1963). Endocrinology 74, 960.

Dukes, P. P., Hammond, D., Shore, N. A., and Ortega, J. A. (1970). J. Lab. Clin. Med. 76, 439.

Erslev, A. J., and Kazal, L. A. (1968). Proc. Soc. Exp. Biol. Med. 128, 845.

Espada, J., and Gutnisky, A. (1970). Acta Physiol. Latinoamer. 20, 122.

Evans, E. S., Rosenberg, L. L., and Simpson, M. E. (1961). Endocrinology 68, 517.

Faura, J., Ramos, J., Reynafarje, C., English, E., Finne, P., and Finch, C. A. (1969). Blood 33, 668.

Feigin, W. M., and Gordon, A. S. (1950). Endocrinology 47, 364.

Feleppa, A. E., Jr., Meineke, H. A., and McCuskey, R. S. (1971). Scand. J. Haematol. 8, 86.

Finne, P. H. (1965). Scand. J. Clin. Lab. Invest. 17, 135.

Fischer, S., and Roheim, P. S. (1963). Nature (London) 200, 899.

Fisher, J. W. (1969). In "The Biological Basis of Medicine" (E. E. Bittar and N. Bittar, eds.), pp. 41–79. Academic Press, New York.

Fisher, J. W., and Roh, B. L. (1970). Blood 36, 847 (abstract).

Fisher, J. W., and Roh, B. L. (1971). In "Renal Pharmacology" (J. W. Fisher, ed.), pp. 167–196. Appleton, New York.

Fisher, J. W., Lajtha, L. G., Buttoo, A. S., and Porteous, D. D. (1965). Brit. J. Haematol. 11, 342.

Fisher, J. W., Hatch, F. E., Roh, B. L., Allen, R. C., and Kelley, B. J. (1968). Blood 31, 440.

Forssell, J. (1954). Acta Med. Scand. 150, 155.

Fox, H. M., Gilchrist, B. M., and Phear, E. A. (1951). Proc. Roy. Soc. B. 138, 514.

Fried, W., Johnson, C., and Heller, P. (1970). Blood 36, 607.

Gallien-Lartigue, O., and Goldwasser, E. (1965). Biochim. Biophys. Acta 103, 319.

Goldwasser, E., and Kung, C. K-H. (1968). Ann. N.Y. Acad. Sci. 149, 49.

Goldwasser, E., and Kung, C. K.-H. (1971a). Proc. Nat. Acad. Sci. 68, 697.

Goldwasser, E., and Kung, C. K.-H. (1971b). Fed. Proc. 30, 1128, (abstract).

Gordon, A. S. (1954). *Rec. Progr. Horm. Res.* 10, 339.

Gordon, A. S. (1957). *Amer. J. Clin. Nutr.* 5, 461.

Gordon, A. S. (1959). *Physiol. Rev.* 39, 1.

Gordon, A. S. (1968). *Proc. Congr. Int. Soc. Hematol., 12th* pp. 288–303.

Gordon, A. S., and Zanjani, E. D. (1970a). *In* "Regulation of Hematopoiesis" (A. S. Gordon, ed.), pp. 413–457. Appleton, New York.

Gordon, A. S., and Zanjani, E. D. (1970b). *In* "Formation and Destruction of Blood Cells" (T. J. Greenwalt and G. A. Jamieson, eds.), pp. 34–64. Lippincott, Philadelphia, Pennsylvania.

Gordon, A. S., LoBue, J., Dornfest, B. S., and Cooper, G. W. (1962). *In* "Erythropoiesis" (L. O. Jacobson and M. Doyle, eds.), pp. 321–327. Grune and Stratton, New York.

Gordon, A. S., Katz, R., Zanjani, E. D., and Mirand, E. A. (1966). *Proc. Soc. Exp. Biol. Med.* 123, 475.

Gordon, A. S., Zanjani, E. D., and Cooper, G. W. (1967). *Seminars Hematol.* 4, 337.

Gordon, A. S., Mirand, E. A., Wenig, J., Katz, R., and Zanjani, E. D. (1968). *Ann. N.Y. Acad. Sci.* 149, 318.

Gordon, A. S., Zanjani, E. D., and Zalusky, R. (1970). *Blood* 35, 151.

Grant, W. C., and Root, W. S. (1947). *Amer. J. Physiol.* 150, 618.

Grant, W. C., and Root, W. S. (1952). *Physiol. Rev.* 32, 449.

Gross, M., and Goldwasser, E. (1969). *Biochemistry* 8, 1795.

Guidi, E. E., and Scaro, J. L. (1970). *Rev. Espan. Fisiol.* 26, 151.

Hammond, D., and Ishikawa, A. (1962). *In* "Erythropoiesis" (L. O. Jacobson and M. Doyle, eds.), pp. 128–133. Grune and Stratton, New York.

Hildemann, W. H., and Keighley, G. (1955). *Amer. Natur.* 89, 169–174.

Hrinda, M. E., and Goldwasser, E. (1969). *Biochim. Biophys. Acta* 195, 165.

Jacobsen, E. M., Davis, A. K., and Alpen, E. L. (1956). *Blood* 11, 937.

Jacobson, L. O., and Goldwasser, E. (1957). *In* "Homeostatic Mechanisms" (*Brookhaven Symposia in Biol.* No. 10), pp. 110–131.

Jacobson, L. O., Goldwasser, E., Gurney, C. W., Fried, W., and Plzak, L. (1959). *Ann. N.Y. Acad. Sci.* 77, 551.

Keighley, G. (1962). *In* "Erythropoiesis" (L. O. Jacobson and M. Doyle, eds.), pp. 106–110. Grune and Stratton, New York.

Krantz, S. B., and Jacobson, L. O. (1970). Erythropoietin and the Regulation of Erythropoiesis. Univ. of Chicago, Chicago, Illinois.

Krantz, S. B., Gallien-Lartigue, L., and Goldwasser, E. (1963). *J. Biol. Chem.* 238, 4085.

Krzymowski, T., and Krzymowska, H. (1962). *Blood* 19, 38.

Kuratowska, Z. (1968). *Ann. N.Y. Acad. Sci.* 149, 128.

Kuratowska, Z., Lewartowski, B., and Lipinski, B. (1964). *J. Lab. Clin. Med.* 64, 226.

LaBella, F. S., Reiffenstein, R. J., and Beaulieu, G. (1963). *Arch. Biochem. Biophys.* 100, 399.

Lange, R. D., McDonald, T. P., and Jordan, T. (1969). *J. Lab. Clin. Med.* 73, 78.

Levere, R. D., and Granick, S. (1967). *J. Biol. Chem.* 242, 1903.

LoBue, J., Monette, F. C., Camiscoli, J. F., Gordon, A. S., and Chan, P-C. (1968). *Ann. N.Y. Acad. Sci.* 149, 257.

Loge, J. P., Lange, R. D., and Moore, C. V. (1950). *J. Clin. Invest.* 29, 830.

Lowy, P. H. (1970). In "Regulation of Hematopoiesis" (A. S. Gordon, ed.), pp. 395–412. Appleton, New York.
Lowy, P. H., Keighley, G., Borsook, H., and Graybiel, A. (1959). Blood 14, 262.
McClugage, S. G., McCuskey, R. S., and Meineke, H. A. (1971). Blood 38, 96.
McCuskey, R. S., Meineke, H. A., Townsend, S. F., and Kaplan, S. M. (1971). Blood 38, 821 (abstract).
McDonald, T. P., Lange, R. D., Congdon, C. C., and Toya, R. E. (1970). Radiat. Res. 42, 151.
McDonald, T. P., Zanjani, E. D., Lange, R. D., and Gordon, A. S. (1971). Brit. J. Haematol. 20, 113.
Medici, P. T., Gordon, A. S., Piliero, S. J., Luhby, A. L., and Yuceoglu, P. (1957). Acta Haematol. 18, 325.
Meyer, O. O., Thewlis, E. W., and Rusch, H. P. (1940). Endocrinology 27, 932.
Mirand, E. A., and Murphy, G. P. (1971). N.Y. State J. Med. 71, 860.
Mirand, E. A., Prentice, T. C., and Slaunwhite, W. R. (1959). Ann. N.Y. Acad. Sci. 77, 677.
Mirand, E. A., Weintraub, A. H., Gordon, A. S., Prentice, T. C., and Grace, J. T., Jr. (1965). Proc. Soc. Exp. Biol. Med. 118, 823.
Mirand, E. A., Gordon, A. S., Zanjani, E. D., Bennett, T. E., and Murphy, G. P. (1972). Proc. Soc. Exp. Biol. Med. 139, 161.
Miyakawa, M., Uno, Y., and Asai, J. (1965). In "The Reticuloendothelial System, Morphology, Immunology and Regulation," p. 132. Nissha, Kyoto, Japan.
Naets, J. P., and Wittek, M. (1968). Acta Haematol. 39, 42.
Naets, J. P., and Wittek, M. (1970). Abstr. Vol. 13th Int. Congr. of Hematol. (Munich), p. 12. J. F. Lehmanns Verlag, München.
Necheles, T. F., and Rai, U. S. (1969). Blood 34, 380.
Necheles, T. F., Sheehan, R. G., and Meyer, H. J. (1968). Ann. N.Y. Acad. Sci. 149, 449.
Paul, J., and Hunter, J. A. (1969). J. Mol. Biol. 42, 31.
Piliero, S. J. (1959). Ann. N.Y. Acad. Sci. 77, 518.
Powsner, E., and Berman, L. (1967). Life Sci. 6, 1713.
Prentice, T. C., and Mirand, E. A. (1957). Proc. Soc. Exp. Biol. Med. 95, 231.
Reissmann, K. R., and Samorapoompichit, S. (1970). Blood 36, 287.
Rosse, W. F., and Waldmann, T. A. (1964). J. Clin. Invest. 43, 1348.
Schooley, J. C., and Garcia, J. F. (1962). Proc. Soc. Exp. Biol. Med. 109, 325.
Schooley, J. C., Zanjani, E. D., and Gordon, A. S. (1970). Blood 35, 276.
Stephenson, J. R., and Axelrad, A. A. (1971). Endocrinology 88, 1519.
Stohlman, F., Jr. (1959). Ann. N.Y. Acad. Sci. 77, 710.
Thorbecke, G. J., and Benacerraf, B. (1959). Ann. N.Y. Acad. Sci. 78, 247.
Thorling, E. B., and Erslev, A. J. (1968). Blood 31, 332.
Thurmon, T. F., Boyer, S. H., Crosby, E. F., Shepard, M. K., Noyes, A. N., and Stohlman, F., Jr. (1970). Blood 36, 598.
Trentin, J. J. (1970). In "Regulation of Hematopoiesis" (A. S. Gordon, ed.), pp. 159–186. Appleton, New York.
Van Dyke, D. C., and Pollycove, M. (1962). In "Erythropoiesis" (L. O. Jacobson and M. Doyle, eds.), pp. 340–350. Grune and Stratton, New York.
Van Dyke, D. C., Nohr, L. M., and Lawrence, J. H. (1966). Blood 28, 535.
Vollmer, E. P., Gordon, A. S., Levenstein, I., and Charipper, H. A. (1939). Endocrinology 25, 970.

Vollmer, E. P., Gordon, A. S., and Charipper, H. A. (1942). *Endocrinology* **31**, 619.

Weintraub, A. H., Gordon, A. S., Becker, E. L., Camiscoli, J. F., and Contrera, J. F. (1964). *Amer. J. Physiol.* **207**, 523.

Whitcomb, W. H., and Moore, M. Z. (1965). *J. Lab. Clin. Med.* **66**, 641.

Zanjani, E. D., Contrera, J. F., Cooper, G. W., Wong, K. K., and Gordon, A. S. (1967). *Proc. Soc. Exp. Biol. Med.* **125**, 505.

Zanjani, E. D., Gordon, A. S., Wong, K. K., and McLaurin, W. D. (1968). *Life Sci.* **7**, 1233.

Zanjani, E. D., McLaurin, W. D., Gordon, A. S., Rappaport, I. A., Gibbs, J., and Gidari, A. S. (1971). *J. Lab. Clin. Med.* **77**, 751.

3

HUMORAL REGULATION OF NEUTROPHIL PRODUCTION
AND RELEASE

*Edward F. Schultz, David M. Lapin, and Joseph LoBue**

I. HISTORICAL BACKGROUND

Increasing evidence indicates that the maintenance of adequate numbers of the various blood cells may be regulated by specific humoral

* Original work reported in this chapter was supported by U.S. Public Health
Service Grants 5-RO1-HL03357-15 and 1-RO1-CA12815-01.

substances. Thus, depletion of a particular blood cell type has been postulated to stimulate the formation of a plasma-borne humor concerned with its regulation. The specific controlling agent, by acting either directly or indirectly upon the blood-forming tissues, would then facilitate production and release of the particular cellular element in question. The subsequent return of the cell type to within normal limits might then serve as a negative feedback on the site(s) of production of the humor leading to decreased plasma levels of the agent.

It has been conclusively demonstrated that erythropoiesis is, in part, controlled by the erythropoiesis stimulating factor, erythropoietin (Gordon, 1959; Jacobson and Doyle, 1963). Furthermore, the mechanism by which the kidney participates in the production of this factor in response to hypoxia has been elucidated (Gordon et al., 1967, 1973). Similarly, thrombopoiesis appears to be regulated by a plasma-borne "thrombopoietin" that is elicited by a reduction in the numbers of circulating platelets or their increased rate of destruction (Abildgaard and Simone, 1967; Ebbe, 1970; Cooper, 1970; Odell, 1973). Evidence further suggests that lymphocytopoiesis may be regulated by principles exerting their effects on lymphoid tissues. Thus, the demonstration of the presence of: a thymic-lymphocytosis promoting factor in mice (Metcalf, 1958); a humoral lymphocyte stimulating factor in the serum of irradiated rats (Ito and Weinstein, 1963); a thymic lymphopoietic factor, "thymosin" (Goldstein et al., 1966); and a lymphocytosis inducing factor in the plasma of rats treated with antilymphocyte serum (Rakowitz et al., 1972) lend support to this hypothesis. Several other reports in the literature suggest the presence of humoral agents regulating leukocyte numbers. Hence, a leukocytosis inducing factor (LIF) has been demonstrated in the plasma of rats subjected to repeated leukocyte withdrawal (leukocytopheresis) (Gordon et al., 1959, 1960a, b, 1964; Katz et al., 1966; Lapin et al., 1969). Menkin (1946, 1955) has reported a leukocytosis-promoting factor that not only caused the release of both mature and immature medullary granulocytes, but also produced marrow granulocytic hyperplasia. Similarly, Steinberg and Martin (1950), and Steinberg et al. (1959, 1965) found leukopoietic activity in the albumin fraction of normal human serum that resulted in the expulsion of marrow granulocytes into the blood. In addition, the extraction of a factor termed "leukopoietin G" from the plasma of leukocytopheresed humans and from various bovine organs has been reported by Bierman and associates (1962, 1964). Reports have also appeared concerning the presence of a granulocytosis-promoting factor extractable from tumor tissue (Delmonte and Liebelt, 1965; Delmonte et al. 1966) and kidneys (Delmonte et al. 1968) and a myelopoiesis stimulating agent from rabbit sera

(Gidali and Feher, 1964; Feher and Gidali, 1965). A leukocytosis inducing substance termed leucogenenal has also been isolated from metabolic products of *Penicillium gilmanii* (Rice, 1966) and from bovine and human liver (Rice and Shaikh, 1970). Similarly, Rytomaa and Kiviniemi (1968a, b) indicate the existence of a specific granulocyte inhibitor (granulocytic chalone) and a stimulatory antichalone in rats. Moreover, the presence of a diffusible granulocytopoiesis stimulating agent in irradiated mice (Rothstein *et al.*, 1971), the appearance of a leukocytosis inducing factor in the plasma of rats injected with typhoid-paratyphoid vaccine (Gordon *et al.*, 1964; Handler *et al.*, 1966), neutrophilia inducing activity in the plasma of dogs recovering from drug-induced myelotoxicity (Boggs *et al.*, 1968a) and man (Boggs *et al.*, 1968b) lend further support to the concept of humoral regulation of leukocyte numbers. Additionally the relatively constant total white blood cell number in normal individuals and animals; the predictable blood cell reactions following the administration of steroids, endotoxin, drugs (e.g., nitrogen mustard, vinblastine sulfate, cytoxan), irradiation, leukocytopheresis, or induction of infection; and the recovery and return of the hematopoietic tissue to a normal steady state following perturbation, all suggest the existence of regulatory mechanisms.

Hematopoietic regulatory mechanisms could involve the control of stem cell pool size, recruitment of precursor cells from the stem cell pool, the proliferation and maturation of cells in the mitotic compartment of the marrow, the release of mature cells from marrow reserves, the distribution of granulocytes between the circulating and marginated pools in the vascular system, and the migration of the mature cells from the vascular system into the tissues. Factors such as adrenal steroids, endotoxin, vasoactive agents, etc., that affect numerous physiological parameters and subsequently alter the rate of utilization, removal, and destruction of blood leukocytes must be considered as secondary regulators of blood homeostasis (Craddock *et al.*, 1959). In order for a neutrophilic factor to be considered as a primary regulator it should meet the following criteria (Dornfest *et al.*, 1962b): (a) the agent cannot exert any nonspecific effect that leads to an inflammatory reaction or change in peripheral utilization and destruction, as these events themselves will increase or decrease hemic cell production; (b) the humoral principle should produce morphological changes in the hematopoietic organ, preferably involving one cell type, as well as peripheral blood changes that are characterized by a sustained leukopenia or leukocytosis; and (c) the factor must act directly on the blood-forming or releasing tissues and not through changes in the body's endocrine and metabolic balance.

II. STEM CELLS

The various lines of evidence that indicate that a pluripotential stem cell serves as a common progenitor for megakaryocytic, myeloid, and erythroid cells include: the presence of the Ph_1 chromosome in marrow megakaryocytes, myeloid cells, and nucleated erythroid elements of patients with chronic myelocytic leukemia, suggesting the chromosome marker was derived from a precursor common to all of these marrow elements (Whang et al., 1963); splenic colonization studies in lethally irradiated recipient mice (Becker et al., 1963), and the observation of granulocytic, megakaryocytic, and erythroid elements in a single colony (Till and McCullock, 1961). Recently evidence has been presented indicating that lymphoid cells may be derived from colony forming units or that both cell types have a common progenitor cell (Wu et al., 1968).

Experiments in mice have shown that stem cells can exist in a transplantable, pluripotential state (colony forming unit, CFU) (Till et al., 1964) with the majority of this population postulated to be in Go (Becker et al., 1965), or they can give rise to secondary stem cell populations. These secondary populations are believed to be responsive to differentiating stimuli (Till et al., 1964) and include, among others, erythropoietin-sensitive cells (Bruce and McCullock, 1964; Lajtha, 1966; Hellman and Grate, 1968; Schooley, 1968; Hellman et al., 1969; Iscove et al., 1970; Lajtha et al., 1971; Stephenson and Axelrad, 1971) and granulocytic stem cells (Pluznik and Sachs, 1965; Bradley and Metcalf, 1966; Bennet et al., 1968; Worton et al., 1969; Chen and Schooley, 1970; Sutherland et al., 1970; Iscove et al., 1970; Morley et al., 1971; Shadduck and Nagabhushanam, 1971).

Recent experiments (Fried and Gurney, 1968; Fried et al., 1971) indicate that the proliferation of hematopoietic precursor cells may be regulated by a humoral agent. This substance is released subsequent to events leading to a reduction in stem cell numbers. Moreover, the proliferative response of hematopoietic precursor cells exposed to this stimulatory principle may, in part, be controlled by the numbers of cells present in the blood cell-forming tissues (Fried et al., 1971).

III. PROLIFERATION AND MATURATION OF NEUTROPHILS

Granulocytic-stem cells that give rise to colonies in agar appear to be rapidly proliferating elements (Hodgson, 1967; Lajtha et al., 1969; Rickard et al., 1970; Iscove et al., 1970). In addition, the numbers of these cells committed to granulopoiesis is reduced under circumstances in which there is an increased demand for erythropoiesis (Hellman and Grate, 1968; Lawrence and Craddock, 1968; Hellman *et al.*, 1969; Rickard *et al.*, 1971). Thus, hematopoietic precursor cell populations committed to granulopoiesis appear to fluctuate in accordance with the demands for other formed elements of the blood. Their survival and subsequent differentiation into recognizable myeloid elements may depend on the availability of essential nutrients (vitamins, minerals, etc.), a suitable cell population density within the blood-forming tissue, and the influence of specific regulatory or microenvironmental factors (Trentin, 1971).

The mechanism whereby leukocytes participate in the regulation of granulopoiesis remains unclear. Some reports have suggested that incubated marrow or peripheral leukocytes release granulocytic stimulators (Bierman and Hood, 1969; Chervenick and Boggs, 1970; Robinson and Pike, 1970; Iscove *et al.*, 1970) while others indicate that leukocytes may inactivate stimulatory factors (Feher and Gidali, 1965) or inhibitors of granulocytopoiesis (Paran *et al.*, 1969; Metcalf, 1970). In this regard, it has been recently demonstrated that the *in vivo* destruction of mature granulocytes results in the release of "colony stimulating factor" CSF (Shadduck and Nagabhushanam, 1971). CSF may act directly on the marrow and enhance granulocytopoiesis, provided that conditions permit expansion of the medullary myeloid compartment in response to this stimulus (Fried *et al.*, 1971). Another relevant finding is that normal marrow cells from mice are capable of adsorbing or inactivating granulocytopoiesis-stimulating agents. Thus, the recruitment of cells from the "granulocytopoietin" (CSF?) sensitive stem cell compartment into recognizable myeloid precursors may depend on the number of cells in this compartment that are capable of responding to a humoral principle, and the time of exposure of the "sensitive" cells to the regulatory substance. The latter could be altered by the rate of marrow cell adsorption or metabolism of the CSF. In addition, the appearance of CSF in normal urine (Stanley and Metcalf, 1969) suggests that the kidney may clear such stimulatory principles from the plasma and, hence, also act as a regulator of the time of exposure of hematopoietic precursor cells to stimulants of granulocytopoiesis.

Successful and sufficient hematopoietic proliferation is also related to marrow structure. The integrity of the stroma of the red marrow, its ability to repair and maintain itself, and the proper vascularity of this gelatinous framework are all essential prerequisites of hemic cell production (Tavassoli and Crosby, 1970; Maniatis *et al.*, 1971).

To summarize, normal marrow function depends upon (1) the

presence of hemic precursor cells of a pluripotential (CFU) and a uni-
potential nature (agar colony forming cell or erythropoietin sensitive
cell); (2) regulatory principles; (3) proper vascularity and availability
of essential nutrients; (4) optimal cell density; and (5) a functional
stroma.

IV. CELLULAR AND HUMORAL REGULATION OF
NEUTROPHIL RELEASE

One problem encountered in the study of granulocyte production and
release is the determination of the mechanism(s) by which plasma-borne
factors effect a change in the numbers of circulating neutrophils. In
any study of such agents it is always important to ascertain whether
test plasmas and their contained "factor(s)" are altering the equilibrium
between the circulating and "marginated" granulocyte pools in the blood,
accelerating the rate of neutrophil release from bone marrow reserves,
or possibly doing both (Boggs, 1967). In addition, the properties exhibited by any proposed regulatory principle should be distinguished
from endotoxin or other vasoactive agents that may be present in the
test material.

A. Perfusion Experiments in Rats

Dornfest et al. (1962a, b) and Gordon et al. (1964) have demonstrated
in rat hindlimb and femur perfusion experiments that the rate of leukocyte release from the marrow reserves was influenced by the blood flow
rate through the marrow and by the numbers of leukocytes present
in the blood. Thus, an increase in blood flow through the marrow led
to an increased release of leukocytes from the marrow reserves. Similarly,
leukocyte depleted blood enhanced leukocyte release in contrast to a
much lower release rate observed when blood with normal levels of
leukocytes ($7\text{--}15 \times 10^3/\text{mm}^3$) was used as perfusate. These investigators
also demonstrated that a plasma-borne leukocytosis-inducing factor
(LIF) obtained from rats subjected to repeated leukocyte withdrawal
(leukocytapheresis, LAP) not only enhanced the rate of blood flow
through the perfused marrow, but also increased the rate of granulocyte
discharge. In addition, tibial marrows obtained from intact recipient
rats injected with the LIF-active plasma contained reduced numbers
of granulocytes and an increased concentration of lymphocytes (Gordon
et al., 1964). These results are in agreement with similar findings in
the perfusion experiments. Moreover, the characteristic changes produced by the action of the LIF on the perfused marrows could be

distinguished from effects produced by endotoxin and various vasoactive agents (Dornfest *et al.*, 1962a, b; Gordon *et al.*, 1964). Thus, these experiments provide good evidence for the marrow leukocyte releasing properties of the LIF.

B. LIF Assay Experiments in Intact Recipient Rats

Many experiments have established the existence of the LIF in the plasma of rats subjected to repeated leukocyte withdrawal (Gordon *et al.*, 1959, 1960a, b, 1964; Katz *et al.*, 1966; Lapin *et al.*, 1969). This factor is obtained in the plasma of the leukocytapheresed donor rats after removal of at least 1 billion leukocytes from the peritoneal cavities of these animals. Administration of LIF-containing plasma to intact rats evokes increased release of mononuclear cells and medullary granulocytes into the peripheral circulation of the recipients. Verification of these results have come from bone marrow studies performed by Gordon *et al.* (1964) and the hindlimb and femur perfusion experiments reported by Dornfest *et al.* (1962a, b) mentioned earlier. Recently, an *in vivo* assay system employing tritiated thymidine ([³H]Tdr) prelabeling of granulocytes was developed to reliably measure the marrow-granulocyte mobilizing activity of LIF-containing plasma (Schultz, 1970). Twenty-four hours after [³H]Tdr administration (1 μCi/gm body weight) labeled neutrophilic granulocytes were observed in the blood of normal recipient rats (150 gm male Long-Evans animals), at which time 1.2–3.7% of the cells were labeled. An increase in the percentage of labeled blood neutrophils was observed at 48 hours after isotope injection (10.3%), and this continued to rise through 72 hours ($\cong$ 22%). On the other hand, the numbers of [³H]Tdr labeled neutrophils/mg of bone marrow was found to be low at 24 hours after isotope administration (4413 $\pm$ 1322 cells/mg), significantly increased by 48 hours (43,633 $\pm$ 5175 cells/mg), and only slightly above these values at 72 hours after [³H]Tdr injection (49,950 $\pm$ 8723 cells/mg). Since the numbers of ³H-containing neutrophils in the marrow appeared to peak at 48 hours and remained there through 72 hours while the numbers of labeled blood neutrophils stayed relatively low at 48 hours only increasing thereafter, it was decided that assay recipient rats should be treated with [³H]Tdr 48 hours prior to their use in an assay for plasma-borne LIF activity. Plasmas tested in this assay system were derived from normal rats, rats subjected to repeated endotoxin injection without the removal of mobilized leukocytes from the peritoneal cavity (sham-LAP rats), and rats subjected to repeated endotoxin injection and peritoneal lavage to remove the leukocytes contained therein (LAP rats). Only LAP-plasma produced significant increases in the numbers of labeled

58   E. F. Schultz, D. M. Lapin, and J. LoBue

neutrophils/mm³ and the neutrophil labeling index in the blood of [³H]Tdr prelabeled assay recipient rats 6 hours after administration of the test plasma. In addition, significantly higher than normal values for total lymphocytes, labeled lymphocytes, and lymphocyte labeling indices were observed in LAP plasma recipients 48 hours after the start of the assay. This appears to provide additional support for the existence of a plasma-borne LIF as originally postulated by Gordon et al. (1960a, b; 1964) and verifies the marrow granulocyte-mobilizing activity manifested by this humoral principle (Dornfest et al., 1962a, b; Gordon et al., 1964).

Recent experiments (Rakowitz et al., 1972) have also demonstrated the presence of the LIF in the plasma of rats rendered leukopenic by treatment with antilymphocyte serum. Administration of this LIF to assay recipient rats produced increases in the total leukocyte counts, total neutrophils/mm³ and total lymphocytes/mm³ in the blood at 4–6 hours after the start of the assay. These experiments provide further evidence supporting a role for the LIF in the regulation of blood leukocyte levels.

C. Experiments in Dogs

Neutrophilia-inducing activity has been found in the plasma of neutropenic dogs recovering from drug-induced myelotoxicity (Boggs et al., 1966). Infusion of this principle into assay recipient dogs demonstrated that the factor specifically increases the total white blood cell counts and numbers of neutrophils/mm³; that it produces its peripheral blood effects by enhancing the rate of neutrophils released from the marrow reserves, and that it may be distinguished from the effects produced by endotoxin, epinephrine, and cortisone. Normal dog plasma could not reproduce effects of the active plasma when infused into control recipient animals. Thus, the neutrophilia inducing activity appears to be equivalent to the leukocytosis inducing factor in that both are evoked when the demand for peripheral blood neutrophils is great; both enhance granulocyte release from the marrow reserves and, subsequently, produce a leukocytosis and a neutrophilia; and, both factors can be distinguished from endotoxin and other vasoactive agents in the respective assay systems.

D. Evidence from Endotoxin Administration in Animals and Man

The demonstration of a leucocytosis-inducing factor in the plasma of rabbits (Fukuda et al., 1960) and rats (Gordon et al., 1964; Handler et al., 1966) following injections of typhoid-paratyphoid vaccine, and neutrophil releasing activity in plasma following injection of endotoxin in dogs (Boggs et al., 1968a) and man (Boggs et al., 1968b) lend further support to the concept of humoral regulation of leukocyte release. These investigations have also demonstrated conclusively that the effects produced by the active principles were not due to vaccine pyrogens. Moreover, Gordon et al. (1964) and Boggs et al. (1968a) showed that plasma activity could not be attributed to increased cortisol levels.

Finally, it is interesting to note that neutrophilia-inducing activity is detectable in the plasma of humans recovering from drug-induced neutropenia (Marsh and Levitt, 1971). This provides additional supporting evidence for the humoral regulation concept.

E. Tissue Extracts

Menkin (1946, 1955) originally reported the isolation of a leukocytosis promoting factor from inflammatory exudates and blood that not only caused the release of mature and immature medullary granulocytes, but also produced marrow granulocytic hyperplasia. Similarly, Steinberg and Martin (1950) and Steinberg et al. (1959, 1965) found leukopoietic activity in the albumin fraction of normal human serum that caused an expulsion of marrow granulocytes into the blood. In addition, the extraction of a factor termed "leucopoietin G" from the plasma of leuko-cytopheresed humans and from various bovine organs has been reported by Bierman et al. (1962) and Bierman (1964). Similarly, reports have appeared concerning the presence of a granulocytosis-promoting factor extractable from tumor tissue (Delmonte and Liebelt, 1965; Delmonte et al., 1966), and kidneys (Delmonte et al., 1968). Reifenstein and associates (1941) observed the presence of leukocytosis-inducing substance obtained from rabbit peritoneal exudates. Likewise, the intravenous injection of starch in rabbits resulted in the production of a myelopoiesis-stimulating agent that was detectable in serum (Gidali and Feher, 1964; Feher and Gidali, 1965). In addition, a substance termed leucogenenol has been isolated from human and bovine liver and from metabolic products of Penicillium gilmanii (Rice et al., 1968, 1971; Rice, 1968; Rice and Darden, 1968) which induces a leukocytosis as well as an increase in the number of marrow myeloid elements in assay recipients. Once again, such observations support the concept of humoral regulation of leukocyte numbers.

F. Steroids

Recent investigations have indicated that two steroid substances, etio-cholanolone and cortisone, enhance the release of granulocytes from

the marrow reserves and produce a peripheral granulocytosis. Etio-
cholanolone was found to produce a granulocytosis via an acceleration
in marrow release that was proportional to the size of the marrow reserves (Godwin *et al.*, 1968a, b). On the other hand, cortisone not only
affected the rate of marrow granulocyte release, but also the rate of
egress of neutrophils from the total blood granulocyte pool (Bishop
et al., 1968).

V. REGULATION WITHIN THE CIRCULATION

It has been clearly established (Cartwright *et al.*, 1964; Boggs *et
al.*, 1965, 1966; Boggs, 1967; Bishop *et al.*, 1968) that the blood
neutrophils are approximately equally divided in numbers between a
circulating pool and a marginated pool. Moreover, the sum of these
two pools, the total blood granulocyte pool, may be considered as a
single kinetic unit due to the rapidity with which cell exchange between
the two occurs. Changes in blood granulocyte concentration may come
about by alterations in the rate of neutrophil release from the marrow,
changes in circulating and marginated pool equilibria, variations in the
rate of cell loss from the blood, or any combination of these factors.
For a thorough review of the possible mechanisms affecting neutropenia
and neutrophilia the reader is referred to the excellent review of Boggs
(1967).

To date, there is no evidence implicating humoral control of the equilibria existing between the circulating and marginated granulocyte pools
of the blood. However, neutrophil adherance qualities, vascular diameter, blood flow rates, or possibly combinations of these as well as other
factors may affect the size of the marginated pools.

It is interesting that clinical studies suggest that marginal pools of
granulocytes may be maintained in preference to the circulating elements
(Boggs, 1967). Since the loss of neutrophils from the blood into the
tissues must occur from the marginated granulocyte pools, the mecha-
nism(s) relating to the maintenance of this pool within the circulation
is of considerable importance.

VI. REGULATORY EVENTS IN AN INFLAMMATORY REACTION

Neutrophils leave the blood by diapedesis (Allison *et al.*, 1955) and
migrate toward sites of inflammation by positive chemotaxis (Keller
and Sorkin, 1968). Increasing evidence indicates that the complement

system plays an important role in the regulation of this leukocyte migration process. Thus, tissue injury (Hill and Ward, 1969; Ward, 1971) apparently leads to the generation of chemotactic substances via the sequential interaction of the first seven of the nine components of complement (C′). The chemotactic substances consist of (1) an activated trimolecular complex of the fifth (C′5), sixth (C′6), and seventh (C′7) complement components designated as C′(5,6,7)a; (2) the plasmin-split fragment of the third component (C′3); and (3) a C′5 fragment (Ward and Becker, 1968; Ward, 1969, 1971). Subsequently, interaction of C′(5,6,7)a with neutrophils leads to the activation of esterases within these cells and the activated enzyme, in turn, presumably participates in the events that guide neutrophil movement (Becker, 1969; Ward, 1971). Although the chemotactic factors generated primarily affect the influx of neutrophilic and eosinophilic granulocytes into an inflammatory exudate, the presence of granulocytes also seems to be related to the subsequent appearance of mononuclear cells (Ward, 1968).

Activation of the complement system may lead to the release of vasoactive amines from platelets and initiation of the coagulation mechanism (Becker, 1969; Henson, 1969). In turn, these events would increase vascular permeability and lead to an intensification of the inflammatory reaction (Barnhart, 1968). Likewise, the phagocytic activity of granulocytes within inflammatory regions may result in further activation of the complement system (Ward, 1971), the kallikrein system (Webster, 1968) and the kinins (Melmon and Cline, 1967; Kellermeyer and Graham, 1968). Thus, additional granulocytes will be drawn into the inflammatory site. Hence the complement system appears to play a substantial role in the complex series of events involving the chemotaxis of neutrophils and the progress of an inflammatory reaction.

VII. POSSIBLE MECHANISMS REGULATING NEUTROPHIL PRODUCTION AND RELEASE

Considering the information discussed in previous papers dealing with regulatory mechanisms (Craddock *et al.*, 1959; Gordon *et al.*, 1964; Boggs, 1966; King-Smith and Morley, 1970) and in the previous sections of this chapter, it is possible to speculate on the mechanisms involved in the regulation of the numbers of circulating neutrophilic granulocytes (Fig. 1). Any inflammatory reaction resulting from tissue damage or infection may lead to the following sequence of events: (1) activation of the complement system with the subsequent production of chemotactic substances; (2) the release of vasoactive amines from platelets

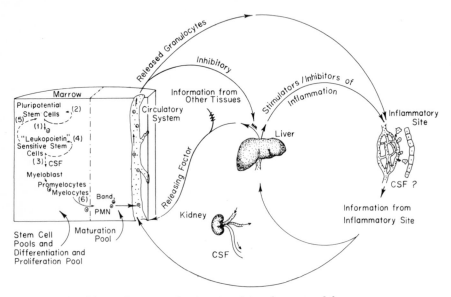

Fig. 1. Possible regulatory mechanisms involving the neutrophil.

Pluripotential stem cells in the blood-forming tissues have the capacity to produce "leukopoietin" sensitive stem cells (1) and more pluripotential stem cells (2). This population of stem cells may respond to local regulators of pool size. Similarly, the "leukopoietin" sensitive stem cells may respond to CSF and become committed to further differentiation and proliferation (3), produce more of their kind in response to local or external regulators (4), or provide local information to the pluripotential stem cell compartment that affects the numbers of cells entering the "leukopoietin" sensitive stem cell compartment (5). Myeloblasts, promyelocytes, and myelocytes that make up the differentiation and proliferation pool provide increasing numbers of the granulocytic elements. Once maturation has proceeded beyond the myelocyte stage (6) the capacity for cell division is lost and cells enter the maturation pool. The release of cells from this pool depends on the blood flow rate through the marrow and information (releasing factors, etc.) emanating from the various tissues of the body. Increases in the total blood granulocyte pool will provide the extra cells necessary to handle infection or trauma and may also act as a feedback control mechanism to decrease the rate of neutrophil release from the marrow, and inhibit the production of release and/or stimulatory principles by the body tissues. In addition, excessive levels of releasing and/or stimulatory principles in the blood may be cleared by the kidney.

resulting in increased vascular permeability; (3) participation of increased numbers of neutrophils in the inflammatory response and, subsequently, activation of the kallikrein and kinin systems, as well as the coagulation mechanism; (4) enhanced neutrophil margination and increased passage of these cells into the inflammatory site; and (5) destruction of cells participating in the inflammatory process. The loss of granulocytes from the blood and their ultimate destruction in the tissues may activate a feedback system in that these events could lead to production of a circulating colony-stimulating factor acting upon "granulopoietin"-sensitive stem cells and thus causing increased myelopoiesis and production and/or release of principles (leukocytosis inducing factor; neutrophil releasing factor) regulating the release of mature neutrophils from marrow reserves.

The degree to which myelopoiesis is enhanced may depend upon the following: the rate of neutrophil destruction in the tissues to the extent that this determines the level of colony-stimulating factor; the number of granulopoietin sensitive stem cells in the marrow; the rate of adsorption or metabolism of this stimulatory principle by the marrow cells; the rate of clearance of this factor from the blood by the kidneys; marrow cell density (since increased cell interaction may counteract the effects of a stimulatory principle); and, the rate of neutrophil release from the marrow reserves (since this would alter the number of cells contained therein).

The rate of granulocyte release from the marrow reserves is enhanced when there is an increase in the blood flow rate through the marrow; when a plasma-borne LIF or NRF is present, and when the blood perfusing the marrow contains a decreased number of mature neutrophils. As the blood neutrophil concentration increases this could serve as part of a feedback mechanism by directly suppressing the rate of granulocyte release from the marrow reserves; by acting indirectly by inhibiting production and/or release of a plasma-borne releasing principle, or possibly doing both.

It also seems relevant to dwell upon the changes in protein synthesis that occur within the liver during an inflammatory reaction. At such times, liver metabolism is characterized by: the appearance of a carbohydrate-containing macroglobulin (α_2-globulin, acute phase) (John and Miller, 1968, 1969; Sarcione and Bohne, 1969; Sarcione, 1970; Menninger *et al.*, 1970); enhanced fibrinogen synthesis (John and Miller, 1968, 1969; Glenn, 1969); increases in α_1-acid glycoprotein (John and Miller, 1968, 1969); and increases in the synthesis of other proteins as well (Glenn *et al.*, 1968; John and Miller, 1969). It would be interesting to determine if the α_2-globulin (acute phase) is in any way related to the 19 S α_2-globulin reported to enhance bone marrow regeneration, cell proliferation (Berenblum *et al.*, 1968; Sontag *et al.*, 1971a, b), and leukocyte release (Sontag *et al.*, 1971b), or the α_1–α_2 globulins possessing properties of the leukocytosis inducing factor (Katz *et al.*, 1966). The α_2-acute phase-globulin may be related to a heat-labile α_2-globulin that inhibits plasmin, kallikreins, plasma permeability factor, and complement

(Ratnoff *et al.*, 1969). Thus, an inflammatory reaction may induce the synthesis of stimulatory proteins that affect bone marrow activity, and inhibitory substances that help to control the biochemical events taking place at the inflammatory site. This liver synthetic activity could be triggered by (a) processed endotoxin, (b) carbohydrate complexes released from cell surfaces, (c) lysosomal enzymes derived from cells participating in the inflammatory reaction, (d) activated complement components, (e) peptides derived from the cleavage of fibrinogen, (f) agar-colony stimulating factor derived from dying neutrophils, inflammed tissues, or other body tissue sites, or (g) any combination of these factors.

VIII. PROBLEMS TO BE RESOLVED

Although considerable information has been gathered concerning the production and release of granulocytes there are many problems requiring further exploration. These include the following: (1) the determination of the nature of the signal that initiates the production of releasing factors; (2) the mechanism whereby the factor is produced and/or released; (3) clarification of whether or not the plasma-borne factor affects neutrophil release alone, or also affects the recruitment of stem cells and proliferation within the myeloid compartment (in this regard, *in vitro* enhancement of DNA synthesis in granulocyte precursors incubated with plasma from leucopheresed animals has been reported by Rothstein *et al.*, 1971); (4) elucidation of any interaction between the feedback systems that appear to be operative in regulation of granulocyte production and release; (5) an evaluation of the regulation of neutrophil exchange between the circulating granulocyte pool and the marginated granulocyte pool; (6) studies to decide if the leucophilic α-globulin is important to any other aspects of neutrophil physiology beside phagocytosis (Fidalgo and Najjar, 1967a, b); (7) determination of the extent to which neuroendocrine effects play a role in leukocyte production and release.

REFERENCES

Abildgaard, C. F., and Simone, J. V. (1967). *Semin. Hematol.* **4**, 424.
Allison, F., Smith, M. R., and Wood, W. B. (1955). *J. Exp. Med.* **102**, 655.
Barnhart, M. I. (1968). *Biochem. Pharmacol. Suppl.* 205–219.
Becker, A. J., McCullock, E. A., and Till, J. E. (1963). *Nature (London)* **197**, 452.

Becker, A. J., McCullock, E. A., Siminovitch, L., and Till, J. E. (1965). *Blood* **26**, 296.

Becker, E. L. (1969). *Fed. Proc.* **28**, 1704.

Bennet, M., Cudkowicz, G., Foster, R. S., and Metcalf, D. (1968). *J. Cell. Physiol.* **71**, 211.

Berenblum, I., Burger, M., and Knyszynski, A. (1968). *Nature* (*London*) **217**, 857.

Bierman, H. R. (1964). *Ann. N.Y. Acad. Sci.* **113**, 753.

Bierman, H. R., and Hood, J. E. (1969). *Blood* **34**, 844.

Bierman, H. R., Marshall, G. J., Maekawa, T., and Kelly, K. H. (1962). *Acta Haematol.* **27**, 217.

Bishop, C. R., Athens, J. W., Boggs, D. R., Warner, H. R., Cartwright, G. E., and Wintrobe, M. M. (1968). *J. Clin. Invest.* **47**, 249.

Boggs, D. R. (1967). *Semin. Hematol.* **4**, 359.

Boggs, D. R., Athens, J. W., Cartwright, G. E., and Wintrobe, M. M. (1965). *J. Clin. Invest.* **44**, 643.

Boggs, D. R., Cartwright, G. E., and Wintrobe, M. M. (1966). *Amer. J. Physiol.* **211**, 51.

Boggs, D. R., Chervenick, P. A., Marsh, J. C., Cartwright, G. E., and Wintrobe, M. M. (1968a). *J. Lab. Clin. Med.* **72**, 177.

Boggs, D. R., Marsh, J. C., Chervenick, P. A., Cartwright, G. E., and Wintrobe, M. M. (1968b). *Proc. Soc. Exp. Biol. Med.* **127**, 689.

Bradley, T. R., and Metcalf, D. (1966). *Aust. J. Exp. Biol. Med. Sci.* **44**, 287.

Bruce, W. R., and McCullock, E. A. (1964). *Blood* **23**, 216.

Cartwright, G. E., Athens, J. W., and Wintrobe, M. M. (1964). *Blood* **24**, 780.

Chen, M. G., and Schooley, J. C. (1970). *J. Cell. Physiol.* **75**, 89.

Chervenick, P. A., and Boggs, D. R. (1970). *Science* **169**, 691.

Cooper, G. W. (1970). *In* "Regulation of Hematopoiesis" (A. S. Gordon, ed.), Vol. II, pp. 1611–1629. Appleton, New York.

Craddock, C. G., Jr., Perry, S., and Lawrence, J. S. (1959). *In* "The Kinetics of Cellular Proliferation" (F. Stohlman, Jr., ed.), pp. 242–259. Grune and Stratton, New York.

Delmonte, L., and Liebelt, R. A. (1965). *Science* **148**, 521.

Delmonte, L., Liebelt, A. G., and Liebelt, R. A. (1966). *Cancer Res.* **26**, 149.

Delmonte, L., Starbuck, W. C., and Liebelt, R. A. (1968). *Amer. J. Physiol.* **215**, 768.

Dornfest, B. S., LoBue, J., Handler, E. S., Gordon, A. S., and Quastler, H. (1962a). *Acta Haematol.* **28**, 42.

Dornfest, B. S., LoBue, J., Handler, E. S., Gordon, A. S., and Quastler, H. (1962b). *J. Lab. Clin. Med.* **60**, 777.

Ebbe, S. (1970). *In* "Regulation of Hematopoiesis" (A. S. Gordon, ed.), Vol. II, pp. 1587–1610. Appleton, New York.

Feher, I., and Gidali, J. (1965). *J. Lab. Clin. Med.* **66**, 272.

Fidalgo, B. V., and Najjar, V. A. (1967a). *Proc. Nat. Acad. Sci. U.S.* **57**, 957.

Fidalgo, B. V., and Najjar, V. A. (1967b). *Biochemistry* **6**, 3386.

Fried, W., and Gurney, C. W. (1968). *J. Lab. Clin. Med.* **71**, 948.

Fried, W., Knospe, W. H., Gregory, S. A., and Trobaugh, F. E., Jr. (1971). *J. Lab. Clin. Med.* **77**, 239.

Fukuda, T., Olde, H., and Miyasaka, A. (1960). *Nature* (*London*) **188**, 860.

Gidali, J., and Feher, I. (1964). *Amer. J. Physiol.* **206**, 585.

Glenn, E. M. (1969). *Biochem. Pharmacol.* **18**, 317.

Glenn, E. M., Bowman, B. J., and Koslowske, T. C. (1968). *Biochem. Pharmacol. Suppl.* 27–49.
Godwin, H. A., Zimmerman, T. S., Kimball, H. R., Wolff, S. M., and Perry, S. (1968a). *Blood* 31, 461.
Godwin, H. A., Zimmerman, T. S., Kimball, H. R., Wolff, S. M., and Perry, S. (1968b). *Blood* 31, 580.
Goldstein, A. L., Slater, F. D., and White, A. (1966). *Proc. Nat. Acad. Sci. U.S.* 56, 1010.
Gordon, A. S., (1959). *Physiol. Rev.* 31, 1.
Gordon, A. S., Dornfest, B. S., Neri, R. O., Eisler, M., and Crusco, A. (1959). *Fed. Proc.* 18, 57.
Gordon, A. S., Neri, R. O., Siegel, C. D., Dornfest, B. S., Handler, E. S., LoBue, J., and Eisler, M. (1960a). *Acta Haematol.* 23, 323.
Gordon, A. S., Siegel, C. D., Dornfest, B. S., Handler, E. S., LoBue, J., Neri, R. O., and Eisler, M. (1960b). *Trans. N.Y. Acad. Sci.* 23, 39.
Gordon, A. S., Handler, E. S., Siegel, C. D., Dornfest, B. S., and LoBue, J. (1964). *Ann. N.Y. Acad. Sci.* 113, 766.
Gordon, A. S., Cooper, G. W., and Zanjani, E. D. (1967). *Semin. Hematol.* 4, 337.
Gordon, A. S., Zanjani, E. D., Gidari, A. S., and Kuna, R. A. (1973). This monograph.
Handler, E. S., Varsa, E. E., and Gordon, A. S. (1966). *J. Lab. Clin. Med.* 67, 398.
Hellman, S., Grate, H. E., and Chaffey, J. T. (1969). *Blood* 34, 141.
Hellman, S., and Grate, H. E. (1968). *J. Exp. Med.* 127, 605.
Henson, P. M. (1969). *Fed. Proc.* 28, 1721.
Hill, J. H., and Ward, P. A. (1969). *J. Exp. Med.* 130, 505.
Hodgson, G. (1967). *Proc. Soc. Exp. Biol. Med.* 125, 1206.
Iscove, N. N., Till, J. E., and McCullock, E. A. (1970). *Proc. Soc. Exp. Biol. Med.* 134, 33.
Ito, Y., and Weinstein, F. B. (1962). *J. Nat. Cancer Inst.* 29, 229.
Jacobson, L. O., and Doyle, M. (1963). "Erythropoiesis." Grune and Stratton, New York.
John, D. W., and Miller, L. L. (1969). *J. Biol. Chem.* 244, 6134.
John, D. W., and Miller, L. L. (1968). *J. Biol. Chem.* 243, 268.
Katz, R., Gordon, A. S., and Lapin, D. M. (1966). *J. Reticuloendothelial. Soc.* 3, 103.
Keller, H. U., and Sorkin, E. (1968). *Experientia* 24, 641.
Kellermeyer, R. W., and Graham, R. D., Jr. (1968). *New England J. Med.* 279, 754.
King-Smith, E. A., and Morley, A. (1970). *Blood* 36, 254.
Lajtha, L. G. (1966). *J. Cell. Physiol. Suppl.* 1 67, 133.
Lajtha, L. G., Pozzi, L. V., Schofield, R., and Fox, M. (1969). *Cell Tissue Kinet.* 2, 39.
Lajtha, L. G., Gilbert, C. W., and Guzman, E. (1971). *Brit. J. Haematol.* 20, 343.
Lapin, D. M., LoBue, J., Gordon, A. S., Zanjani, E. D., and Schultz, E. F. (1969). *Proc. Soc. Exp. Biol. Med.* 131, 756.
Lawrence, J. S., and Craddock, C. G., Jr. (1968). *J. Lab. Clin. Med.* 72, 731.
Maniatis, A., Tavassoli, M., and Crosby, W. H. (1971). *Blood* 37, 581.
Marsh, J. C., and Levitt, M. (1971). *Blood* 37, 647.

Melmon, K. L., and Cline, M. J. (1967). *Amer. J. Med.* **43**, 153.
Menkin, V. (1946). *Arch. Pathol.* **41**, 376.
Menkin, V. (1955). *Ann. N.Y. Acad. Sci.* **59**, 956.
Menninger, F. F., Jr., Esber, H. J., and Bogden, A. E. (1970). *Clin. Chim. Acta* **27**, 385.
Metcalf, D. (1958). *Ann. N.Y. Acad. Sci.* **73**, 113.
Metcalf, D. (1970). *J. Cell. Physiol.* **76**, 89.
Morley, A., Richard, K. A., Guesenberry, P., Garrity, M., and Stohlman, F., Jr. (1971). *J. Cell. Physiol.* **77**, 301.
Odell, T. T., Jr. (1973). This monograph.
Paran, M., Ichikawa, Y., and Sachs, L. (1969). *Proc. Nat. Acad. Sci. U.S.* **62**, 81.
Pluznik, D. H., and Sachs, L. (1965). *J. Cell. Comp. Physiol.* **66**, 319.
Rakowitz, F., Schultz, E. F., D'Onofrio, S. E., Jr., Siegel, C. D., and Gordon, A. S. (1972). *J. Lab. Clin. Med.* **78**, 363.
Ratnoff, O. D., Pensky, J., Ogston, D., and Naff, G. B. (1969). *J. Exp. Med.* **129**, 315.
Reifenstein, G. H., Ferguson, J. H., and Weiskotten, H. G. (1941). *Amer. J. Pathol.* **17**, 233.
Rice, F. A. H. (1966). *Proc. Soc. Exp. Biol. Med.* **123**, 189.
Rice, F. A. H. (1968). *J. Infect. Dis.* **118**, 76.
Rice, F. A. H., and Dander, J. H. (1968). *J. Infect. Dis.* **118**, 289.
Rice, F. A. H., and Shaikh, B. (1970). *Biochem. J.* **116**, 709.
Rice, F. A. H., Lepick, J., and Darden, J. H. (1968). *Radiat. Res.* **36**, 144.
Rice, F. A. H., McCurdy, J. D., and Aziz, K. (1971). *Proc. Soc. Exp. Biol. Med.* **136**, 56.
Rickard, K. A., Shadduck, R. K., Howard, D., and Stohlman, F., Jr. (1970). *Proc. Soc. Exp. Biol. Med.* **134**, 152.
Rickard, K. A., Rencricca, N. J., Shadduck, R. K., Monette, F. C., Howard, D. E., Garrity, M., and Stohlman, F., Jr. (1971). *Brit. J. Haematol.* **21**, 537.
Robinson, W. A., and Pike, B. L. (1970). *In* "Hemopoietic Cellular Proliferation" (F. Stohlman, Jr., ed.), pp. 249–259. Grune and Stratton, New York.
Rothstein, G., Hugl, E. H., Bishop, C. R., Athens, J. W., and Ashenbrucker, H. E. (1971). *J. Clin. Invest.* **50**, 2004.
Rytomaa, T., and Kiviniemi, K. (1968a). *Cell Tissue Kinet.* **4**, 329.
Rytomaa, T., and Kiviniemi, K. (1968b). *Cell Tissue Kinet.* **4**, 341.
Sarcione, E. J. (1970). *Biochemistry* **9**, 3059.
Sarcione, E. J., and Bohne, M. (1969). *Proc. Soc. Exp. Biol. Med.* **131**, 1454.
Schooley, J. C. (1968). Semiannual rep. Lawrence Radiat. Lab., Univ. Calif., p. 99.
Schultz, E. F. (1970). Ph.D. Thesis, New York Univ.
Shadduck, R. K., and Nagabhushanam, N. G. (1971). *Blood* **38**, 559.
Sontag, J. M., Trainin, N., and Berenblum, I. (1971a). *Radiat. Res.* **45**, 499.
Sontag, J. M., Berenblum, I., and Trainin, N. (1971b). *Radiat. Res.* **45**, 511.
Stanley, E. R., and Metcalf, D. (1969). *Aust. J. Exp. Biol. Med. Sci.* **47**, 467.
Steinberg, B., and Martin, R. A. (1950). *Amer. J. Physiol.* **161**, 14.
Steinberg, B., Dietz, A. A., and Martin, R. A. (1959). *Acta Haematol.* **21**, 78.
Steinberg, B., Cheng, F. H. F., and Martin, R. A. (1965). *Acta Haematol.* **33**, 279.
Stephenson, J. R., and Axelrad, A. A. (1971). *Blood* **37**, 417.

68 — *E. F. Schultz, D. M. Lapin, and J. LoBue*

Sutherland, D. J. A., Till, J. E., and McCullock, E. A. (1970). *J. Cell. Physiol.* **75**, 267.

Tavassoli, M., and Crosby, W. H. (1970). *Science* **169**, 291.

Till, J. E., and McCullock, E. A. (1961). *Radiat. Res.* **14**, 213.

Till, J. E., McCullock, E. A., and Siminovitch, L. (1964). *Proc. Nat. Acad. Sci. U.S.* **51**, 29.

Trentin, J. J. (1971). *Amer. J. Pathol.* **65**, 621.

Ward, P. A. (1968). *J. Exp. Med.* **128**, 1201.

Ward, P. A. (1969). *Amer. J. Pathol.* **54**, 121.

Ward, P. A. (1971). *Amer. J. Pathol.* **64**, 521.

Ward, P. A., and Becker, E. L. (1968). *J. Exp. Med.* **127**, 693.

Webster, M. E. (1968). *Fed. Proc.* **27**, 84.

Whang, J., Frei, E., III, Tjio, J. H., Carbone, P. P., and Brecker, G. (1963). *Blood* **22**, 664.

Worton, R. G., McCullock, E. A., and Till, J. E. (1969). *J. Cell. Physiol.* **74**, 161.

Wu, A. M., Till, J. E., Siminovitch, L., and McCullock, E. A. (1968). *J. Exp. Med.* **127**, 455.

HUMORAL REGULATION OF EOSINOPHIL
PRODUCTION AND RELEASE

*Natalie S. Cohen, Joseph LoBue, and Albert S. Gordon**

I. INTRODUCTION

An understanding of the complex processes involved in cellular differentiation is an important goal of modern biological research. The actions
of hormones on their target tissues and the control of cell numbers

* The authors' original investigations reported in this chapter were supported
by research grants 5-RO1-HL03357-15 and 1-RO1-CA12815-01 from the United
States Public Health Service.

in proliferating cell systems represent fundamental aspects of differentiation and functional homeostasis. In many cell lines, including those of the blood and bone marrow, there is every indication that both proliferation and cytodifferentiation are closely interrelated. Erythropoietin, which controls the production of erythrocytes, has been demonstrated to initiate the synthesis of RNA and δ-aminolevulinic acid synthetase and, consequently, the production of hemoglobin (Goldwasser, 1966; Bottomley et al., 1971). Factors which specifically stimulate granulocyte (primarily neutrophil) proliferation have also been proposed (Bierman, 1964; Craddock, 1960; Rytomaa and Kiviniemi, 1968a, b; Yoffey, et al., 1964; Metcalf, 1973). Of course, another type of regulatory mechanism, one which involves the direct inhibitory influence of the mature cell on its own precursors (i.e., a negative feedback mechanism) has also been implicated in the control of erythropoiesis (Whitcomb and Moore, 1965), thrombopoiesis (Linman and Pierre, 1963), and neutrophilic granulocyte production (Craddock, 1960; Gordon et al., 1964; Rytomaa and Kiviniemi, 1968a, b).

On the other hand, with the sole exception of Komiya's work on "eosinopoietin" (1956), no inhibitory or stimulatory humoral agents specific for eosinophils have been described. There is, however, a vast literature on the various functional responses of eosinophils, such as the eosinophilia accompanying certain foreign protein responses (Hudson, 1968; Litt, 1960), and the reactions of these cells to various physiologically active substances such as histamine (Archer, 1968; Hudson, 1968), and adrenal hormones (Hudson, 1968). The present chapter will be concerned with possible cell-specific humoral factors involved in the control of eosinopoiesis, i.e., factors similar to those mentioned above for the erythrocyte and neutrophil. Work in our laboratory has suggested the existence of such humoral controls for the eosinophil.

II. EOSINAPHERESIS AS A METHODOLOGICAL APPROACH FOR THE STUDY OF THE CONTROL OF EOSINOPHIL PRODUCTION

The investigations described herein were designed to determine the effects of intensive eosinophil withdrawal on the bone marrow, and to explore the possibility of the existence of a humoral factor(s) controlling some aspect(s) of eosinophil production.

The technique of leukapheresis (LAP), in which substantial numbers of white blood cells are withdrawn from an organism, thus disturbing the normal leukocyte balance, has constituted a successful approach to many problems of neutrophil production and release (Craddock et al.,

1955; Gordon *et al.*, 1960; Schultz *et al.*, 1973). We have developed an effective method of eosinapheresis (EAP) comparable to the LAP techniques. A brief discussion of this method has already been presented in an autoradiographic study of eosinophil kinetics in normal and EAP rats (Cohen *et al.*, 1967).

A. The Technique of EAP

Rats of a modified Long-Evans strain, receiving Purina Chow and tap water *ad libitum* were used. EAP rats and their controls weighed 200–270 gm. Plasma recipients and their controls were 135–170 gm.

The method of peritoneal lavage was patterned after the LAP technique of Gordon *et al.* (1960). In this procedure the animals are lightly anaesthetized with ether and are each injected intraperitoneally with 10 ml of isotonic saline. After rotating the rat several times to allow the saline to wash through the cavity thoroughly, as much fluid as possible is withdrawn and transferred to a graduated centrifuge tube containing a small amount of heparin. A sample of the fluid is removed for a total cell count, the tube is centrifuged (4.5 minutes, 2000 rpm), and the sedimented cells are resuspended in a few drops of rat serum for smearing.

The procedure for EAP involved lavage as described, followed by the intraperitoneal injection of a 0.1% suspension of asbestos fibers in isotonic saline to mobilize eosinophils (as described by Speirs, 1955). Abdomens of EAP rats were shaved and swabbed with 70% ethanol prior to each insertion of a needle. All treatments of the peritoneal cavity were pyrogen-free, since the presence of pyrogen results in an increase in the numbers of neutrophils entering the peritoneal cavity. Syringes, needles, and injection bottles were sealed and heated to 180°–210°C for at least 3 hours. The rubber caps for injection bottles were boiled for 20 minutes in a 1% solution of NaOH in pyrogen-free distilled water, and then thoroughly rinsed with such water. The saline used for lavage and for the asbestos suspension was also pyrogen-free (McGaw Laboratories, Inc.).

Asbestos for injection was prepared from Seitz filters (Speirs, 1955) which were shredded and ashed. The ashes were ground, washed, dried, and weighed into injection bottles which were then heated as described above. Pyrogen-free isotonic saline was added to each bottle to make a 0.1% suspension. Preliminary studies indicated that daily EAP using 2 ml of this asbestos suspension produced cell harvests containing relatively large numbers of eosinpophils. More frequent lavage and/or more asbestos enhanced the neutrophil response at the expense of eosinophils.

B. Initial Studies in EAP Rats

1. MARROW STUDIES

The bone marrow was studied in 67 rats subjected to varying EAP schedules (i.e., daily for 5 days, or 2 or 3 times a day for 4 or 3 days, respectively). As controls, 17 normal rats received a single peritoneal lavage with no asbestos injection. All animals were sacrificed within an hour after lavage. The control procedures were performed at the same time of day at which the EAP rats were sacrificed in order to eliminate any complications of diurnal variations in leukocyte values.

Bone marrow was obtained from all EAP donor and plasma recipient rats after exsanguination by aortic puncture. A femur was removed, cleaned of adhering tissue, and cracked open longitudinally with a sharp scalpel and the marrow was drawn into a weighed, heparinized capillary tube. After reweighing, the entire sample was expelled into a measured amount of rat serum, and all visible cell clumps were dispersed by agitation with a Pasteur pipette. Total nucleated cell and total eosinophil counts were performed on a sample of the suspension. After mild centrifugation (3 minutes, 2000 rpm) the sedimented cells were resuspended in a few drops of rat serum. Thin smears were made, air-dried, and stained with May-Grunwald. Differential counts were done on 1000 cells; for mitotic counts 5000 cells were enumerated on each smear.

2. CELL HARVESTS

The peritoneal cell harvests in a typical experiment in which EAP was performed daily for 5 days are given in Table I. A mean of 21.4×10^6 eosinophils were removed per day. This represents approximately 7 times the number of eosinophils present in the circulation of the rat at any one time (Rytomaa, 1960). Thus, with the present EAP technique, approximately 18–21 times the number of eosinophils in the circulation could be removed in 3 days (Table I). The withdrawal of such quantities of eosinophils would thus seem to constitute a considerable drain on those mature components of the hemopoietic system which might be involved in the negative feedback control of eosinophil production and release.

According to the leukapheresis (LAP) studies of Gordon et al. (1960, 1964), removal of a billion or more neutrophils in 3 days from a rat sufficed to elicit the production of a leukocytosis-inducing factor (LIF). When injected into rats, this LIF produces a peripheral leukocytosis as well as a depletion of bone marrow granulocytes. One billion neutro-

TABLE I

CELLS WITHDRAWN DURING 5 DAILY EAP: NUMBER AND PERCENTAGE OF EACH CELL TYPE[a]

Rat No.	Total cells	Mean ± sem	Eosinophils	Mean ± sem	Neutrophils	Mean ± sem	Mononuclear cells	Mean ± sem
53	277		73		79		125	
54	346	310.6 ±19.9	156	107.0 ±15.9	64	86.8 ±9.2	126	116.8 ±19.7
56	292		123		85		84	
58	267		82		112		73	
58	371		101		94		176	

Number withdrawn per day:

	Average:	Range:
Total cells	62.1	58.1–66.1
Eosinophils	21.4	18.2–24.6
Neutrophils	17.4	15.5–19.2
Mononuclear cells	23.4	19.4–27.3

Percentage of total cells attributable to each type:

	Average:	Range:
Eosinophils	34.3	27.6–42.3
Neutrophils	28.0	23.5–33.0
Mononuclear cells	37.0	29.4–47.0

[a] All values other than percentages are expressed in millions.

phils is approximately 13 times the number of circulating neutrophils. No significant bone marrow alterations were observed in the LAP rats.

3. EFFECTS ON EAP RATS AND THE RECIPIENTS OF THEIR PLASMA

There were definite bone marrow alterations in the EAP rats. A positive correlation was found between the numbers of eosinophils withdrawn during EAP and the eosinophil percentage and the numbers of eosinophils per milligram (both based on hemocytometer counts) in the bone marrow. The correlation values, calculated by the rank difference method, were $r = +0.6$ (S.E. $r = \pm0.079$) and $r = +0.5$ (S.E. $r = \pm0.092$), respectively. Both correlations were highly significant ($p < 0.001$). Mature eosinophils accounted for most of the increases observed.

Plasma of EAP rats was tested for possible effects on the peripheral blood of 7 intact and 6 adrenalectomized rats and on both the blood and bone marrow of 8 intact animals. Adrenalectomy was performed 2 days prior to the plasma assay and the animals were maintained in 1.0% NaCl solution. Peripheral blood samples were taken at 0 hours and at frequent intervals thereafter for 5 to $7\frac{1}{2}$ hours. The plasma was administered subcutaneously as a single 10-ml dose to all but 2 animals. These received 3 injections within a half-hour period, the total volume injected being 18.5 and 20.0 ml of EAP plasma. These 2 rats were killed at $7\frac{1}{2}$ hours postinjection for marrow analysis. The bone marrow of the remaining rats was sampled 5 hours postinjection. As controls, the blood and bone marrow of 3 rats were studied at similar times after the subcutaneous administration of 10 ml of normal plasma from the EAP control animals. In addition, the marrow of 6 uninjected rats was examined. All control marrows were taken at the same time of day as were those of the experimental groups.

There were no consistent alterations in peripheral leukocyte levels of either the EAP rats or the EAP-plasma recipients, nor did the EAP plasma produce any changes in the bone marrow of the recipients.

These studies thus presented an entirely different picture from that observed in the LIF work of Gordon et al. (1960, 1964). In the latter, blood and bone marrow changes were observed in the LAP plasma recipients but not in the donors. The EAP data nonetheless suggested that definite effects on the eosinopoietic system were elicited by the experimental treatment, and indicated that further study of the observed bone marrow alterations and their development would be in order. Thus a more intensive investigation of EAP marrow at various times after different numbers of EAP was initiated. Concomitantly the plasma of these animals was assayed for its effect on recipient rat marrow sampled at 24 hours, rather than at 5–7½ hours after injection.

C. Studies of EAP-Elicited Bone Marrow Alterations

1. Experimental Schedules

A large group (104) of rats were subjected to different numbers (1–5) of daily EAP and sacrificed at 6 or 12 hours after EAP. Immediately before exsanguination a lavage was performed to sample the peritoneal fluid. Marrow was collected for total, differential, and mitotic counts. In 2 control groups (6 animals each) each rat received a single lavage (no asbestos) followed immediately by exsanguination and marrow removal. These groups were treated at the same times of day at which the EAP rats were killed. Two groups of "sham EAP" animals were also studied. The sham A group (4 rats) received the usual 10 ml of lavaging asbestos-saline but only 0.2 ml was removed. This was for peritoneal fluid analysis. In the other group, sham B (4 rats), the lavaging asbestos-saline was withdrawn but was immediately reinjected. Only a few drops were retained for examination. The procedure was carried out daily for 3 days. Sacrifice was at 6 hours after the final sham EAP.

The 6- and 12-hour EAP donor and recipient groups and the peritoneal cell harvests are presented in Table II. In the 4-EAP, 12-hour group the eosinophil harvest was relatively poor. This seems to have been reflected in some of the bone marrow parameters discussed below.

Finally, 4 rats were given 3 daily EAP but were not killed until 24 hours (2 animals) or 36 hours (2 animals) after the final EAP.

2. Bone Marrow Alterations

The bone marrow of 6- and 12-hour EAP rats differed widely from that of control animals in many respects (Figs. 1 and 2). It should be noted that wide individual variations both in the peritoneal and bone marrow response to EAP were often encountered, thus lowering the incidence of attaining statistical significance in many cases, even when differences from controls was substantial. This difficulty was intensified for recipient data since smaller numbers of rats were involved.

Slight increases in marrow cellularity were found in all of the 6-hour groups. At 6 hours the marrow mitotic index (MI) was elevated, often significantly, after 2, 3, 4, and 5 EAP; increases were also found at 12 hours after EAP but individual variation was large. The mean percentage of large mononuclear cells (including myeloblasts) was higher in all EAP groups than in controls.

TABLE II

EAP RATS AND PLASMA RECIPIENTS: TIMES OF MARROW SAMPLING AND PERITONEAL CELL HARVESTS

| | | EAP rats | | | Plasma recipients | |
| | | Numbers of cells withdrawn (Mean ± sem) | | | | |
Group	No. rats	Total cells × 10⁻⁶	Eosinophils × 10⁻⁶	Neutrophils × 10⁻⁶	No. rats	Time of sacrifice (hrs)
6-Hour exsanguination						
C (control)	6	....	....	....	3	24
SA (sham A)	4	7.65 ± 1.03	1.23 ± 0.58	2.52 ± 0.57	2	24
SB (sham B)	4	141.05 ± 19.71	9.63 ± 1.53	84.83 ± 14.41	2	24
1 EAP	7	59.16 ± 13.17	19.83 ± 8.65	20.00 ± 5.78	4	24
2 EAP	11	174.00 ± 10.71	36.51 ± 8.84	46.51 ± 6.14	4	24
3 EAP	26	233.64 ± 15.00	46.61 ± 14.37	67.60 ± 6.44	6	24
4 EAP	8	424.38 ± 40.03	74.03 ± 13.10	154.74 ± 15.90	4	24
5 EAP	6	404.87 ± 34.59	93.73 ± 22.85	109.08 ± 8.56	3	24
12-Hour exsanguination						
C (control)	6	....	....	....	3	24
1 EAP	8	67.78 ± 8.55	9.93 ± 2.04	23.19 ± 4.72	4	24
2 EAP	8	190.56 ± 24.14	34.15 ± 7.49	52.51 ± 7.16	4	24
3 EAP	8	203.54 ± 28.49	58.29 ± 12.74	59.88 ± 13.20	4	24
4 EAP	14	276.74 ± 21.99	55.82 ± 9.48	81.03 ± 7.80	4	24
5 EAP	8	460.81 ± 27.75	110.20 ± 14.09	118.44 ± 7.39	3	6–7

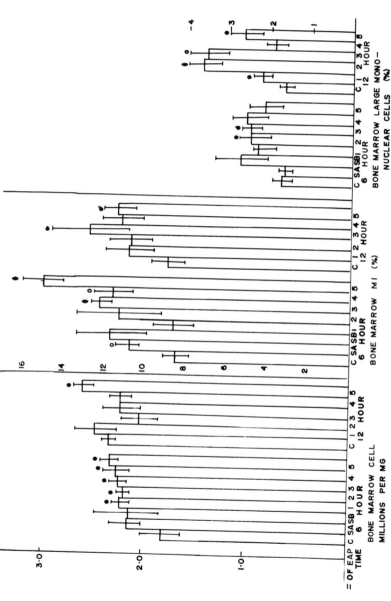

Fig. 1. Bone marrow cells per milligram, mitotic index, and large mononuclear cells in rats and 6 and 12 hours after EAP. The EAP groups designated C, SA, SB, 1, 2, 3, 4, 5 are the same as those described in Table II. Mean values ±1 standard error of the mean (vertical lines) are presented. Symbols above the bars represent the following p values: ● = p = .05, ○ = p = .01, ● = p = .001. The mitotic index indicates the numbers of mitotic figures per 1000 nuclear cells, based on counts of 5000 nucleated cells.

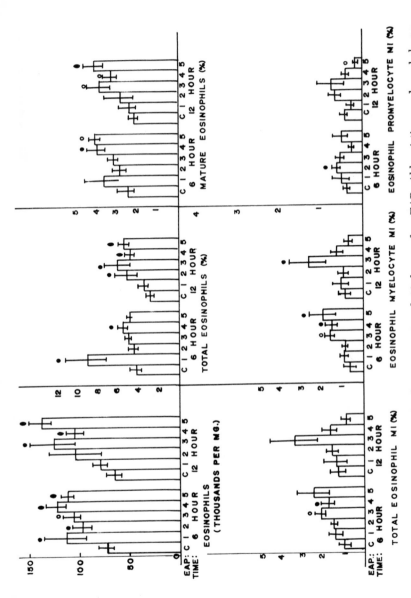

Fig. 2. Bone marrow eosinophil values in rats 6 and 12 hours after EAP. Abbreviations and symbols are used as in Fig. 1. The values for eosinophils per milligram and total eosinophil percent are based on hemocytometer counts. The latter category would probably include all mature cells and myelocytes. Mature eosinophil percentages were derived from smears.

Additional marrow eosinophil data are given in Fig. 2. Total eosinophil percentages were significantly elevated in the 2, 3, 4, and 5 EAP 12-hour animals. Increases in eosinophils per milligram were evident in both groups, and became more pronounced with increasing numbers of EAP, especially in the 12-hour rats. This would seem to represent an increase in the marrow reserve of eosinophils similar to that described in the foreign protein response in guinea pigs (Hudson, 1966). There were also rises, often significant, in the percent mature eosinophils. Eosinophil myelocyte percentages were slightly higher than the control mean in the 12-hour EAP groups, and the promyelocyte percent was somewhat elevated in some groups in both series (data not given in Fig. 2). At 6 hours after EAP there were very high mitotic indices (MI) for total eosinophil and eosinophil myelocyte populations. Although increases were also found in the 12-hour series, considerable individual variation existed. Values for eosinophil promyelocyte MI displayed no consistent alterations, although some substantial elevations were found.

While mature neutrophil and band cell percentages were not consistently affected by EAP, significant increases in neutrophilic myelocyte and promyelocyte percentages and in total neutrophil, neutrophil myelocyte, and neutrophil promyelocyte MI were found in all groups except the 6-hour and 1-EAP rats. In all neutrophil categories the sham A group displayed values intermediate between control and EAP levels while the sham B group was identical to EAP marrow. In contrast, the changes in eosinophil categories were unique to the EAP groups.

Twenty-four hours after 3 EAP, marrow alterations were still evident. After 36 hours, however, the bone marrow MI, eosinophil MI, and eosinophil promyelocyte percentages returned to normal and most other eosinophil parameters were similar in value to those found in controls. One notable exception was the eosinophils per milligram value which remained well above control levels even after 36 hours.

3. Discussion

The very early response of the bone marrow to EAP is readily apparent from the alterations observed in neutrophil counts. The MI, for example, was already significantly increased in almost all animals by 12 hours (but not by 6 hours) after only 1 EAP. This is all the more striking because during the first EAP virtually no neutrophils were yet mobilized into and removed from the peritoneal cavity. Thus the bone marrow changes after 1 EAP would seem to represent an immediate and direct reaction of the leukopoietic system to the exodus of mature neutrophils from the marrow. The fact that the sham B rats displayed the same marrow neutrophil changes as the EAP rats supports this concept. Other investigators (Craddock, 1960; Yoffey *et al.*, 1964; Yoffey, 1966) have also noted the relative rapidity with which the bone marrow can respond to a sudden increase in peripheral utilization of neutrophils. For example, Yoffey *et al.* (1964) observed that the MI in typhoid vaccine challenged animals was increased as early as 4 hours after treatment, peaking at 12 hours. This is in good agreement with the present data on neutrophil MI.

Alterations in eosinophil parameters were more variable and less extensive than those observed for neutrophils. The response of the eosinopoietic system was slower, also more sluggish. Thus, even though many eosinophils were withdrawn in the first EAP, increases in eosinophil MI, when present, were found only after 3, 4, and 5 EAP.

Several factors may have contributed to these differences in neutrophil and eosinophil bone marrow reactions. EAP probably created a far greater drain on neutrophil resources than the present data indicate, perhaps even exceeding that exerted on the eosinopoietic system, especially since peritoneal neutrophils are known to be destroyed at a rapid rate (Speirs, 1955). Moreover, if as Speirs has found (1955), eosinophils enter the peritoneal cavity more slowly than neutrophils, then the effects of the initial EAP on the eosinophil population would be expected to be less intense than those seen for neutrophils. The comparatively late appearance of the marrow eosinophil alterations might also have been a reflection of the longer eosinophil mitotic and intermitotic intervals (Alexander *et al.*, 1969; Andersen and Bro-Rasmussen, 1968; Bro-Rasmussen *et al.*, 1967; Cohen *et al.*, 1967; Dustin, 1959). In fact, the 36-hour EAP bone marrows had attained or were approaching normal levels in all parameters except eosinophils per milligram, which was still very high.

Litt (1960) found that intraperitoneal injection of foreign protein into guinea pigs elicited a peripheral eosinophilia in 2 to 4 hours which peaked at 4 to 9 hours. There was an influx of eosinophils into the peritoneal cavity which was followed by elevations in total marrow eosinophils 48 hours after injection. However, he found no consistent short-term alterations while the cells were accumulating in the peritoneal fluid, nor were the percentages of the different maturational stages changed according to any definitive pattern. In the course of studies of the proliferation of lymphoid and plasmacytoid cells during the secondary response to tetanus toxoid in mice, Cottier *et al.*, (1964) found 2 peaks of peripheral eosinophilia, one at 24 hours and the second 6 days after challenge. Thus an increase in eosinophil production may require some days. In extensive studies of eosinophil responses in foreign protein reactions in the guinea pig, Hudson (1966) observed a peripheral eosinophilia 24 to 30 hours after antigenic challenge. This represented a mobilization of part of the marrow eosinophil reserve. Additional work (Hudson, 1968) suggests that the eosinophil reserve depletion is followed by an increase in proliferative activity. There also appeared to be an increase in the marrow reserve of mature and nearly mature eosinophils in sensitized guinea pigs.

The present data indicate that the efflux of large numbers of mature eosinophils from the marrow is followed by enhanced proliferative activity in eosinophil precursor cells. Autoradiographic studies of eosinophil kinetics in normal and EAP rats (Cohen *et al.* 1967) provide additional indications that granulocytopoiesis was accelerated in the EAP animals. The eosinophil myelocyte cell cycle apparently was shortened in these animals.

There were several differences in bone marrow eosinophil values between the 6- and 12-hour EAP rats. Increased eosinophils per milligram of marrow were found at all time intervals studied, but such increases were attributable mainly to promyelocytes and mature cells at 6 hours after EAP, whereas at 12 hours elevations in eosinophil myelocyte percent were also contributory. Mature cell percentages were enhanced in many more animals at 12 than at 6 hours, and at 1 hour far more than at either of these times. Evidently by 1 hour the large stores of mature marrow eosinophils had not yet been released to replace those cells withdrawn by EAP. By 6 hours many mature cells had left the marrow and had not yet been replaced although promyelocyte percentages had risen. Replacement of mature eosinophils by maturation was probably well underway by the time 12 hours had elapsed.

D. The Effects of EAP Plasma on Recipients

1. Recipient Animals

Plasma from the EAP animals sacrificed at 6 and 12 hours after various numbers of daily EAP was administered subcutaneously to 37 recipients in approximately 10 ml amounts (Table II). After injection, the peripheral blood was sampled at various intervals up to 24 hours, at which time the bone marrow was removed for total, differential, and mitotic counts. There were also 3 recipient animals sacrificed 6 to 7 hours after injection of 5-EAP 12-hour plasma. Plasma from control groups was administered to 6 recipients in a manner identical to that described for EAP plasma; blood and bone marrow were studied at the same time as those of the EAP recipients.

2. Bone Marrow Alterations

In 12-hour plasma recipients, total marrow cells per milligram and mononuclear cell MI were generally higher than in controls (Fig. 3). The total marrow MI and the relative numbers of large mononuclear

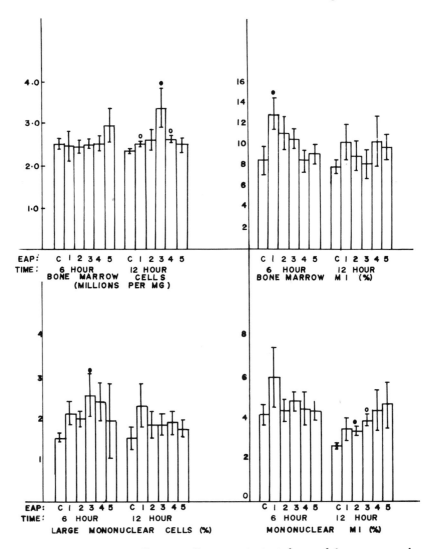

Fig. 3. Bone marrow cells per milligram, mitotic index, and large mononuclear cells in recipients of EAP plasma. The plasma recipient groups designated C, 1, 2, 3, 4, 5 are the same as those described in Table II. Symbols are used as in Fig. 1.

cells also tended to be increased (Fig. 3). On the other hand, total eosinophil percentages, eosinophils/mg., and mature eosinophil and myelocyte counts were decidedly reduced in 12-hour plasma recipients, (Fig. 4). Elevated eosinophil promyelocyte percentages, however, were observed in most 6-hour and 12-hour groups. The total eosinophil MI and the promyelocyte MI tended to be depressed in the 12-hour recipients, although the myelocyte MI was increased in most of these animals. All of the modifications observed in the 12-hour recipients except the latter were already present in the marrow of 3 recipients (5-EAP plasma) sacrificed at 6–7 rather than at 24 hours after injection. No consistent trends in any eosinophil MI categories were apparent in the 6-hour plasma-recipient animals.

Changes in recipient marrow neutrophil parameters were also observed (data not shown). Total neutrophil and mature neutrophil numbers were not consistently altered, although some low values were found. There were decreases in the band cell percent in all 12-hour EAP and in the 2- and 3-EAP 6-hour plasma recipients, however, and rises in percent myelocytes in most 12-hour groups. Substantial increases in neutrophilic promyelocytes were found in all recipient groups, and in both the 6- and 12-hour series there were striking elevations in total neutrophil and neutrophil myelocyte MI. Neutrophil promyelocyte MI, however, was not consistently altered. The data obtained in sham EAP plasma recipients did not differ appreciably from controls.

3. DISCUSSION

As in the donors, changes in neutrophil and eosinophil counts in EAP-plasma recipients did not parallel each other. The differences between the two probably are a reflection of the basically slower eosinophil response and lower production rate (see Section II,C, 3).

Craddock (1960) has noted that the release of granulocytes in response to increased peripheral removal, as in LAP, accelerated granulocytopoiesis. Also, as indicated earlier, Yoffey *et al.* (1964) observed that a rise in neutrophil MI was evident as early as 4 hours after the injection of TAB vaccine which elicited a massive neutrophil release.

It is apparent that the increased numbers of immature neutrophils found in EAP-plasma recipients as early as 6 hours after plasma injection might be explained as the natural sequelae of increased marrow neutrophil release. The primary effect of the LIF of Gordon *et al.* (1960, 1964) has definitely been shown to be enhanced marrow leukocyte release. The peripheral leukocytosis caused by the LIF was not found in EAP-plasma recipients, even though marrow neutrophil band cell num-

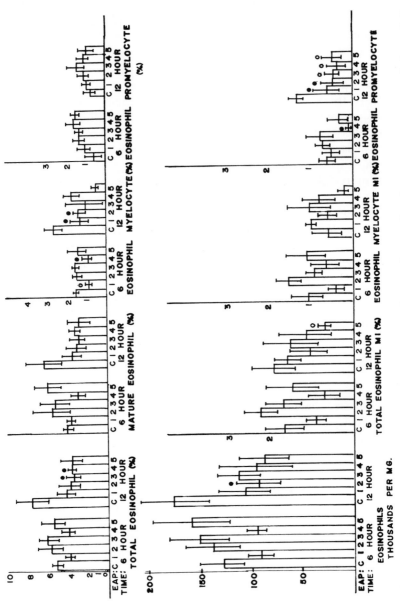

Fig. 4. Bone marrow eosinophil values in recipients of EAP plasma. The plasma recipient groups designated C, 1, 2, 3, 4, 5 are the same as those described in Table II. The values for eosinophils per milligram and total eosinophil percent are based on hemocytometer counts. The latter category would probably include all mature cells and myelocytes. Other eosinophil percents were derived from smears. Symbols are used as in Fig. 1.

bers were depressed in some recipients. Of course, the numbers of neutrophils removed by EAP may not have been sufficient to produce a peripherally detectable quantity of LIF. Yoffey *et al.* (1964), noted that the marrow released cells in response to vaccine within 2 hours but that most of these cells also left the circulation during this period. Thus, rapid sequestration, margination, or enhanced efflux from the circulation could have masked any leukocytosis which may have occurred in response to EAP-plasma. It has been aptly pointed out by Osgood (1954) and others that granulocytes are not primarily blood cells; the circulation serves merely as a means of transporting these cells from sites of production to the tissues where they perform many of their physiological functions.

However, the proportion of neutrophilic myelocytes in mitosis was high, suggesting that production of these cells was occurring at an increased rate so that released cells would be rapidly replenished. Thus, analysis of the bone marrow would appear to provide more sensitive means of ascertaining possible regulatory potentialities of test material on leukocyte populations.

Many of the eosinophil marrow changes could probably be best attributed to enhanced eosinophil release and maturation. It is evident, however, that the plasmas of the 2 donor groups were not identical in their effects. The bone marrow picture in the 12-hour recipients were more extensively modified than in the 6-hour animals (Fig. 4), undoubtedly reflecting donor differences in the degree of marrow compensation to EAP-induced eosinophil exodus. Total eosinophil percent and eosinophils per milligram, for example, showed many more decreases among 12-hour recipients. It has been previously observed (Litt, 1960; Samter *et al.*, 1953) that under certain conditions the bone marrow can release mature eosinophils very quickly. Although no rise in circulating eosinophil levels was found in the present studies, the same arguments, as were presented above in connection with detection of neutrophil release, could be reasonably applied to eosinophils. In this regard, Hudson (1966) observed that 29 hours after an antigenically induced peripheral eosinophilia, 90% of the released eosinophils were gone from the blood.

In the 6-hour recipients there was no reduction in the numbers of mature eosinophils, yet the eosinophil promyelocyte percentages were increased. Nevertheless, a relatively high rate of eosinophil release may also have occurred in these animals but it may have been slow enough for maturation of the eosinophil myelocyte population to keep the supply of mature eosinophils at normal levels. The eosinophil myelocytes did decrease substantially in some 6-hour groups. Possibly, eosinophil release was a stimulus for an increase in maturation rate. Studies of eosinophil kinetics have indicated that eosinophil maturation can be increased by EAP (Cohen *et al.*, 1967).

If EAP plasma effected a release of both neutrophils and eosinophils, why were immature neutrophil parameters increased in recipients while most eosinophil values decreased? The decreases in eosinophil compartments could conceivably have been the result of the rapid release of eosinophils combined with a relatively slow response of the eosinopoietic elements to this stimulus and/or the normally longer eosinophil precursor cell cycle (see Section II,C,3).

Eosinophil myelocyte numbers were reduced in many groups of recipients and eosinophil myelocyte MI was correspondingly altered in the 6-hour groups. In the recipients of 12-hour plasma, however, the myelocyte MI tended to be high, although their numbers were low. In effect, then, the proportion of eosinophil myelocytes in division was relatively higher than normal in the 12-hour animals.

The eosinophil promyelocyte percentages were high in many of the 6-hour and 12-hour recipients. On the other hand, only a few of the former had elevated promyelocyte MI and in all of the latter this measurement was extremely low (actually reduced substantially below normal). Perhaps once again the long intermitotic time of the eosinophil was the reason for this low MI. The most accurate measurement of cell cycle time in the rat eosinophil (Alexander *et al.*, 1969) places it at 22–25 hours, which is about double the duration found for neutrophilic granulocyte precursors (LoBue *et al.*, 1970).

III. THE HUMORAL CONTROL OF LEUKOPOIESIS: GENERAL MECHANISMS AND SOME SPECULATIONS ON FACTORS CONTROLLING EOSINOPOIESIS

There are probably a number of factors at work in the control of leukopoiesis. The presence of the mature cell may result in the inhibition of cell release while its absence may stimulate release (see Schultz *et al.*, 1973) with increased cell release eventually enhancing proliferative activity in the cell lines involved. Finally, Lajtha *et al.* (1962) have pointed out that there must be some mechanism whereby the mitotable cell compartment, upon sufficient stimulation, can counteract the stimulatory signal produced by release.

The present data suggest that the release effect for eosinophils may be mediated via a humoral factor but the precise nature of this regulatory system has not yet been determined. If regulation does involve a circulating factor it is tempting to interpret the puzzling decrease in eosinophil promyelocyte MI in the 12-hour EAP recipients alluded to earlier as circumstantial evidence for a specific inhibitory agent present in this EAP plasma and acting upon myelocyte proliferation in the manner of a counter-stimulation type of feedback signal.

A. Releasing Factors

The nature or source of the humoral factor stimulating eosinophil release is not clear. It may involve the mature cell. Gordon et al. (1960) in their discussion of the LIF have pointed out that the mature cell may produce an inhibitor which actively prevents release or a substance which destroys or inactivates an agent normally present in the circulation responsible for stimulating release.

There is little evidence regarding the nature of the mechanism whereby release triggers enhanced proliferative activity. There seems to be a direct effect of decreased mature cell numbers within the marrow. Such cells apparently produce a substance which inhibits cell proliferation. Thus, Craddock (1960) noted that mature granulocytes inhibit the uptake of tritiated thymidine and ^{32}P into lymphocyte DNA in vitro. Such a system could be contained locally within the marrow, although there is some evidence that a distinct circulating factor also exists. It would appear to be extremely difficult to distinguish, in vivo, a direct proliferative response from that caused indirectly by release. However, Bierman et al. (1959) has claimed that "Leukopoietin G" affects neutrophil production and maturation as well as release, and the tissue culture data of Nowell (1959) demonstrated leukopoietic activity in serum from rats with peripheral granulocytosis caused by bacterial infection, and, of course, the "leukopoietic" effects of "colony-stimulating factor" are well documented (Metcalf, 1973).

B. Chalones and Antichalones

The inhibitory effect of the mature granulocyte on leukopoiesis, as well as the presence of direct stimulatory agents, has recently been investigated in detail by Rytömaa and Kiviniemi (1968a, b). These workers have separated and partially purified a chalone (see Bullough, Chapter 1) which specifically inhibits granulocytopoiesis in vitro and a serum antichalone which is stimulatory. The antichalone did not appear to originate from the blood cells, unlike the chalone, which could be purified from serum, cell extracts, or from leukocyte tissue culture media.

The authors state that during increased functional demand for granulocytes, the antichalone appears in the serum and that it enhances DNA synthesis in precursor cells.

The antichalone levels were particularly high and the chalone levels low in the serum of "leukocytapheresis" rats in which large quantities of mature granulocytes had been mobilized into the peritoneal cavity in response to an aseptic inflammatory stimulus. Such experimental animals would correspond to the EAP rats, and this could explain the direct stimulatory effect on neutrophil production which may have been elicited by EAP plasma.

C. Suggestions and Speculations

The existence of a possible chalone or antichalone substance specific for eosinophils has not yet been investigated. Methods are available for obtaining relatively pure eosinophil preparations (Archer, 1968) and thus direct effects of eosinophil extracts on eosinophil production can be investigated. Another approach might be to prepare antisera to eosinophils and assay the plasma of eosinophil-depleted animals for its effect on eosinopoiesis. A methodology of this kind was applied by Lawrence et al. (1967) to a study of neutrophil kinetics.

It has become rather well established that many hormones produce their effects on target cells via cyclic-3',5'-adenosine monophosphate (Robison et al., 1968) and there are indications that this nucleotide may be involved in the action of erythropoietin on bone marrow both in vitro and in vivo (Bottomley et al., 1971; Cohen and Keighley, unpublished observations; Gidari et al., 1971; Winkert and Birchette, 1970). This substance may play an important role in the control of the differentiation of many cell lines. It would be of considerable interest to examine the levels of cAMP or of adenyl cyclase activity in a chalone assay system. Of course, the establishment of a dependable eosinophil assay system in itself is a formidable hurdle to be overcome, although the spleen colony techniques of Fowler et al. (1967) might be utilized with profit.

The eosinophils present many challenges to studies of their specific humoral control because of their relatively small numbers and because of the varied functional controls which are involved in mediating the requirements of the tissues for these cells. These controls include the adrenal hormones, and substances such as histamine, antigens, and antigen-antibody complexes (reviewed by Archer, 1968, and Hudson, 1968). Despite these complexities, however, it appears that the eosinophil, like the red cell, the neutrophil, and the platelets, is subject to specific humoral control. The nature and precise mode of action of this control has yet to be resolved.

IV. SUMMARY

A large body of literature has substantiated the existence of cell-specific humoral factors which control the proliferation and differentiation of various cell lines, including those of the blood. Little work has been done on possible humoral factors specifically concerned with eosinophil production. The existence of such a factor(s) has been suggested by the investigations discussed in this chapter.

A study was made of the effects of intensive eosinophil removal (eosinapheresis) (EAP) on the bone marrow of the rat. During a typical EAP experiment 21.4×10^6 eosinophils were withdrawn per day and definite bone marrow alterations occurred in rats subjected to EAP. These have been interpreted as an acceleration of granulocyte production in response to the increased peripheral utilization of leukocytes incurred by EAP.

The plasma of rats exsanguinated at 6 or 12 hours after EAP elicited striking alterations in the numbers and mitotic indices of eosinophils, neutrophils, and large mononuclear cells in the bone marrow of recipient animals. This plasma may contain a humoral agent(s) which produce(s) enhanced eosinophil and neutrophil release. There were also suggestions of the existence of another factor in 12-hour plasma inhibiting proliferation in immature eosinophils. These findings are discussed in relation to theories of cell-specific control of cellular proliferation.

REFERENCES

Alexander, P., Jr., Monette, F. C., LoBue, J., Gordon, A. S., and Chan, P.-C. (1969). *Scand. J. Haematol.* **6**, 319.
Andersen, V., and Bro-Rasmussen, F. (1968). *Ser. Haematol.* **1** (4), 33.
Archer, R. K. (1968). *Ser. Haematol.* **1** (4), 3.
Bierman, H. R. (1964). *Ann. N.Y. Acad. Sci.* **113**, 753.
Bierman, H. R., Marshall, G., Maekawa, T., and Kelly, K. H. (1959). *Clin. Res.* **7**, 93.
Bottomley S. S., Whitcomb, W. H., Smithee, G. A., and Moore, M. F. (1971). *J. Lab. Clin. Med.* **77**, 793.
Bro-Rasmussen, F., Andersen, V., and Henriksen, O. (1967). *Scand. J. Haematol.* **4**, 81.
Cohen, N. S., LoBue, J., and Gordon, A. S. (1967). *Scand. J. Haematol.* **4**, 339.

Cottier, H., Odartchenko, N., Keiser, G., Hess, M., and Stoner, R. D. (1964). *Ann. N.Y. Acad. Sci.* 113, 612.

Craddock, C. G., Adams, W. S., Perry, S., Skoog, W. A., and Lawrence, J. S. (1955). *J. Lab. Clin. Med.* 45, 881.

Craddock, C. G. (1960). *In* "Ciba Found. Symp. on Haemopoiesis" (G. E. W. Wolstenholme and M. O'Connor, eds.), pp. 237–261. Little, Brown, Boston, Massachusetts.

Dustin, P., Jr. (1959). *In* "The Kinetics of Cellular Proliferation" (F. Stohlman, Jr., ed.), pp. 50–56. Grune and Stratton, New York.

Fowler, J. H., Wu, A. M., Till, J. E., McCulloch, E. A., and Siminovitch, L. (1967). *J. Cell Physiol.* 69, 65.

Gidari, A. S., Zanjani, E. D., and Gordon, A. S. (1971). *Life Sci.* 10 (part II), 895.

Goldwasser, E. (1966). *Curr. Topics Develop. Biol.* 1, 173.

Gordon, A. S., Neri, R. O., Siegel, C. D., Dornfest, B. S., Handler, E. S., LoBue, J., and Eisler, M. (1960). *Acta Haematol.* 23, 323.

Gordon, A. S., Handler, E. S., Siegel, C. D., Dornfest, B. S., and LoBue, J. (1964). *Ann. N.Y. Acad. Sci.* 113, 766.

Hudson, G. (1966). *In* "Bone Marrow Reactions" (J. M. Yoffey), pp. 86–104. Williams and Wilkins, Baltimore, Maryland.

Hudson, G. (1968). *Semin. Hematol.* 5, 166.

Komiya, E. (1956). Thieme, Stuttgart.

Lajtha, L. G., Oliver, R. G., and Gurney, C. W. (1962). *Brit. J. Haematol.* 8, 442.

Lawrence, J. S., Craddock, C. G., Jr., Campbell, T. N. (1967). *J. Lab. Clin. Med.* 69, 88.

Linman, J. W., and Pierre, R. V. (1963). *J. Lab. Clin. Med.* 62, 374.

Litt, M. (1960). *Blood* 16, 1318.

LoBue, J., Monette, F. C., Gordon, A. S., Chan, P-C., and Alexander, P., Jr. (1970). *In* "Myeloproliferative Disorders of Animals and Man" (W. J. Clarke, E. B. Howard, and P. L. Hachett, eds.), pp. 84–102. U.S.A.E.C. Div. of Tech. Inform., Oak Ridge, Tennessee.

Metcalf, D. (1973). This monograph.

Nowell, P. C. (1959). *Proc. Soc. Exp. Biol. Med.* 101, 347.

Osgood, E. E. (1954). *Blood* 9, 1141.

Robison, G. A., Butcher, R. W., and Sutherland, E. W. (1968). *Ann. Rev. Biochem.* 37, 149.

Rytomaa, T. (1960). *Acta Pathol. Microbiol. Scand., Suppl.* 140 50.

Rytomaa, T., and Kiviniemi, K. (1968a). *Cell Tissue Kinet.* 1, 329.

Rytomaa, T., and Kiviniemi, K. (1968b). *Cell Tissue Kinet.* 1, 341.

Samter, M. M., Kofoed, A., and Pieper, W. (1953). *Blood* 8, 1078.

Schultz, E. F., Lapin, D. M., and LoBue, J. (1973). This monograph.

Speirs, R. S. (1955). *Ann. N.Y. Acad. Sci.* 59, 706.

Winkert, J. and Birchette, C. (1970). *Fed. Proc.* 29, 843.

Whitcomb, W. H., and Moore, M. Z. (1965). *J. Lab. Clin. Med.* 66, 641.

Yoffey, J. M. (1966). "Bone Marrow Reactions." Williams and Wilkins, Baltimore, Maryland.

Yoffey, J. M., Makin, G. S., Yates, A. K., Davis, C. J. F., Griffiths, D. A., and Waring, I. S. (1964). *Ann. N.Y. Acad. Sci.* 113, 790.

5

THE COLONY STIMULATING FACTOR (CSF)

Donald Metcalf

The proliferation of any cell population *in vitro* is dependent on and therefore regulated by a multiplicity of factors ranging from temperature and pH to simple organic building blocks. It is also quite clear that specific cell types must have specific growth regulators which have no, or insignificant, effects on other cell types. It is the goal of those working on the culture of specific cell types to identify and characterize factors which appear to have cell-specific regulatory properties. Furthermore, to establish the biological validity of conclusions drawn from *in vitro* studies, it is mandatory to show that these factors exist and have an essentially similar function in the intact animal.

This chapter will describe an *in vitro* system for growing clones of granulocytes and/or macrophages in which colony growth is dependent on stimulation by one such growth regulator, colony stimulating factor (CSF). In line with the above generalizations, our studies on CSF have attempted (a) to establish the chemical nature and action of this factor using the *in vitro* culture assay system, and (b) to document the production and action of CSF in the body, both in health and disease.

The phenomenon of colony growth in semisolid agar cultures of hematopoietic cells was discovered independently by Bradley and Metcalf (1966) and Pluznik and Sachs (1966). Both groups showed that when mouse bone marrow or spleen cells were cultured in agar medium, discrete colonies of cells developed which were composed of populations of granulocytic and/or macrophage cells (Bradley and Metcalf, 1966; Ichikawa *et al.*, 1966; Metcalf *et al.*, 1967). Colonies were formed by the proliferation of a small number of the hematopoietic cells which had been plated, and colony formation only occurred if an underlayer of feeder cells, or medium conditioned by such cells, was used in the cultures. Subsequent studies showed that certain mouse sera would also stimulate colony formation (Robinson *et al.*, 1967). The operational term colony stimulating factor, or CSF was applied to the active factor released by the feeder layer cells and present in serum or conditioned medium which stimulated colony formation.

I. GRANULOCYTIC AND MACROPHAGE COLONY FORMATION

Hematopoietic cells from most animal species will form colonies of granulocytes or macrophages if cultured in semisolid agar or methylcellulose, but most work so far has used mouse, monkey, or human cells.

In the simplest culture techniques, dispersed cell suspensions are prepared and added to an equal parts mixture of double strength medium (containing a high content of fetal calf or horse serum) and 0.6% agar solution held at 37°C (Metcalf, 1970). Portions of this mixture are then pipetted into glass or plastic petri dishes to form a 1-mm thick layer (1 ml volumes for 35 mm dishes or 2 ml volumes for 50 mm dishes). The agar medium is allowed to gel at room temperature, then the cultures are incubated at 37°C for 7–10 days in a fully humidified atmosphere of 10% CO_2 in air.

Within 24 hours, small aggregates of cells can be seen in the cultures and by 3–4 days, developing colonies are quite obvious. For most mouse work, colonies are counted and cytologically classified after 7 days of incubation without further refeeding. Human and monkey colonies grow somewhat more slowly than do mouse colonies and often are not scored until 10–14 days. Depending on the type of culture used, colonies may vary widely in size but maximum colony size is about 5000–10,000 cells.

One of the early facts established about colony formation *in vitro* was that colonies did not arise by aggregation of the cells originally plated, but were formed by cell proliferation. The semisolid nature of the culture medium virtually precludes any significant cell migration, and analysis of developing colonies showed no significant migratory movement of cells within the agar (Metcalf *et al.*, 1967). The relationship between the number of colonies obtained and the number of cells plated suggested strongly a single-cell origin of colonies and this was supported by experiments in which plastic rings were inserted in the agar to isolate developing colonies (Pluznik and Sachs, 1965). Formal proof of the clonal origin of colonies from single cells has recently been achieved by obtaining colony formation with single monkey bone marrow cells isolated by micromanipulation and implanted individually into agar cultures (Moore *et al.*, 1972).

Most colonies in agar therefore appear to be clones derived from individual cells and the cells initiating such colonies were given the term *in vitro* colony-forming cell or *in vitro* CFC, to distinguish them from the multipotential hematopoietic stem cells (CFU) which form colonies in the spleen of irradiated mice. Recent evidence suggests that the cloning efficiency of the agar culture system is unusually high being at least 30–40% and probably much higher. Because of the ease of scoring and the high culture efficiency, the agar culture system is particularly useful in the detection and enumeration of *in vitro* CFC's. In the adult animal, *in vitro* CFC's are found only in three locations—the bone marrow, spleen, and blood—and the incidence of these cells in the mouse is approximately $200:10^5$ bone marrow cells; $4:10^5$ spleen cells and $0.5:10^5$ nucleated peripheral blood cells.

Studies on the nature of the *in vitro* CFC's have indicated that these cells are heterogeneous with respect to buoyant density (Haskill *et al.*, 1970; Metcalf *et al.*, 1971), adherence to glass bead columns (Metcalf *et al.*, 1971), and size (Worton *et al.*, 1969). Because of this heterogeneity, attempts to use these separative procedures to purify and identify *in vitro* CFC's have not been successful using mouse cells. However the *in vitro* CFC's in adult monkey bone marrow have proved to be much more uniform with respect to buoyant density and separation on continuous albumin density gradients has achieved fractions in which 1 in 3 cells is a colony-forming cell (Moore *et al.*, 1972). Combination of this procedure with thymidine suiciding and autoradiography has shown that most *in vitro* CFC's are medium-sized cells, 9–11 μm in diameter, with a round leptochromatic nucleus and a scanty nongranulated rim of cytoplasm. This cell corresponds to the cell classified by Yoffey as a "transitional lymphocyte" although use of the term "lymphocyte" in this context is probably misleading as the cells have no direct relationship with immunologically competent cells. A small proportion of *in vitro* CFC's may have the morphology of blast cells or promyelocytes. However the information available so far indicates that the majority of cells initiating granulocyte and macrophage colonies *in vitro* are undifferentiated cells with no specific morphological features, indicating their differentiating potentiality.

A number of cell cycle killing agents—titrated thymidine, vinblastine and hydroxyurea—have been used to demonstrate that in the mouse, monkey, and human, most *in vitro* CFC's are in active cell cycle (Lajtha *et al.*, 1969; Iscove *et al.*, 1970; Rickard *et al.*, 1970; Metcalf, 1972; Moore *et al.*, 1972a) although there may be a minority population which could be in G_0 in adult bone marrow.

There is a striking parallelism between *in vitro* CFC's and CFU's in their location and relative frequency in various organs. Analysis of spleen colonies for their content of CFU's and *in vitro* CFC's also indicated some correlation between both cells and the suggestion was made that they might be the same cell population (Wu *et al.*, 1968). Although there is still some difference of opinion (Dicke *et al.*, 1971), most evidence at present suggests that both populations are distinct with possibly a minor population of cells having properties of both cell types. The likely relationship between these two cells was indicated in a series of experiments in which marrow cells were studied from mice previously injected with large doses of vinblastine. These bone marrows contain few *in vitro* CFC's but since most CFU's are in a noncycling G_0 state, they survive vinblastine treatment. Certain density fractions from such marrow populations contained no detectable *in vitro* CFC's but were able to produce spleen colonies in irradiated recipients. These spleen colonies contained large numbers of *in vitro* CFC's (Haskill *et al.*, 1970; Metcalf and Moore, 1971), indicating clearly that most *in vitro* CFC's are the progeny of CFU's, and explaining why so often both cell populations can be shown to have some relationship with each other.

In vitro CFC's also have a relationship with erythropoietic precursor cells. In mice made anemic by bleeding, the incidence of *in vitro* CFC's in the bone marrow is sharply reduced. Conversely, in mice made polycythemic by hypertransfusion, erythropoiesis is suppressed and the incidence of *in vitro* CFC's is high (Bradley *et al.*, 1967; Metcalf, 1969). These observations are compatible with the interpretation that erythropoietin-sensitive cells and *in vitro* CFC's are the immediate progeny of the CFU and, depending on demand, available CFU's are directed into one or other pathway.

The development of colonies in the solid-state culture system should lend itself admirably to time-lapse photographic analysis of the initial stages in colony formation, but the incubating conditions are stringent, making microscopy difficult, and so far no such analysis has been reported. However, a static analysis of events in the first 48 hours of colony growth has revealed an interesting situation which is worthy of further study. Analysis of the size frequency distribution of developing colonies in cultures of mouse bone marrow cells has strongly suggested not only a surprising degree of symmetry but also of synchrony in early cell divisions. Peak frequencies of 2, 4, 8, 16, and 32 cells were observed in growing colonies and all colony cells at this stage appeared to have a uniform morphology—that of early granulocytic cells (Metcalf, D., unpublished data). Because no cell death occurs in colonies in the first 72 hours, colony growth is exponential with a mean doubling time of 12–15 hours under conditions of maximal stimulation.

Subsequent events in colony development are more difficult to analyze because of the occurrence of two important processes: (a) some colony granulocytic cells begin to differentiate to nondividing metamyelocytes and polymorphs, some of which then disintegrate, and (b) colony granulocytes transform into macrophages.

The capacity of colony granulocytes to differentiate *in vitro* to mature polymorphs makes this system of unusual interest as a model not only for analyzing granulopoiesis but also for investigating basic processes in differentiation. However, this phenomenon does complicate the analysis of colony growth rates since an increase in growth rate can be due either to shorter cell cycle times or to reduction in the proportion in differentiating (nondividing) cells.

From the earliest work on colony formation *in vitro*, it was evident that two distinct types of colonies were present in cultures after 7–10 days of incubation: compact colonies which were usually composed of pure populations of granulocytic cells, and loose colonies of widely dispersed cells which were often, but not invariably, composed of mononuclear cells with bulky vacuolated cytoplasm and active phagocytic capacity. Subsequent investigation of these monocytic cells in mouse colonies has shown that they have a full complement of IgG_{2a} receptors typical of other macrophages and characterizing them as a well-differentiated example of macrophages (Cline *et al.*, 1972). It was initially assumed that two distinct populations of *in vitro* CFC's existed, producing either granulocytic or macrophage colonies (Ichikawa *et al.*, 1966). However, the occurrence in all cultures of large numbers of mixed colonies containing both granulocytic and macrophage cells casts doubt on this interpretation (Metcalf *et al.*, 1967). Complicating the investigation of the nature of mixed colonies was the fact that macrophages are present in the agar between developing colonies and might possibly become incorporated as a second population in expanding granulocytic colonies.

Analysis of developing colonies by plate mapping and colony transfer showed that the majority of colonies commenced as aggregates of granulocytic cells and that most macrophage colonies could be shown to have originated from such granulocytic populations (Metcalf, 1969a). Subsequently it was demonstrated that individual granulocytic cells from early colonies could be transferred to new cultures where they transformed without further division to macrophages or generated small clones of macrophages (Metcalf, 1971a). More recently, the clonal origin of mixed colonies has been proved by culturing single monkey *in vitro* CFC's and producing single colonies containing populations of both granulocytic and macrophage cells (Moore *et al.*, 1972).

Some *in vitro* CFC's may only be capable of forming colonies which remain purely granulocytic, become composed of mature polymorphs, and then finally disintegrate. However the majority of granulocytic colonies, regardless of the culture system used, will ultimately transform to macrophage colonies. This transformation is more rapid in cultures of cells from some species (e.g., the mouse) than from others (e.g., the monkey or human) and can be accelerated using various culture systems or by pretreatment of the *in vitro* CFC's (see later). *In vitro* CFC's can be partially separated by buoyant density according to their capacity to form colonies exhibiting early or late macrophage transformation (Janoshwitz *et al.*, 1971) indicating that the timing of this process is related in part to intrinsic differences between individual *in vitro* CFC's.

In vitro colonies are characteristically heterogeneous in size and in any one culture can range from a size of 50–100 cells up to large aggregates of 5000–10,000 cells. The reason for this heterogeneity is not known, but in part may be due to some *in vitro* CFC's being more ancestral than others and having a greater capacity for cell division before differentiating to nondividing cells. A similar comment can be made with respect to the origin of the small aggregates of cells which are present in all cultures between the large aggregates scored as colonies. These small aggregates (clusters) may be initiated by cells which are the progeny of *in vitro* CFC's and therefore have a more limited capacity for division. Cluster-forming cells have essentially similar physical properties to *in vitro* CFC's although some separation has been achieved by physical methods (Metcalf *et al.*, 1971).

Hematopoietic colony growth also shows several other phenomena of general biological interest. In cultures of normal cells using submaximal concentrations of CSF, colony growth rates can vary according to the number of colonies present in the culture dish, colony crowding potentiating colony growth until intense overcrowding with confluence of colonies obscures further observations (Metcalf, 1968). Potentiation of normal colony growth can also be achieved by increasing the number of hematopoietic cells cultured and this phenomenon can be reproduced by adding additional nonviable bone marrow cells, or lymphoid cells which survive only for short periods in culture (Metcalf, 1968). Autoradiographic analysis of colonies developing in cultures where labeled lymphoid or nonviable bone marrow cells had been added showed that all colony cells exhibited nuclear labeling. This indicates that extensive reutilization of nuclear material occurs in developing colonies and this may in part be the mechanism by which disintegrating cells potentiate colony growth. However, as shall be discussed later, some hematopoietic cells can also produce CSF and thus potentiate colony growth by this mechanism.

II. NATURE AND ORIGIN OF COLONY STIMULATING FACTOR

In cultures of mouse bone marrow cells, colony growth as described in the preceding section is entirely dependent on the presence in the culture medium of an adequate concentration of a specific factor—colony stimulating factor (CSF). In the absence of added CSF, some abortive proliferation occurs, but the developing aggregates soon stop increasing in size and within a few days disintegrate. In the culture systems originally developed, CSF was supplied to the hematopoietic cells by using an underlayer of agar in which were 10^6–10^7 feeder layer cells per ml. Cell types successfully used as feeder layers were mouse, embryo, fibroblasts, trypsinised embryo, kidney, spleen, and thymic cells (Pluznik and Sachs, 1965; Bradley and Metcalf, 1966). As a simplification of this method, fluid harvested from cultures of such cells (conditioned medium) was substituted in the underlayer or mixed directly in the hematopoietic cell layer and shown to have colony stimulating activity (Pluznik and Sachs, 1966; Ichikawa et al., 1966; Bradley and Sumner, 1968).

Because of the complexity of biological processes occurring in underlayers and in an attempt to obtain evidence that CSF existed in vivo in significant concentrations, the colony stimulating activity of mouse sera was tested by mixing the serum directly in with the agar medium containing in vitro CFC's. It was found (Bradley et al., 1967a; Robinson et al., 1967) that the addition of a small volume of mouse serum to cultures of mouse bone marrow cells stimulated colony formation, the number and growth rate of colonies increasing with the volume of serum used. Similar colony stimulating activity was observed with human serum

(Foster *et al.*, 1968) and with dialyzed, but unconcentrated, human urine (Robinson *et al.*, 1969; Metcalf and Stanley, 1969).

Attempts were made to purify CSF from normal and leukemic mouse serum (Stanley *et al.*, 1968) but, with the recognition that urine exhibited equally high biological activity, efforts were concentrated on purifying CSF from normal human urine. Initial results made it probable that CSF was a protein, and chromatographic and gel filtration procedures were developed which resulted in a 500- to 1000-fold purification of CSF with respect to protein and with an overall recovery of 50% (Stanley and Metcalf, 1969). The molecular weight of CSF in this material was established as 45,000 using gradient gel electrophoresis or sedimentation in sucrose gradients and as 60,000 by gel filtration (Stanley and Metcalf, 1971). Studies on the response of this material to digestion with a variety of proteolytic enzymes showed its relative resistance to peptidases but confirmed the peptide content of CSF, biological activity being destroyed following incubation with α-chymotrypsin and subtilisin (Stanley and Metcalf, 1971a). A further purification procedure was based on the relative resistance of human urine CSF to proteolytic enzymes. Semipurified material was digested with an insolubilized papain preparation which broke down much of the remaining contaminating protein in these preparations. Electrophoresis of this digested material in polyacrylamide gels localized all colony stimulating activity to a single band which stained positively for protein and carbohydrate (Stanley and Metcalf, 1972). Digestion of this material with purified neuraminidase altered the electrophoretic mobility of CSF, presumably by removing charged neuraminic acid residues but did not destroy biological activity. It has been concluded that CSF in normal human urine is a neuraminic acid-containing glycoprotein of MW, 45,000–60,000.

The material resulting from papain digestion and electrophoretic separation was between 100- 200,000-fold purified with respect to protein, when compared with starting material, and the overall recovery was 10–20%. This material stimulated colony formation at concentrations of less than 100 pg/ml (10^{-12} M) and calculations from these data indicated that the concentration of CSF in normal human serum was approximately 1 ng/ml, similar to that of other glycoprotein regulators, e.g., growth hormone. Of some interest was the fact that this highly purified material stimulated the development and growth of both granulocytic and macrophage colonies and no evidence has been obtained for two separate CSF's in human urine which might be specific for each cell class.

There is a general chemical similarity of CSF to erythropoietin (Goldwasser and Kung, 1971) which suggests that these hematopoietic regulators may belong to a family of molecules with broadly similar chemical properties and mechanisms of action, although there are likely to be many differences in detail.

The sedimentation coefficient of CSF in human urine is considerably lower than that obtained for CSF in mouse serum (Stanley et al., 1968) which raises the possibility that the CSF molecule might be partially degraded before being cleared by the kidney. The difficulty encountered in demonstrating an action of neuraminidase on urine CSF compared with its action on CSF from other sources (see later) also suggests that urine CSF may differ in having a lower carbohydrate content or less accessible carbohydrate residues.

Parallel studies on the chemical nature of CSF from other murine sources have also indicated that not all CSF's are identical in size or composition. Analysis of CSF in medium conditioned by mouse fibroblasts or embryo cells has confirmed the general properties of CSF although this material seems less heat-labile than serum or urine CSF. Again the molecular weight ranged from 40,000–70,000 according to the techniques used (Stanley et al., 1971; Austin et al., 1971) and a carbohydrate content was demonstrable by neuraminidase digestion (Stanley et al., 1971). However analysis of CSF in embryo-conditioned medium indicated that as incubation proceeded, CSF released by the embryo cells developed progressively greater electrophoretic heterogeneity with mobilities ranging from the α post-albumin region to the globulin region. The biological activity of these different fractions appeared to be similar.

Evidence of more obvious differences between CSF's from different sources was obtained from an analysis of CSF extractable from different organs. Here CSF from various organs differed in its elution characteristics after adsorption on calcium phosphate gels, CSF in organs like the submaxillary gland differing sharply from serum CSF (Sheridan and Stanley, 1971). Even greater differences were observed in CSF extracted from various mouse tissues after injection of endotoxin into the mouse. CSF extracted from the submaxillary gland differed in elution profiles and electrophoretic mobility from CSF extracted from the normal organ and from other organs. Furthermore, in sucrose gradients of CSF produced by lung tissue from endotoxin-injected mice, at least 3 peaks of active material were obtained with apparent molecular weights of 15,000, 30,000, and 45,000 (Sheridan, J., and Metcalf, D., unpublished data). The significance of these more recent observations has not yet been established. It may be that not the whole of CSF molecule is required for biological activity, although careful analysis of enzyme-digested urinary CSF has failed to reveal breakdown products with biological activity. However the existence of active subunits has not yet been disproved and clearly there can be considerable variation in chemical structure without gross changes in biological activity.

In this context the CSF released by lung tissue in liquid culture stimulates an abnormally high proportion of granulocytic colonies of characteristic morphology, possibly indicating a slightly different mode of action. However with this exception, little difference has been noted between colonies stimulated by equivalent concentrations of CSF from different sources.

Some species specificity has been observed in the activity of CSF in different combinations of CSF and target *in vitro* CFC's. Mouse *in vitro* CFC's appear to be unusual in being responsive to stimulation by CSF from a wide variety of mammalian species. However, with this exception, *in vitro* CFC's from various species appear to be most responsive to CSF of homologous origin and can be quite unresponsive to some foreign CSF's (Bradley *et al.*, 1969; Moore, M. A. S., unpublished data). Further evidence of species specificity was obtained when rabbits were immunized with human urinary CSF. High titers of antibody were produced which blocked or inactivated human CSF in culture but these antisera either had no, or only weak, inhibitory activity on a variety of mouse CSF's (Stanley *et al.*, 1970).

III. MECHANISM OF ACTION OF CSF *IN VITRO*

A number of investigations have been made to determine the possible mode of action of CSF in stimulating colony formation. In general terms, the activity of highly purified CSF at a concentration of 10^{-12} M suggests that it may act mainly or exclusively at the membrane of the target *in vitro* CFC's. Although it has been possible to radio-iodinate CSF without loss of biological activity (Stanley and Metcalf, 1972) no studies have yet been made to identify possible receptor sites on the *in vitro* CFC membrane.

When standard numbers of bone marrow cells are cultured, the number and growth rate of the colonies which develop are related to the concentration of CSF in the culture medium (Robinson *et al.*, 1967; Metcalf and Stanley, 1969; Metcalf, 1970). A sigmoid dose-response relationship exists between CSF concentration and colony numbers and the linear portion of this curve can be used as a sensitive bioassay system for determining CSF concentration in biological fluids.

CSF is not simply a trigger molecule initiating the first division in the *in vitro* CFC. CSF is required continuously throughout colony growth and when developing colonies are transfered to cultures lacking CSF, colony growth ceases immediately and most colonies disintegrate (Metcalf and Foster, 1967; Paran and Sachs, 1968). Assays of the agar medium in which active colony formation has occurred have indicated that CSF concentrations are reduced (Metcalf, 1970). This does not necessarily indicate utilization by dividing colony cells, as CSF might be nonspecifically broken down by enzyme action or be inactivated or bound to cells in the cultures. When *in vitro* CFC's are incubated in agar medium in the absence of added CSF, most CFC's either rapidly die or lose their capacity for proliferation (Metcalf, 1970). This finding raises the possibility that CSF may merely be a survival factor, allowing cells to survive and express their intrinsic capacity for proliferation. If this were the case, CSF could not be regarded as a bona fide growth stimulator and might not necessarily have a significant role *in vivo*.

However recent observations on the growth rate of granulocytic colonies early in the incubation period have indicated that CSF has a concentration-dependent effect on the mean cell cycle time of colony cells—the higher the concentration the shorter the cell cycle time. Reduction of the CSF concentration over a 36-fold range progressively lengthened the mean cell cycle time from 15 to 18.6 hours (Metcalf, D., unpublished data).

Further evidence that CSF actively stimulates cellular proliferation was provided by analysis of the events occurring in cultures during the early part of the incubation period. There is a variable lag period before individual *in vitro* CFC's or cluster-forming cells begin proliferation *in vitro*. Because of this, the total number of discrete aggregates (developing colonies and clusters) increases linearly throughout the first 5–7 days of the incubation period (Metcalf, 1969a). The length of the lag period does not appear to be related to the cell cycle status of the cells at the time of initiation of the cultures (Metcalf, 1972). Why some *in vitro* CFC's should remain dormant in culture for up to 4–6 days is obscure since the majority of these cells were in active cell cycle. Increasing the concentration of CSF in these cultures shortens the mean lag period before the cells commence proliferation, although this effect has an upper limit and no matter how high the concentration of CSF, not all *in vitro* CFC's can be made to initiate proliferation simultaneously (Metcalf, 1970).

Because some colony cells differentiate to nondividing cells, colony growth rates are not only dependent on mean cell cycle times but also on the proportion of colony cells remaining capable of division. The influence of CSF on the differentiation of colony cells has not been analyzed but there is some suggestion that with very high concentrations of CSF some colonies develop which appear to be composed solely of very primitive granulocytic cells. This suggests that CSF may have an influence on cellular differentiation but there is such variation in pattern of colony differentiation, even between adjacent colonies in the same culture, that intrinsic differences between colony-forming cells and their immediate progeny may be the primary factor determining the pattern of cellular differention in individual colonies.

The fact that some colonies are only initiated by very high concentrations of CSF whereas others proliferate with low concentrations, raises the intriguing possibility that there may be marked heterogeneity in *in vitro* CFC's with respect to their responsiveness to stimulation by this regulator. This type of possibility exists with respect to the responsiveness of all target cell populations to specific regulators but cannot readily be tested *in vivo*. The agar culture system should permit answers to be obtained to this question, at least for the *in vitro* CFC population. An alternative possibility is that the higher number of colonies obtained with increasing CSF concentrations may reflect stochastic processes and it must be admitted that computer-generated dose-response curves, based on models in which *in vitro* CFC's are assumed to have uniform thresholds for stimulation, fit many observed titration curves (Maritz *et al.*, 1972). However a strong argument against the stochastic model of uniform CFC's is the fact that colonies also vary greatly in size. On a simple stochastic model, most colonies would tend to be uniform in size since daughter cells in all colonies would have an equal probability of making contact with suprathreshold numbers of CSF molecules. In fact, however, gross differences in colony size are noted and although other variables could be operative, these findings suggest strongly that not only do CFC's vary in their responsiveness to CSF but that the responsiveness of individual CFC's may be inherited by the immediate progeny of that CFC. These possibilities should be capable of investigation by reciprocal colony transfer experiments in which colonies initiated by high or low concentrations of CSF are transfered to cultures with varying concentrations of CSF.

Cultures stimulated by CSF in mouse or human serum tend to contain many macrophage colonies in distinction to the large numbers of granulocytic or mixed colonies stimulated by feeder layers or conditioned medium (Robinson *et al.*, 1967; Metcalf and Foster, 1967a; Foster *et al.*, 1968). Furthermore, when large doses of mouse or human serum are employed, often inhibition of colony formation is observed. Investigation of such sera revealed the presence of a complex of lipoproteins which was responsible for the inhibition of colony formation and for the premature transformation of granulocytic to macrophage colonies

(Stanley *et al.*, 1968; Chan *et al.*, 1971). These inhibitors were heat-labile and were removable by centrifugation, ether extraction, or by precipitation following dialysis of the serum against water. Fractionation of serum by gel filtration or electrophoresis indicated that more than one type of inhibitory lipoprotein was present in normal serum and an assay system was developed for determining inhibitor levels in sera (Chan *et al.*, 1971). The inhibitory activity of these lipoproteins is relatively species-specific, specificity being determined by the source of CSF rather than the species of CFC. The inhibitors do not appear to be cytotoxic for *in vitro* CFC's or colony cells. Being of large molecular weight, the inhibitors are poorly diffusible and underlayers containing both CSF and inhibitors stimulate colony formation in overlayers containing target cells, presumably because of the inability of the inhibitors to diffuse upwards into the target cell layer.

The mechanism by which these inhibitors block colony formation is not known but if complexing occurs with CSF, then presumably such complexes must be readily dissociable to allow CSF in underlayers to diffuse readily and stimulate colony formation in overlayers of agar.

Preincubation in liquid culture of *in vitro* CFC's with inhibitors does not reduce the total number of these cells but when such cells generate colonies, premature transformation to a macrophage composition occurs (Chan, 1971). This interesting example of "directed" differentiation warrants further exploration.

Although exhaustive tests have not been carried out using semipurified CSF, evidence so far available suggests that the stimulating action of CSF is specific and limited to cells of the granulocytic and macrophage series. CSF in human urine or mouse serum did not stimulate the transformation of mouse lymphocytes, or potentiate the generation of antibody-producing cells in antigen-stimulated cultures of mouse spleen cells (Byrd, W. and Metcalf, D., unpublished data). Similarly, CSF-containing conditioned medium did not significantly stimulate erythroid colony formation in plasma gel cultures (Stephenson *et al.*, 1971) and erythropoietin has no effect on granulocyte and macrophage colony formation in agar cultures (Bradley *et al.*, 1969). CSF in conditioned medium also did not influence the growth rate of P388 or HeLa cells in liquid or agar (Bradley *et al.*, 1969) or the growth of colonies of mouse plasma cell tumor cells or mastocytoma cells in agar (Metcalf, D., unpublished data).

As discussed earlier, the probable morphology of most *in vitro* CFC's is that of a medium-sized cell with no obvious cytoplasmic granularity and no IgG receptor sites. In the case of the mouse, when such cells divide following stimulation by CSF, the two daughter cells invariably have the morphology of early granulocytic cells with ring-shaped nuclei. Since bone marrow fractions containing CFU's but no *in vitro* CFC's cannot be induced to form colonies by the action of CSF, it can be concluded that the CFU to CFC transition is not under direct influence of CSF. General evidence indicates that the stem cell to progenitor cell sequence is regulated by the tissue microenvironment and is tantamount to derepression of a specific portion of the genome, committing the progenitor cell to a restricted pathway of differentiation. However, the progenitor cell does not itself phenotypically express this derepression of specific genetic material until activated by the appropriate humoral regulator, e.g., erythropoietin or CSF. Precommitment however probably results in the generation by the progenitor cell of specific molecular configurations at the cell membrane (receptor sites) which subsequently allows specific contact to be made between target cell and humoral regulator.

From the evidence so far available, this aspect of the mode of action of CSF appears to be essentially similar to that documented for erythropoietin. The primary target cell for CSF—the *in vitro* CFC—seems precommitted to the granulocyte and macrophage pathway of differentiation but exhibits no morphological evidence of this until stimulated to divide by CSF, following which both progeny are clearly granulocytic in nature. Contact between CSF and specific receptor sites presumably allows an RNA synthesis to be initiated leading to granule formation and changes in nuclear morphology; but analysis of the details of such events must await the preparation of pure CSF and purified populations of *in vitro* CFC's. The fact that both daughter cells are identifiable as granulocytes clearly speaks against any self-replicative capacity of CFC's as does the symmetry of subsequent divisions which was referred to earlier. This description of the action of CSF on *in vitro* CFC's has glossed over certain important questions about which no information is as yet available. Does granulocyte differentiation becomes evident in the *in vitro* CFC *before* its first division *in vitro?* Do CFC's self-replicate *in vivo?* In the absence of CSF does the *in vitro* CFC die or simply lose its capacity for division and/or differentiation? Is the active cell cycle status of most *in vitro* CFC's in the body due to stimulation by CSF or regulated by other control mechanisms?

At the present time the action of CSF on *in vitro* CFC's can be summarized as (a) the induction of proliferation and differentiation of the *in vitro* CFC's which then generate two daughter granulocytic cells, (b) the stimulation of the proliferation and subsequent differentiation of these granulocytic cells, and (c) the stimulation of the proliferation of macrophages derived from these colony granulocytes. The granulocyte to macrophage transformation may not actually be under CSF regulation, and may involve the action of the lipoprotein inhibitors identified in serum.

IV. ORIGIN AND METABOLISM OF CSF

CSF is detectable in the unconcentrated serum and urine of normal mice (Robinson *et al.*, 1967; Metcalf and Foster, 1967a; Chan, S. H., unpublished data) and man (Chan *et al.*, 1971; Robinson *et al.*, 1969; Metcalf and Stanley, 1969). In view of the fact that the standard bioassay involves the addition of 0.1 ml of serum or urine to a 1 ml culture, the concentration of CSF potentially impinging on the corresponding target cells *in vivo* is at least 10 times that producing detectable stimulation *in vitro*.

CSF levels in the serum appear to be highly labile. Very rapid rises in CSF levels occur following infections (Foster *et al.*, 1968; 1968a) of the injection of bacterial antigens (Metcalf, 1971c) and these elevated levels decline quickly, indicating a probable half-life of less than 4 hours. The serum half-life of injected human urine CSF was studied in mice using a discriminatory antibody to distinguish human for mouse CSF (Metcalf and Stanley, 1971). The data indicated a half-life of appoximately 3 hours and human CSF has been detected in significant quantities in the urine of such mice.

In view of the fact that kidney cells can release CSF *in vitro*, the CSF in urine potentially represents either CSF cleared from the plasma or produced by renal or other genitourinary tract cells. Bilateral nephrectomy in mice led to a rise in serum CSF levels (Chan, 1970), and elevated serum CSF levels have been observed in anephric patients (Chan, S. H., unpublished data). Nephrectomy in mice tends to prevent the fall in serum CSF levels following the injection of large doses of cortisone and analysis of the urine from cortisone-injected mice has indicated that elevated CSF levels are present roughly corresponding to the CSF lost in the serum (Metcalf, 1969b; Chan, 1970). These observations suggest that clearance by the kidney is a major metabolic fate of serum CSF, although this is unlikely to be the sole pathway of CSF breakdown or elimination.

The tissue origin of CSF has not been determined. Many tissues are able to provide cells which, in feeder layers, can stimulate colony formation. Among the most active in this regard are trypsinized embryo, kidney, and thymus cells (Pluznik and Sachs, 1965; Bradley and Metcalf, 1966). Neoplastic hematopoietic cells can also serve as underlayers

(Bradley, T. R. and Metcalf, D., unpublished data ; Paran *et al.*, 1968; Metcalf *et al.*, 1969). Similarly, long-term *in vitro* cell lines derived from normal lymphocytes or fibroblasts have also been used to stimulate colony formation.

An alternative approach has been to preincubate cell suspensions in liquid medium for varying periods and to assay the conditioned medium for colony stimulating activity. Pluznik and Sachs (1966) and Ichikawa *et al.* (1966) reported that embryo cells conditioned medium and Bradley and Sumner (1968) showed strong colony stimulating activity using medium conditioned by kidney or embryo cells. Active conditioned medium has also been prepared from suspensions of bone marrow, spleen, thymus, and lymph node cells (Metcalf, 1971c). As before, many neoplastic hematopoietic populations have been shown to actively condition medium (Paran *et al.*, 1968; Pluznik, 1969; Metcalf, 1971d).

A more direct attempt to determine the tissue or tissues producing CSF has been the chemical extraction of CSF from various organs. Studies in adult C57BL mice indicated that CSF was extractable in large amounts from submaxillary gland, lung, thymus, kidney, and spleen and in lesser amounts from lymph node, pancreas, bone marrow, brain, heart, small intestine, testis, skeletal muscle and liver. On a wet weight basis, all organs extracted contained a higher concentration of CSF than was present in whole blood. The highest CSF concentrations were from salivary gland and the male salivary gland contained higher concentrations than the female (Sheridan and Stanley, 1971). This is of interest in view of the high content and known sex difference in the submaxillary gland content of both nerve growth factor and epidermal growth factor (Levi Montalcini and Angeletti, 1968; Turkington *et al.*, 1971).

Resection of the spleen, kidneys, thymus, or 75% of the liver did not cause appreciable falls in the serum CSF levels and these individual organs do not appear to be major contributors to serum CSF pools.

The data so far available are consistent with two general interpretations: (a) many different cell types in the body are able to produce or release CSF, or (b) CSF is produced in different tissues by a specific cell, e.g., macrophage, endothelial cell or fibroblast, which is common to all tissues. The capacity of neoplastic cells or established cell lines to produce CSF may not be relevant in this context since this capacity may represent an abnormal derepression of the genome of such cells.

So far the only relatively pure population of normal cells tested directly from the animal has been glass-adherent peritoneal macrophages. These produced highly active conditioned medium, strengthening the possibility that the tissue macrophage might be a significant source of CSF in various organs (Sheridan, J. W., unpublished data).

Against the concept of a dispersed population of one morphological cell type being the origin of CSF is the finding of physicochemical differences between CSF's extracted from different normal organs. The CSF from the male submaxillary gland is outstanding as being very different from other tissue CSF's (Sheridan and Stanley, 1971). Unless cells like macrophages differ significantly from one organ to another, such diversity in CSF's is strongly suggestive that quite different cell types can produce CSF. A solution to this question may come from the use of fluorescein-conjugated antisera prepared against purified CSF to stain tissue sections. None of the techniques used so far is capable of answering the problem of which organs are the main contributors to the serum CSF pool, since apparent function *in vitro* depends so much on selective cell survival and the assessment of tissue CSF pools does not in itself indicate the turnover of CSF within the cells.

One favored suggestion has been that the mature polymorph is the major source of CSF (Robinson and Pike, 1970). This simple situation would provide a direct feedback regulation of granulocyte production, adjusted to balance polymorph breakdown. Such a system would not of course explain regeneration of granulopoiesis after neutropenia. Some support for the polymorph origin of CSF has been the observation that serum colony stimulating activity is elevated by whole body irradiation and endotoxin, both of which cause polymorph breakdown. Furthermore, peripheral white cells but not peripheral blood lymphocytes strongly stimulated colony formation when used as underlayers (Robinson and Pike, 1970).

While granulocytes may be a good source of CSF they seem unlikely to be a major source. Fractionation of blood cells on adherence columns and by buoyant density separation has produced pure fractions of granulocytes which have no measurable feeder layer activity (Moore, M. A. S., and Williams, N., unpublished data). Colony stimulating activity was exhibited by those fractions containing monocytes and adherent lymphocytes although neither of these cell types has yet been tested in a pure population. Additional evidence against a major role of polymorphs in CSF production has been the observation that extracts of polymorph-rich tissues, e.g., bone marrow, are not notably active in stimulating colony formation whereas highly active extracts can be prepared from organs like the submaxillary gland which contain relatively few granulocytes (Sheridan and Stanley, 1971).

One outstanding problem in relation to this general question is the origin of the CSF actually impinging on the target *in vitro* CFC's in the bone marrow. With a single organ source of a humoral regulator, the local concentration of the regulator at the surface of the target cells can be monitored satisfactorily by measuring plasma or serum levels of the regulator. Where the possibility exists that a local cellular source of CSF may exist adjacent to the target cell, the local concentration of CSF around target cells need not necessarily parallel serum CSF levels. This is a particular problem for studies in humans where extensive investigation of local bone marrow production of CSF may not be feasible. Studies using mouse bone marrow shafts substantially free of hematopoietic cells have shown that such shafts have a 10- to 20-fold higher capacity to conditioned medium than have the hematopoietic populations contained within the shaft (Chan, S. H., and Metcalf, D., unpublished data). Furthermore, as shall be discussed later, significant fluctuations in the functional activity of stromal cells occur in relation to hematopoietic regeneration, which are not paralleled by similar changes in the capacity of the hematopoietic cells to produce CSF or in serum CSF levels.

Since CSF in the plasma or nonhematopoietic tissues cannot be of relevance in regulating hematopoiesis unless it reaches the target cells, the role and the extent of local CSF production within the marrow is at present a major complication in attempting to assess the significance of the CSF extractable from nonhematopoietic tissues. In short, does the marrow compartment import or export CSF?

V. EFFECTS OF CSF *IN VIVO*

The history of attempts to alter granulopoiesis and macrophage formation in intact animals by injecting material of various kinds has made it evident that many foreign substances can affect the formation or blood levels of these cells. With this confused background, and the appreciation that blood white cell levels are a very unreliable index of white cell production, attempts to test suspected regulators of white cell formation *in vivo* need to be approached with some circumspection. It is desirable that only highly purified material, preferably of isologous origin, be tested and this together with adequate control preparations known to have no *in vitro* activity. A second general approach to establishing an *in vivo* function for a regulator is a correlative one, comparing regulator levels with blood cell production in a variety of abnormal situations. Both approaches have been used to determine whether CSF has a function *in vivo* comparable with its observed action *in vitro*.

Observations have been made on *in vivo* effects of medium conditioned by syngeneic embryo cells on granulopoiesis in mice. Serum-free material with high *in vitro* activity was given as a single or multiple

injections and recipient mice were found to slowly develop a granulocytosis together with a rise in the frequency and total number of *in vitro*
CFC's in the spleen but not the marrow (Bradley *et al.*, 1969). Kinetic
studies indicated that the granulocytosis was due to an increased production of polymorphs rather than the release of preformed cells. Injections
of control medium, conditioned by syngeneic thymic cells and having
no *in vitro* colony stimulating activity, did not produce these changes.

　　Somewhat similar responses were observed in $C_{57}BL$ mice following
the injection of 1000-fold purified CSF prepared from normal human
urine. Following a single injection of 10–20,000 units (10 units equals
concentration of CSF required to stimulate the formation of 10 colonies
in cultures of 75,000 bone marrow cells) neonatal mice developed a
granulocytosis and monocytosis which were maximal at 48 hours. Adult
mice developed a monocytosis, again maximal at 24–48 hours and passing
off by 72 hours, but did not develop a polymorph leukocytosis. However
analysis of the accumulation of labeled polymorphs in the blood of
such mice following pulse labeling with tritiated thymidine indicated
an increased rate of appearance of labeled cells, suggesting an increased
rate of production of polymorphs not evident in absolute polymorph
levels (Metcalf and Stanley, 1971a). In these experiments, control mice
received human serum or urine fractions with no significant *in vitro*
colony stimulating activity. Mice injected with active human urine preparations also exhibited at 24 hours a slight increase in bone marrow levels
of *in vitro* CFC's, cluster-forming cells, and immature granulocytes. Of
considerable interest was the fact that the bone marrow cells from these
mice commenced proliferation *in vitro* earlier than did cells from control
mice. As was discussed earlier, the length of the lag period before proliferation commences *in vitro* can be shortened by increasing the concentration of CSF and the above effects noted in mice injected with CSF
suggest that CSF may have a similar action *in vivo*. In these experiments,
no influence was noted of injected CSF-containing material on erythropoiesis or on the levels of lymphocytes and eosinophils and no histological changes in hematopoietic organs were apparent at 48 or 72 hours.

　　In view of the likelihood that CFC's are not self-replicating and that
CSF has no action on the CFU to CFC sequence, it is not to be expected
that injected CSF would increase the number of CFC's. Indeed it might
well deplete the total CFC numbers, at least temporarily. What would
be anticipated is a moderate increase in the probable progeny of
CFC's—cluster-forming cells, immature granulocytes, and eventually
polymorphs and monocytes. Although these changes were observed in
mice following the injection of human urine CSF (Metcalf and Stanley,
1971a), the magnitude of the changes was relatively small (2- to 4-fold).

Two possible explanations can be advanced for this: (a) urine CSF may not be fully active *in vivo* if some degradation of the molecule occurred prior to excretion, or (b) human CSF might not be as efficient in stimulating mouse CFC's *in vivo* as it is *in vitro*. An alternative is that CSF may be only one of a number of regulators operating *in vivo* on granulopoiesis and that, in the presence of a balanced set of such regulators, perturbations of granulopoiesis may not be readily achieved by single pulse injections.

Although the control materials used in the experiments were reasonable, it must be reemphasised that there are many pitfalls in interpreting changes induced *in vivo* in granulopoiesis and monocyte formation. The materials used were *not* pure preparations of CSF and may have contained extraneous material responsible for the changes observed. For example, CSF in human urine has now been purified in excess of 100,000-fold, which means that 99% of the protein material in the 1000-fold purified material tested in the above experiments must have represented contaminating protein.

Insofar as the results go, the changes observed are consistent with CSF having effects *in vivo* comparable with the known action of CSF *in vitro,* but the results can only be accepted as initial observations and much further work is required using more highly purified material.

The indirect approach to obtain evidence of the action of CSF *in vivo* is to correlate changes in urine, serum, or tissue CSF levels with changes in granulopoiesis or macrophage formation in a variety of experimental or disease states. While this is technically feasible it must be recognized that this type of approach does not readily distinguish cause from effect and, above all, cannot eliminate the possibility that changes in other regulators as yet unidentifiable are occurring in parallel with fluctuations in CSF levels and are actually responsible for the altered hematopoiesis.

The simplest experimental system analyzed has been the response of mice to the injection of foreign antigens. The injection of a number of antigens has been observed to increase the number of CFC's in the bone marrow, blood, and particularly the spleen (Bradley *et al.,* 1969; McNeill, 1970; Pluznik and Scheinman, 1970; Metcalf, 1971b). Similar changes have been observed in the number of cluster-forming cells and in peripheral levels of granulocytes and macrophages. Material inducing the largest changes were bacterial antigens, endotoxin, Freund's adjuvant or sheep red cells while foreign serum proteins were essentially inactive. When bacterial antigens, e.g., *Salmonella* flagellin, are added to cultures of bone marrow cells from normal or immunized mice, no colony formation is stimulated. However it is of some interest that these antigens have been observed to potentiate colony formation stimulated by low concentrations of CSF (McNeill, 1970a; Metcalf, 1971b). This effect was related to antigen dose, being only seen with intermediate dose ranges of antigen and was also seen in cultures of fetal liver CFC's taken from embryos before the development of the thymus or other lymphoid organs (McNeill, 1970a). Analysis of this phenomenon indicated that a serum α-macroglobulin was also needed for the response and antigen-induced potentiation was not observed if the CSF source was urine or serum-free conditioned medium (McNeill, 1970b; Metcalf, 1971b). It was shown that antigens, when added to liquid cultures of bone marrow cells, increased the capacity of bone marrow cells to form CSF and the potentiating effect of antigens in culture may be due to stimulation of an increased level of endogenous CSF formation, supplementing the low levels of CSF used in the experimental procedure.

These observations raised the possibility that the stimulating effects of antigens *in vivo* on the number of CFC's and progeny cells may be indirect and be mediated by CSF. In fact, the injection of bacterial antigens or endotoxin into mice produced spectacular rises in serum CSF levels (Metcalf, 1971c). Following a single IV injection of endotoxin, serum levels commenced rising within 1 hour and peaked at 3–6 hours at which time levels were 50- to 100-fold elevated above normal. The timing and magnitude of this response makes it quite possible for the subsequent cellular proliferation to have been mediated by this increased level of CSF (Metcalf, 1971b, c).

The serum CSF response to bacterial antigens is radioresistant, but is specifically depressed in mice preimmunized against the particular antigen being used (Metcalf, 1971c). Depressed reactivity begins to develop within 2 days of the immunization and is transferable to normal recipients by immune serum (Metcalf, D., unpublished data).

Earlier studies indicated that the serum CSF levels are high in patients in the acute stages of bacterial and viral infections (Foster *et al.*, 1968; Metcalf and Wahren, 1968) and in mice in the acute stages of viral infections (Foster *et al.*, 1968a). Studies on patients with infectious mononucleosis showed that most exhibited elevated serum CSF levels and that serum CSF levels also rose in family contacts of these patients, the rises preceding detectable rises in E-B antibodies (Metcalf and Wahren, 1968; Wahren *et al.*, 1970). The small percentage of these patients who failed to exhibit serum CSF rises in the acute stages of mononucleosis were those who tended to have a more severe clinical illness, although here CSF may merely have been paralleling changes in other components of the host resistance system and not necessarily have been directly responsible for the increased resistance.

Somewhat similar observations have been made in patients with acute granulocytic leukemia where serum and urine CSF levels tended to rise following the development of intercurrent infections. It was observed that patients in relapse tended to be unable to elevate CSF levels in response to these infections and again this correlates with the known high susceptibility to infections of leukemic patients in relapse (Metcalf *et al.*, 1971a).

These observations on the responses of granulocyte and macrophage precursors to bacterial antigens and infections has reawakened interest in the importance of the granulocyte-macrophage (G-M) system in infections and strongly suggests that CSF in the mediator of the prompt responses. In line with this conclusion is the low level of serum CSF .and spleen *in vitro* CFC's in germfree animals (Metcalf *et al.*, 1967a; Metcalf, 1972). The host defense system against bacterial infections needs to be operative within hours of the onset of significant infections. It is unlikely that the lymphoid-plasma cell system in a person not previously exposed to the antigens concerned can provide an effective defense system, since cellular responses and antibody production are relatively slow in onset and do not reach significant levels for several days after antigenic challenge. This suggests that the CSF-mediated G-M response may be of paramount importance in providing the initial defense until the subsequent elaboration of specific antibody and reactive cells. It is of interest in this context that the CSF response is actually depressed in immune animals, suggesting that the two systems, if not mutually exclusive, at least phase in with one another to provide a continuing flexible cover for the body.

One of the most intriguing aspects of the CSF response to bacterial products is the multiplicity of organs responding to the challenge and the speed of their response. Studies on CSF extractable from various organs after the injection of endotoxin has shown that within minutes, levels of CSF rise in all tissues and that these levels exceed CSF levels in the serum. These changes are seen most dramatically in the submaxillary gland, where not only is there a rapid rise in the total content of CSF but also clear evidence of the elaboration of a chemically different type of CSF, more nearly related to the CSF seen in the serum of these animals (Sheridan, J. and Metcalf, D., unpublished data). Many questions remain to be resolved concerning this response. Do such cells as submaxillary gland parenchymal cells have the capacity to recognize and respond to bacterial antigens? Is this a nonspecific response to injury mediated by some intermediate humoral factor formed by the action of endotoxin or antigens with cells elsewhere in the body? Are the responses observed due to the activity of macrophages, endothelial cells, or fibroblasts within these organs? Until these questions can be resolved, the biological significance of the changes cannot be fully appreciated. It is clear however that the CSF response is not restricted in origin to a limited number of tissues but appears to involve tissues throughout the body in an extremely rapid and widespread change in cellular metabolic activity.

A number of studies have been made on response of *in vitro* CFC's and CSF levels to whole-body irradiation *in vitro*. CFC's are radiosensitive with a D_{37} of approximately 85–95 R (Robinson *et al.*, 1967; Chen and Schooley, 1970). Regeneration of cluster-forming cells occurs more rapidly than that of colony-forming cells but *in vitro* CFC's regenerate within 2–3 weeks of a 250 R whole-body irradiation (Hall, 1969). Serum CSF levels rise abruptly in the first 6–8 hours following irradiation, the magnitude of the rise paralleling the dose of irradiation (Hall, 1969; Sheridan, J. and Metcalf, D., unpublished data). Subsequent changes in CSF levels are more complex. In mouse strains with low serum inhibitor levels, e.g., $C_{57}BL$, serum CSF levels do not usually fluctuate significantly during the regeneration of *in vitro* CFC's (Hall, 1969). Other workers have reported a rise in serum colony stimulating activity during hematopoietic regeneration (Morley *et al.*, 1971). However reinvestigation of this response in mice given 250 R whole-body irradiation has shown that the apparent increase in colony stimulating activity is due to an abrupt fall in inhibitor levels in the serum coinciding with the onset of CFC regeneration at 10 days (Chan, S. H., unpublished data). This lowering of inhibitor levels gives a higher net colony stimulating activity for unfractionated serum. This observation underlines the possible importance of CSF inhibitors as a modulating influence in regulating granulopoiesis. Of even more interest is the fact that bone marrow stromal production of CSF rises abruptly immediately before regeneration commences, while the CSF-producing ability of the hematopoietic population within the shaft remains unchanged at a low level throughout regeneration (Chan, S. H., unpublished data). These results strongly imply that the regeneration of *in vitro* CFC's and their progeny following irradiation may be regulated by increased local CSF production in combination with a fall in serum inhibitor levels. Further studies are needed in other experimental systems involving perturbation of *in vitro* CFC levels to determine whether a similar pattern is observable but these observations reemphasise the problems of assessing CSF action *in vivo*, if serum CSF levels do not necessarily reflect local concentrations of CSF at the membrane of the target cells.

Prolonged periods of hormone imbalance favoring proliferation of target cells have been shown to result in neoplasia development in such target cells (Furth, 1969). Because of the analogies between hematopoietic regulator-target cell and hormone-target cell interactions, a major subject of interest has been a possible role of CSF and inhibitors in the development and progression of granulocytic or myelomonocytic leukemia. Are abnormal levels of CSF and inhibitors demonstrable in patients with leukemia and does the progression of the disease correlate with such abnormalities? This subject has recently been reviewed elsewhere (Metcalf, 1971e) but in the present context, certain points can be summarized:

a. In mice with spontaneous, viral-induced or transplanted leukemia, serum CSF levels are elevated compared with levels in control mice (Robinson *et al.*, 1967; Metcalf and Foster, 1967a; Metcalf *et al.*, 1969; Metcalf and Bradley, 1970; Hibberd and Metcalf, 1971).

b. Germfree mice with leukemia also exhibit elevated serum CSF levels, making it unlikely that the elevated levels are necessarily due to secondary bacterial or protozoal infections (Metcalf *et al.*, 1967a).

c. Elevated serum levels of CSF are observed in both lymphoid leukemia and erythroleukemia in mice but in both cases, excess granulopoiesis is known to occur (Metcalf *et al.*, 1959; Hibberd and Metcalf, 1971).

d. In humans with leukemia or lymphoma, elevated serum and urine levels of CSF have been observed and often parallel the stage and clinical activity of the disease (Foster *et al.*, 1968, 1971).

e. In human acute granulocytic leukemia, elevated serum and urine CSF levels have been observed at some stage in all patients (Robinson and Pike, 1970a; Metcalf *et al.*, 1971a). In these patients inhibitor levels were subnormal in 60% of the sera and low inhibitor levels correlated with shortened survival time and were more common in relapse and in patients with remission who had low granulocyte levels. Serum inhibitor levels have also been observed to fall terminally in mice with myelomonocytic leukemia (Chan, 1971).

f. Elevated CSF levels and low inhibitor levels have also been observed in some patients with aplastic anemia, polycythemia vera and neutropenia—conditions associated with an increased risk of granulocytic leukemia development (Metcalf *et al.*, 1971b).

g. Mouse myelomonocytic leukemic cells can form colonies *in vitro* and the growth of these leukemic colonies was observed to be responsive to CSF (Metcalf *et al.*, 1969). Similarly, human leukemic cells from patients with chronic and acute granulocytic leukemia have been shown to form colonies *in vitro* and in every case the growth of these colonies has been shown to be responsive to stimulation by feeder layers or conditioned medium (Brown and Carbone, 1971; Greenberg *et al.*, 1971;

Iscove *et al.*, 1971; Moore *et al.*, 1972; Moore, M. A. S. and Metcalf, D., unpublished data).

The significance of these observations for the development and progression of granulocytic leukemia in mice and humans cannot yet be fully assessed. However they constitute clear evidence of abnormal regulator levels before leukemia development and during the course of the disease and strongly indicate that the leukemic cells are still responsive to these regulators. It is likely therefore that the regulators are responsible for at least some aspects of the disturbed hematopoiesis in leukemia.

The foregoing evidence from normal animals and man and from changes in CSF levels during infections and leukemia together constitute both direct and indirect evidence that CSF and possibly CSF inhibitors are involved significantly in the regulation of granulopoiesis and monocyte formation. Although only fragmentary evidence exists at present, it seems possible that the action of CSF *in vivo* is essentially similar to that observed *in vitro* during the stimulation of granulocyte and macrophage colony formation.

VI. SUMMARY

The colony stimulating factor (CSF) is a glycoprotein of molecular weight 45,000–60,000 which stimulates and proliferation and differentiation *in vitro* of clones of granulocytic and/or macrophage cells. It is produced in most tissues in the body and is cleared by the kidney from the plasma to the urine. Levels of CSF fluctuate sharply in response to antigenic stimulation and infections and are elevated in leukemia. CSF appears to be a major humoral regulator of granulopoiesis and monocyte formation. A complex of lipoproteins in the serum blocks *in vitro* action of CSF and may serve *in vivo* to modulate the action of CSF and to direct granulocytic cells to a macrophage pathway of differentiation.

Progress in the preparation of purified CSF and of pure populations of primary target cells (granulocyte and macrophage progenitor cells) have made this agar culture system ideal for the analysis of many general aspects of the regulation of cellular proliferation and differentiation.

REFERENCES

Austin, P. E., McCulloch, E. A., and Till, J. E. (1971). *J. Cell. Physiol.* **77**, 121.
Bradley, T. R., and Metcalf, D. (1966). *Aust. J. Exp. Biol. Med. Sci.* **44**, 287.

Bradley, T. R., and Sumner, M. A. (1968). *Aust. J. Exp. Biol. Med. Sci.* **46**, 607.

Bradley, T. R., Robinson, W., and Metcalf, D. (1967). *Nature (London)* **213**, 511.

Bradley, T. R., Metcalf, D., and Robinson, W. (1967a). *Nature (London)* **213**, 926.

Bradley, T. R., Metcalf, D., Stanley, E. R., and Sumner, M. A. (1969). In "In Vitro 4" (P. Farnes, ed.), p. 22. Williams and Wilkins, Baltimore, Maryland.

Brown, C. H., and Carbone, P. P. (1971). *J. Nat. Cancer Inst.* **46**, 989.

Chan, S. H. (1970). *Proc. Soc. Exp. Biol. Med.* **134**, 733.

Chan, S. H. (1971). *Aust. J. Exp. Biol. Med. Sci.* **49**, 553.

Chan, S. H., Metcalf, D., and Stanley, E. R. (1971). *Brit. J. Haematol.* **20**, 329.

Chen and Schooley (1970). *J. Cell. Physiol.* **75**, 89.

Cline, M. J., Warner, N. L., and Metcalf, D. (1972). *Blood* **39**, 326.

Dicke, K. A., Platenburg, M. G. C., and Van Bekkum, D. W. (1971). *Cell Tissue Kinet.* **4**, 463.

Foster, R., Metcalf, D., Robinson, W. A., and Bradley, T. R. (1968). *Brit. J. Haematol.* **15**, 147.

Foster, R., Metcalf, D., and Kirchmyer, R. (1968a). *J. Exp. Med.* **127**, 853.

Foster, R. S., Metcalf, D., and Cortner, J. (1971). *Cancer* **27**, 881.

Furth, J. (1969). *Harvey Lectures* **63**, 47.

Goldwasser, E., and Kung, C. K. H. (1971). *Proc. Nat. Acad. Sci.* **68**, 697.

Greenberg, P. L., Nichols, W., and Schrier, S. L. (1971). *New England J. Med.* **284**, 1225.

Hall, B. M. (1969). *Brit. J. Haematol.* **17**, 553.

Haskill, J. S., McNeill, T. A., and Moore, M. A. S. (1970). *J. Cell. Physiol.* **75**, 167.

Hibberd, A. D., and Metcalf, D. (1971). *Israel J. Med. Sci.* **7**, 202.

Ichikawa, Y., Pluznik, D. H., and Sachs, L. (1966). *Proc. Nat. Acad. Sci.* **56**, 488.

Iscove, N. N., Till, J. E., and McCulloch, E. A. (1970). *Proc. Soc. Exp. Biol. Med.* **134**, 33

Iscove, N. N., Senn, J. S., Till, J. E., and McCulloch, E. A. (1971). *Blood* **37**, 1.

Janoshwitz, H., Moore, M. A. S., and Metcalf, D. (1971). *Exp. Cell. Res.* **68**, 220.

Lajtha, L. G., Pozzi, L. V., Schofield, R., and Fox, M. (1969). *Cell Tissue Kinet.* **2**, 39.

Levi-Montalcini, R., and Angeletti, P. V. (1968). In *CIBA Symp.* "Growth of the Nervous System" (G. E. W. Wolstenholme and M. O'Connor, eds.), p. 126. Churchill, London.

McNeill, T. A. (1970). *Immunology* **18**, 61.

McNeill, T. A. (1970a). *Immunology* **18**, 39.

McNeill, T. A. (1970b). *Immunology* **18**, 49.

Maritz, J. S., Stanley, E. R., Yeo, G. F., and Metcalf, D. (1972). *Biometrics* **28**, 801.

Metcalf, D. (1968). *J. Cell. Physiol.* **72**, 9.

Metcalf, D. (1969). *Brit. J. Haematol.* **16**, 397.

Metcalf, D. (1969a). *J. Cell. Physiol.* **74**, 323.

Metcalf, D. (1969b). *Proc. Soc. Exp. Biol. Med.* **132**, 391.

Metcalf, D. (1970). *J. Cell. Physiol.* **76**, 89.
Metcalf, D. (1971a). *J. Cell. Physiol.* **77**, 277.
Metcalf, D. (1971b). *Immunology* **20**, 727.
Metcalf, D. (1971c). *Immunology* **21**, 427.
Metcalf, D. (1971d). *Aust. J. Exp. Biol. Med. Sci.* **49**, 351.
Metcalf, D. (1971e). *Advan. Cancer Res.* **14**, 181.
Metcalf, D. (1972). *Proc. Soc. Exp. Biol. Med.* **139**, 511.
Metcalf, D., and Bradley, T. R. (1970). In "Regulation of Hematopoiesis" (A. S. Gordon, ed.), p. 187, Appleton. New York.
Metcalf, D., and Foster, R. (1967). *Proc. Soc. Exp. Biol. Med.* **126**, 758.
Metcalf, D., and Foster, R. (1967a). *J. Nat. Cancer Inst.* **39**, 1235.
Metcalf, D., and Moore, M. A. S. (1971). "Haemopoietic Cells." North-Holland Publ., Amsterdam.
Metcalf, D., and Stanley, E. R. (1969). *Aust. J. Exp. Biol. Med. Sci.* **47**, 453.
Metcalf, D., and Stanley, E. R. (1971). *Brit. J. Haematol.* **20**, 549.
Metcalf, D., and Stanley, E. R. (1971a). *Brit. J. Haematol.* **21**, 481.
Metcalf, D., and Wahren, B. (1968). *Brit. Med. J.* **3**, 99.
Metcalf, D., Furth, J., and Buffett, R. F. (1959). *Cancer Res.* **19**, 52.
Metcalf, D., Bradley, T. R., and Robinson, W. (1967). *J. Cell. Physiol.* **69**, 93.
Metcalf, D., Foster, R., and Pollard, M. (1967a). *J. Cell. Physiol.* **70**, 131.
Metcalf, D., Moore, M. A. S., and Warner, N. L. (1969). *J. Nat. Cancer Inst.* **43**, 983.
Metcalf, D., Moore, M. A. S., and Shortman, K. (1971). *J. Cell. Physiol.* **78**, 441.
Metcalf, D., Chan, S. H., Gunz, F. W., Vincent, P., and Ravich, R. M. B. (1971a). *Blood* **38**, 143.
Metcalf, D., Chan, S. H., Gunz, F. W., and Vincent, P. (1971b). *Proc. Int. Conf. Leukemia, 5th,* Padua. Karger-Basel (in press).
Moore, M. A. S., Williams, N., and Metcalf, D. (1972). *J. Cell. Physiol.* **79**, 283.
Moore, M. A. S., Williams, N., Metcalf, D., Garson, M. O., and Hurdle, A. D. F. (1972a). In "Cell Differentiation" (R. Harris and D. Viza, eds.). Munksgaard. Copenhagen, p. 108.
Morley, A., Rickard, K. A., Howard, D., and Stohlman, F. (1971). *Blood* **37**, 14.
Paran, M., and Sachs, L. (1968). *J. Cell. Physiol.* **72**, 247.
Paran, M., Ichikawa, Y., and Sachs, L. (1968). *J. Cell. Physiol.* **72**, 251.
Pluznik, D. (1969). *Israel J. Med. Sci.* **5**, 306.
Pluznik, D., and Sachs, L. (1965). *J. Cell. Comp. Physiol.* **66**, 319.
Pluznik, D. H., and Sachs, L. (1966). *Exp. Cell. Res.* **43**, 553.
Pluznk, D. H., and Scheinman, (1970). *Israel J. Med. Sci.* **6**, 456.
Rickard, K. A., Shadduck, R. K., Howard, D. E., and Stohlman, F. (1970). *Proc. Soc. Exp. Biol. Med.* **134**, 152.
Robinson, W. A., and Pike, B. L. (1970). In "Hematopoietic Cellular Differentiation" (F. Stohlman, Jr. ed.), p. 249. Grune and Stratton, New York.
Robinson, W. A., and Pike, B. L. (1970a). *New England J. Med.* **282**, 1291.
Robinson, W., Metcalf, D., and Bradley, T. R. (1967). *J. Cell. Physiol.* **69**, 83.
Robinson, W. A., Stanley, E. R., and Metcalf, D. (1969). *Blood* **33**, 396.
Sheridan, J. W., and Stanley, E. R. (1971). *J. Cell. Physiol.* **78**, 451.
Stanley, E. R., and Metcalf, D. (1969). *Aust. J. Exp. Biol. Med. Sci.* **47**, 467.
Stanley, E. R., and Metcalf, D. (1971). *Proc. Soc. Exp. Biol. Med.* **137**, 1029.

Stanley, E. R., and Metcalf, D. (1971a). *Aust. J. Exp. Biol. Med. Sci.* **49**, 281.

Stanley, E. R., and Metcalf, D. (1972). In "Cell Differentiation" (R. Harris and D. Viza, eds.). Munksgaard, Copenhagen, p. 149.

Stanley, E. R., Robinson, W. A., and Ada, G. L. (1968). *Aust. J. Exp. Biol. Med. Sci.* **46**, 715.

Stanley, E. R., McNeill, T. A., and Chan, S. H. (1970). *Brit. J. Haematol.* **18**, 585.

Stanley, E. R., Bradley, T. R., and Sumner, M. A. (1971). *J. Cell. Physiol.* **78**, 301.

Stephenson, J. R., Axelrad, A. A., McLeod, D. L., and Shreeve, M. M. (1971). *Proc. Nat. Acad. Sci.* **68**, 1542.

Turkington, R. W., Males, J. L., and Cohen, S. (1971). *Cancer Res.* **31**, 252.

Wahren, B., Lantorp, K., Sterner, G., and Espmark, A. (1970). *Proc. Soc. Exp. Biol. Med.* **133**, 934.

Worton, R. G., McCulloch, E. A., and Till, J. E. (1969). *J. Cell. Physiol.* **74**, 171.

Wu, A. M., Siminovitch, L., Till, J. E., and McCulloch, E. A. (1968). *Proc. Nat. Acad. Sci.* **59**, 1209.

HUMORAL REGULATION OF THROMBOCYTOPOIESIS *

T. T. Odell, Jr.

I. INTRODUCTION

Investigation into the humoral regulation of thrombocytopoiesis has
had a fairly long though intermittently active history, but our under-

* Research sponsored by the U.S. Atomic Energy Commission under contract
with the Union Carbide Corporation.

standing is still rather rudimentary. That humoral agents have an active part seems certain. One of the greatest needs is a sensitive, reproducible, and reasonably easy method for assaying possible regulatory substances. There is promise that such assays may soon be available.

Recently acquired knowledge about the process of megakaryocyte maturation has provided a necessary foundation for unraveling the mechanisms that maintain homeostatic levels of circulating platelets by regulating platelet production.

This chapter emphasizes findings of approximately the last decade with the objective of outlining our present knowledge of the regulation of megakaryocytopoiesis and platelet production. Some reviews are included in the list of references (Abildgaard and Simone, 1967; Cooper, 1970; Ebbe, 1970; Ebbe and Phalen, 1970; Levin, 1970; Paulus, 1971a).

II. ALTERATIONS IN THROMBOCYTOPOIESIS UNDER CHANGING DEMANDS FOR CIRCULATING PLATELETS

When the need for platelets changes, many alterations in megakaryocytopoiesis and platelet production can be observed. Investigation of these changes has promoted the understanding of normal thrombocytopoiesis, has indicated various maturation processes or stages where regulation may occur, and has suggested methods that can be used to further elucidate the mechanisms and agents involved in the regulation of platelet production. Experiments in which animals have been made thrombocytopenic (platelet-poor) or thrombocytotic (platelet-rich) have clearly demonstrated that the number of circulating platelets has a regulatory influence on the rate of production of blood platelets. Platelet production is stimulated in thrombocytopenic individuals and inhibited in animals made thrombocytotic by transfusion of fresh, viable platelets.

A. Thrombocytopenia

Thrombocytopenia has commonly been produced experimentally by injections of antiplatelet antiserum (APS), by exchange transfusion with platelet-poor blood, or by exposure of animals to whole-body radiation. Various chemicals that interfere with cell proliferation, such as chemotherapeutic agents and alcohol, have also occasionally been used to reduce platelet counts. In addition, some studies of congential or accidental thrombocytopenia in man have been sources of useful and interesting observations.

1. PLATELETS

a. *Platelet Production.* The peripheral platelet counts of rats and other
experimental animals can be reduced to zero by injection of APS. The
APS appears to remove circulating platelets without affecting megakaryocytes when a suitable dose is administered, although "large" doses produce a more extended platelet depression, either because of the action
of residual antiserum on circulating platelets or because of an effect
of APS on maturing megakaryocytes (Odell *et al.*, 1969). After treatment
with APS, the recovery of circulating platelets to normal levels as well
as changes in the megakaryocytic cell line can be studied (Witte, 1955;
Ebbe *et al.*, 1968a; Odell *et al.*, 1969; Rolovic *et al.*, 1970).

The platelet count of rats begins to rise within 24 hours after a moderate reduction (e.g., 50%) brought about by a single dose of APS (Fig.
1). Platelet return is postponed, but the rate is accelerated after more
severe thrombocytopenia (less than 3% of the normal platelet count)
(Odell, Jackson and Murphy, in preparation). Such results indicate a
quantitative relation between the mass of circulating platelets and the
rate of platelet production. Platelet return is accelerated even more after
prolonged platelet depression brought on by repeated injections of APS
or by a single large dose of APS (Ebbe and Stohlman, 1970).

The peak number of platelets in the peripheral circulation of rats
is usually reached about 5 days after production of acute thrombocytopenia and is an important consideration in the planning and interpretation of experiments with potential thrombocytopoietic agents. The maxi-

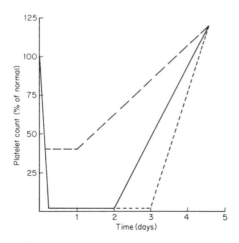

Fig. 1. Graph shows platelet return after induced thrombocytopenia in relation
to degree and duration of depletion.

mum increase often does not exceed 50–75% above the normal platelet count, although increases greater than 100% have been observed after severe thrombocytopenia. After reaching its peak, the count then returns to normal within about 4 days.

The platelet count has also been reduced by exchange transfusion with platelet-poor blood (Craddock et al., 1955; Matter et al., 1960; Odell et al., 1962; Ebbe et al., 1968b). The degree of platelet reduction is exponentially related to the blood volumes exchanged. The rate of platelet return is proportional to the degree of platelet depression (Odell et al., 1971b).

Platelet return has also been studied after whole-body exposure of rats and mice to radiation (Ebbe and Stohlman, 1970; Odell, 1971; Odell et al., 1971a). The time between the nadir and restoration of normal platelet levels is longer after radiation than after exchange transfusion or moderate doses of APS, because the radiation produces a hypoplastic or aplastic marrow that must be rebuilt before megakaryocytopoiesis and platelet production can proceed.

b. Platelet Size. New platelets produced under conditions of increased demand are larger than the average platelet in unperturbed individuals. It has been observed that the average size of platelets is greater in rats and dogs recovering from thrombocytopenia induced by APS or by radiation than in control animals (Detwiler et al., 1962; Spertzel et al., 1963; McDonald et al., 1964; Batikyan, 1965). Observations in human beings with various platelet disorders have demonstrated a positive correlation between the percentage of large platelets in the peripheral circulation and the number of megakaryocytes in the marrow (Garg et al., 1971). The foregoing experiments as well as data on functional and metabolic characteristics of platelet subpopulations have suggested that platelets become smaller as they age (Detwiler et al., 1962; Booyse et al., 1968; Steiner and Baldini, 1969; Karpatkin and Charmatz, 1970; Amorosi et al., 1971). However, other results have indicated that the large, dense platelets produced under conditions of increased platelet production remain large throughout their life span (Minter and Ingram, 1971).

2. CHANGES IN MEGAKARYOCYTES IN THROMBOCYTOPENIA

Several changes in megakaryocytes have been observed after induction of thrombocytopenia, either by injection of APS or by exchange transfusion with platelet-poor blood (Fig. 2). The number of megakaryocytes begins to increase above normal about 20 hours after induction of thrombocytopenia by injection of APS (Odell et al., 1969). The increase in

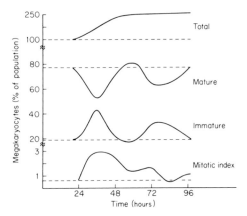

Fig. 2. Changes in megakaryocytes with time after induction of acute thrombocytopenia.

megakaryocytes is greater after a prolonged than after a brief platelet depression (Harker, 1968). When the megakaryocyte number increases, the age composition of the population also changes (Craddock *et al.*, 1955; Odell *et al.*, 1969). The proportion of immature megakaryocytes increases first. As the added cells pass through the stages of megakaryocytopoiesis, an increase in the mature cells is seen. The mitotic index of the population of recognizable megakaryocytes begins to increase about 18 hours after injection of APS, rises rapidly to a peak several times that of controls at around 36 hours, and then declines (Odell *et al.*, 1969). In addition, average megakaryocyte size is increased by 24 hours. The size increase is observed first in the immature and later in the mature cells (Witte, 1955; Ebbe *et al.*, 1968a).

The average ploidy level of megakaryocytes also increases after thrombocytopenia (Penington and Olsen, 1970; Odell, Jackson, and Murphy, unpublished). An increased number of nuclear lobes accompanies the increase in DNA (Harker, 1968), although it has been clearly demonstrated that there is not a regular relation within individual cells between ploidy value, as determined by measurement of relative amount of DNA, and lobe number (Paulus, 1971b). The complement of DNA in each lobe is usually 4N or more and often does not correspond to an integral power of the diploid value.

The 24-hour labeling index of megakaryocytes in rats exchange-transfused with platelet-poor blood was proportional to the remaining platelet mass (Odell *et al.*, 1971b). [³H]TdR was administered 24 hours after the exchange transfusion.

All of the foregoing results demonstrate that a reduction in the mass of circulating platelets causes marked changes in megakaryocytopoiesis and platelet production.

B. Thrombocytosis

Transfusion of freshly collected, viable platelets also produces changes in the megakaryocyte-platelet system (Cronkite et al., 1961; de Gabriele and Penington, 1967a; Odell et al., 1967; Harker, 1968; Evatt and Levin, 1969; Penington and Olsen, 1970). When the circulating platelet count was raised to three or more times normal, the platelet count 10 days later was below normal. In addition, the production rate of platelets was reduced, as judged by the amount of label in the circulating platelet mass after injection of $Na_2[^{35}S]O_4$ or [^{75}Se]methionine. The number of megakaryocytes in the marrow was significantly reduced 3 and 4 days after platelet transfusion. It was also shown that megakaryocyte size, DNA content, and number of lobes in megakaryocyte nuclei decline after thrombocytosis (Harker, 1968; Ebbe, 1970; Penington and Olsen, 1970).

These studies in thrombocytotic rats, like the studies in thrombocytopenic rats, demonstrate that the size of the circulating platelet mass affects thrombocytopoiesis.

III. THROMBOPOIETIN AND ITS ASSAY

Some of the changes in platelets and megakaryocytes observed after thrombocytopenia have also been seen after injections of various test materials, such as plasma, serum, and spleen extracts, usually obtained from thrombocytopenic donors or patients with platelet disorders. It is thought that these test materials contain an agent that increases the rate of platelet production. The agent is generally called thrombopoietin.

A. Platelet Number

An increase in the number of circulating platelets after injection of a test material has often been used as an assay of thrombopoietin. Such an increase indicates either an increased rate of platelet production, release from a platelet pool, or a lengthened platelet survival. The lag period prior to a detectable increase in the number of circulating platelets tends to rule out platelet release from a pool as the cause of the increase. In addition, the pool of platelets in the spleen of rats is relatively small, about 12% of the platelet mass (Aster, 1967). Neither has an increase in platelet survival been reported.

An increase in the number of circulating platelets has been reported in rabbits, rats, and mice injected with homologous plasma or serum of platelet-deficient donors (Spector, 1961; Odell et al., 1961; Kelemen et al., 1963). The peak value occurred 4 or 5 days after injections of the test material were begun, a timing similar to that seen after experimental induction of acute thrombocytopenia. Interspecies injections of serum or plasma have had a similar effect (Chat and Ebbe, 1962; Linman and Pierre, 1963; Schulman et al., 1965; McClure and Choi, 1968). Injections of spleen extracts also stimulated new production of platelets, as indicated by an increase in the number of circulating platelets (Cooney et al., 1965). Injections of normal human plasma produced marked increases in the platelet count of a thrombocytopenic patient with an apparent deficiency in megakaryocyte maturation (Schulman et al., 1965). In addition, plasma from a man recovering from alcohol-induced thrombocytopenia exhibited thrombopoietic activity. Six days after the plasma was reinfused, his platelet count more than doubled (Sullivan, 1969).

An advantage of measuring the number of circulating platelets as an assay of potential thrombopoietic agents is the direct relation between increased platelet production and platelet number. An increase could, of course, occur in the absence of increased production if exit from the circulating compartment decreased as a result either of increased platelet survival or of decreased platelet utilization; however, lengthened platelet survival is not known, and decreased platelet utilization is probably not a frequent occurrence.

A disadvantage is that the variability of the results introduced by inherent errors in making platelet counts and by the real variation in platelet counts among animals reduces the resolution of the method. Variation among animals can be eliminated by expressing platelet counts as a percentage of an initial count taken before starting an experiment. Another disadvantage of this method is that the maximum increase (about 50–100%) is relatively small, making it difficult to measure a dose-response relation. Some evidence for a dose-response relation has, however, been reported (Odell et al., 1961).

The observation that new platelets produced in response to platelet demand are larger and heavier than the average platelet in a control population (Section II) suggests that platelet size or density might be employed to assay thrombopoietic agents, although their use has not yet been reported.

B. Platelet Labeling with Radioisotopes

In the past few years, the uptake of radioisotopes by platelets has been used to assay potential thrombocytopoietic agents. ^{35}S-labeled sodium sulfate and ^{75}Se-labeled selenomethionine ($[^{75}$Se]M) both label platelets while they are still part of the cytoplasm of megakaryocytes, and the label is not eluted from the platelets when they enter the circulatory system (Odell and McDonald, 1964; Evatt and Levin, 1969). In animals with heightened platelet production during recovery from thrombocytopenia, the radioactivity per average platelet 2 or 3 days after injection of the labeling compound is greater than in controls (Evatt and Levin, 1969; Najean and Ardaillou, 1969; Penington, 1969; Cooper et al., 1970; Harker, 1970). A greater platelet radioactivity in animals injected with a test substance than in controls has been interpreted as evidence for the presence of a thrombopoietic agent (Evatt and Levin, 1969; Cooper et al., 1970; Harker, 1970; Penington, 1970; Shreiner and Levin, 1970). The sensitivity of assay animals to thrombopoietic substances has been enhanced by suppressing endogenous thrombocytopoiesis by means of repeated transfusions of fresh platelets (Harker, 1970; Shreiner and Levin, 1970). Some evidence for a dose-response relation between the amount of thrombopoietic plasma injected and platelet radioactivity has been observed in hypertransfused recipients. In one set of experiments, radioactivity of platelets of hypertransfused mice injected with $[^{75}$Se]M was 1.2–3.2 times greater at the higher dose when the amount of test plasma was quintupled; the plasma donors were patients with various platelet disorders (Penington, 1970). In another experiment that utilized platelet-hypertransfused rabbits as the test animals, platelet radioactivity was 1.4–2.2 times greater in recipient rabbits when the amount of rabbit plasma injected was increased 4–7 times (Shreiner and Levin, 1970). Both thrombocytopenic and normal plasmas produced a dose-response effect. However, a larger amount of normal than thrombocytopenic plasma was needed for an equivalent response in the intermediate dose range.

It has been suggested that the method of measuring the radioactivity of labeled platelets after injection of $Na_2[^{35}S]O_4$ or $[^{75}$Se]M is a more sensitive technique for assaying thrombopoietin than counting the number of platelets in the peripheral circulation, because in some experiments injection of serum from thrombocytopenic donors resulted in an increased labeling of platelets without an increase in platelet count (Evatt and Levin, 1969; Cooper et al., 1970; Penington, 1970). What are the implications of these findings? What does the platelet radioactivity method measure? In these experiments platelet radioactivity was expressed as counts per minute per average platelet (usually translated to percent incorporation of injected dose of radioactivity). Therefore, greater platelet radioactivity in animals given a test substance than in controls indicates that the amount of activity per platelets is, on the average, greater. This can occur by an increase in the concentration of radioactive compound within individual platelets, by an increase in the size of platelets without a change in labeled compound concentration, or by an increase in the proportion of labeled cells in the total population. Therefore, this method does not necessarily measure an increased production of platelets but may measure a change in the kind of platelets that are produced. The latter seems to be the case in experiments where the radioactivity per average platelet increased while the platelet population was not enlarged. Perhaps platelet function is improved in such cases without a change in platelet number. It has indeed been shown that new platelets in animals with stimulated thrombocytopoiesis are larger than those in controls (Section II,A,1,b), and that megakaryocyte size and ploidy are greater. Perhaps the increase in platelet size accounts in part for the higher radioactivity per platelet in the labeling experiments. Nevertheless, a true thrombopoietic agent should increase production of new platelets, as well as size or protein concentration, and thereby the number of circulating platelets in assay animals.

When the responses of platelet number and platelet radioactivity are being compared in assay experiments (Table I), it is important to keep the time of sampling in mind, because the maximum response will occur at different times for the two endpoints. The maximum rate of platelet output in rats generally occurs sometime between 48 and 96 hours after stimulation, whereas the maximum platelet count occurs at 5 or 6 days. Indeed, the time of maximum rate of platelet production varies with the potency of the stimulus. The relation between the platelet labeling and the peripheral platelet count methods of assay needs further definition.

C. Measurement of Megakaryocytopoiesis

Since it has been amply demonstrated that a variety of changes in megakaryocytes are seen in individuals stimulated to produce platelets at an increased rate (Section II), potential methods for assaying thrombopoietic agents include measurement of such changes. The changes include number of megakaryocytes, mitotic index, age composition of the megakaryocyte population, size of megakaryocytes, ploidy of the megakaryocyte population, and DNA synthesis in the megakaryocyte

TABLE I

HYPOTHETICAL COMPARISON OF THE PLATELET COUNTING AND
LABELING METHODS OF ASSAY[a]

	New platelets[b]	Platelets lost[b]	Platelet count[c]	Platelet labeling index[d]
Control	250,000	250,000	100	25
Experimental	500,000	250,000	125	40
Ratio (experimental/control)	2	1	1.25	1.60

[a] Calculations are for a 24-hour period after injection of a radioisotopic compound such as [^{75}Se]M or $Na_2[^{35}S]O_4$, specifically the 24-hour period between 48 and 72 hours after injection of thrombopoietin into the experimental group and an inactive material into the controls. It is assumed that the rate of platelet production in the experimental group is doubled during this 24-hour period, and that all newly produced platelets are equally labeled (the error introduced by the latter assumption may be similar in both experimental and control groups). This exercise illustrates that new platelets replace old platelets, but that new labeled platelets do not replace old labeled platelets under the stipulated conditions.

[b] Platelets/mm³ of blood/day.

[c] Percent of normal. Normal platelet count is approximately 1×10^6.

[d] Estimated.

population. So far, only megakaryocyte number has been examined after injection of potential thrombopoietic materials. The number of megakaryocytes was increased 2 and 3 days after initiating injections of serum from platelet-depleted donors (Odell *et al.*, 1964; Krizsa *et al.*, 1968). However, this is not a sensitive endpoint.

In addition to measuring these various parameters *in vivo*, it may be possible to develop techniques for measuring megakaryocytopoiesis and platelet production in tissue culture.

D. Immunologic Measurement of Thrombopoietin

An exciting recent development is the report of a hemagglutination-inhibition assay for sheep thrombopoietin (McDonald, 1971). To stimulate production of a thrombopoietic agent, sheep were made thrombocytopenic by injection of APS. Their serum was collected and fractionated on DEAE-phosphate cellulose columns. To measure thrombopoietic activity, the fractions were tested for their ability to increase radioactivity in the circulating platelet mass of thrombocytotic mice injected with $Na_2[^{35}S]O_4$. The most active fractions were then used as antigens to produce antisera against "thrombopoietin." Mixing of these antisera with tanned red blood cells that had been coated with the thrombopoietically active serum fractions produced agglutination of the red blood cells. Specificity of the antiserum for "thrombopoietin" was indicated in tests in which the hemagglutinating activity of the antiserum was removed completely when it was mixed with an active fraction separated from serum of platelet-poor sheep but only partially when it was mixed with a similar fraction from serum of normal sheep. To test unknowns for thrombopoietic activity, the test material is first incubated with the antiserum, and then this mixture is added to the tanned and "thrombopoietin"-coated red cells. Thrombopoietin in the test material ties up the antiserum, thereby inhibiting the hemagglutination reaction. By serial dilutions of the test material, the thrombopoietin can be quantified. Problems concerned with the impurity of the antigen used to make antisera must be considered. This immunoassay technique may, however, provide a useful tool in future investigations of thrombopoietin.

E. Other Assay Recipients

Splenectomized animals have also been used as recipients in assays of potential thrombopoietic agents, but the results did not differ from those in untreated recipients (Odell *et al.*, 1961; Spector, 1961).

It has been suggested that an iron-deficient thrombocythemic rat would be a sensitive animal for thrombopoietin assay, because such rats have an increased megakaryocyte pool that is responsive to thrombopoietin (Reed and Penner, 1970).

F. Important Considerations in the Design of Assays

It is necessary in *in vivo* assays of thrombopoietic agents to make platelet counts during the first 24 hours after administration of the test agent to determine whether the agent itself produces a thrombocytopenia that provides the stimulus for thrombocytopoiesis. High platelet counts at 4 or 5 days or increased platelet radioactivity around 3 days can be erroneously attributed to thrombopoietic activity of the test material when they are actually a reaction to a marked early thrombocytopenia induced by the test material.

It can be questioned whether platelet number in the peripheral circulation has resulted from accelerated platelet production or a release of platelets from a storage pool, such as the spleen. Attention to the time of the response will distinguish between these two possibilities. A lag period occurs between stimulation and increased production, whereas release from a pool would be rapid. Release from a pool is of relatively little consequence in the rat, a commonly used subject in these investigations, because comparatively few platelets are stored in its spleen, not more than 12% (Aster, 1967). It may, however, be a consideration in other species.

The specificity of the test material is also a matter of concern. Stimulation of thrombocytopoiesis has been observed after administration of some nonspecific substances, such as egg albumin (Odell *et al.*, 1964). Transfer studies indicated that such substances may stimulate the production of thrombopoietin, which in turn increases platelet production. Inflammation, as well as thrombocytopenia, may provide a stimulus that promotes production of thrombopoietin. It should also be noted here that several investigators have reported successful testing of human thrombopoietic agents by animal assays (including Schulman *et al.*, 1965; McClure and Choi, 1968; Penington, 1970; Penner, 1970; Ardaillou *et al.*, 1971). This appears to be a valid assay system when equivalent materials from normal humans do not produce a positive response when administered to control recipients.

In tests of materials from thrombocytopenic donors, the time after induction of thrombocytopenia when the test substance is collected is undoubtedly important. Collection has usually been made some time during the first 24 hours and often within 5 hours. A systematic study has not yet been made, but assay experiments and studies of megakaryocytopoiesis suggest that the thrombopoietin titer in blood increases rapidly after induction of thrombocytopenia.

The time when the test animal is sampled is, of course, also important, the optimum time depending on the assay method being used. A source of potential error is to restrict sampling to a single time period before it has been determined that it is the best time or the only time needed.

G. Source of Thrombopoietin

The site of production of thrombopoietin is not known. The spleen and kidney have each been proposed as possible locations, but the evidence is inconclusive.

A particular fraction of bovine spleen extracts regularly caused an increase in the number of circulating platelets of rabbits, beginning at about 3 days after administration and reaching a peak on the fourth, fifth, or sixth day, a time schedule consistent with a stimulation of megakaryocytopoiesis to produce new platelets (Cooney *et al.*, 1965). However, the authors cautioned against definitely attributing this effect to the presence of a thrombopoietic agent in their extracts, pending additional study. On the other hand, it has been shown that the replenishment of platelets in the peripheral circulation after induction of thrombocytopenia with APS was similar in splenectomized and normal rats (de Gabriele and Penington, 1967b), implying that the spleen is not the source of the thrombopoietin that experimental results have shown to be present in the blood of thrombocytopenic hosts.

It was recently reported that nephrectomized rats failed to produce thrombopoietin in response to thrombocytopenia induced by bleeding (about 20% of blood volume), whereas thrombopoietic activity was present in the serum of donors that were missing other organs, such as spleen or adrenal. The serum of the organectomized donor rats was tested in mice, using the 5-day platelet count as the assay (Krizsa, 1970). It was concluded that the kidney may have a part in the elaboration of thrombopoietin. In another report, recovery from APS-induced thrombocytopenia was normal in nephrectomized rats, implying that the kidneys are not essential to thrombopoietic activity (de Gabriele and Penington, 1967a). In conclusion, the source of thrombopoietin remains to be clarified.

H. Chemical and Physical Properties

The chemical and physical nature of thrombopoietin is little known; collection of clearer information again depends in part on the development of a more satisfactory assay system. Various investigations have indicated that it is relatively heat stable, nondialyzable, and cold resistant. It may migrate with the α- or β-globulins of blood plasma, and it has been suggested that it is a glycoprotein. In fact, the characteristics described for it resemble those of erythropoietin. However, various experimental results make it clear that it is not identical to erythropoietin.

IV. MODE OF ACTION OF THROMBOPOIETIN

Studies of megakaryocytopoiesis and platelet production rates in normal and perturbed states are providing the foundation for investigation of platelet regulation. Questions before us are: (A) Where in the system does thrombopoietin act? (B) What is its action? (C) How does it act? Hypotheses about the action of thrombopoietin include: (1) increased differentiation of cells from a pluripotential stem cell pool, (2) increased proliferation of cells of a committed stem cell compartment, (3) increased polyploidization of megakaryocytes by additional DNA replication accompanied by increased synthesis of cytoplasm, and (4) increased rate of maturation of cells in the megakaryocytic series. What have studies of normal and perturbed megakaryocytopoiesis told us about these questions and hypotheses?

A. Where Thrombopoietin Acts

The fact that the rate of platelet production does not usually begin to increase until about 2 days after a stimulus, such as bleeding, treatment with antiplatelet antiserum, or injection of a presumed thrombopoietic agent, suggests that thrombopoietin acts on an early cell in the megakaryocytic series, which must then pass through a maturation process before it is able to produce platelets. Data concerned with transit times in megakaryocytopoiesis indicate what may be happening during this 2-day period. The generation cycle time (T_{GC}) of recognizable megakaryocytes (actually endoreduplication of DNA without cell division) is about 9.3 hours (Odell et al., 1968). Assuming that the T_{GC} of immediate precursor cells is similar in duration, the transit time from a diploid precursor in G_1 or G_0 to a recognizable cell approaching 8N would take about 14–18 hours.* This is consistent with the observation that the number of megakaryocytes and their mitotic index begin to increase about 20 hours after induction of thrombocytopenia (Odell et al., 1969). Between 20 and 48 hours, the 8N megakaryocytes undergo additional DNA replication and cytoplasmic maturation. The foregoing results are consistent with the idea that thrombopoietin acts on a diploid precursor of megakaryocytes and with the notion that thrombopoietin is released in greater quantity soon after induction of thrombocytopenia.

An alternate possibility is that it takes many hours after stimulation for new thrombopoietin to be released and to begin to take effect. However, thrombopoietin activity has been found in serum of rabbits within 2 hours after induction of thrombocytopenia (Spector, 1961).

B. What Thrombopoietin Does

1. POLYPLOIDIZATION, CELL PROLIFERATION AND DNA SYNTHESIS

The megakaryocyte population has more DNA per cell than diploid somatic cell populations. Microspectrophotometric measurements of Feulgen-stained nuclei have demonstrated that most cells fall in ploidy

* Recognizable megakaryocytes are defined here as cells that can be readily identified as megakaryocytes with a light microscope in bone marrow smears or squash preparations stained with conventional hematologic stains, such as Giemsa. Morphological identification of cells in marrow smears and squashes stained with Giemsa followed by microspectrophotometric determination of the relative amounts of DNA in the same cells revealed that megakaryocytes are not usually recognized by us in conventional preparations until they approach a ploidy value of 8N, at which time they are distinguishable from tetraploid cells in other hemopoietic cell lines.

classes having DNA values that are some integral power of two (Garcia, 1964). The frequency distribution among ploidy classes is characteristic of the species. The average ploidy value of recognizable megakaryocytes increases when platelet production is stimulated (Penington and Olsen, 1970; Odell, Jackson, and Murphy, unpublished). Present evidence suggests that, on the average, a megakaryocyte in a stimulated animal undergoes one more DNA replication than it ordinarily would; the ploidy class with the highest frequency (about 70% of the population of recognizable megakaryocytes) in control rats is 16N, whereas 32N megakaryocytes are most frequent in post-thrombocytopenic rats. Since the shift in frequency distribution of cells among ploidy classes was seen at 24 hours but not at 12 hours (Odell, Jackson, and Murphy, unpublished), and since the cell generation cycle time of megakaryocytes is about 9–10 hours, the programming for the increased DNA synthesis responsible for the ploidy shift may take place in cells of the unrecognized precursor population (before 8N) and become evident only when these cells reach the recognizable population.

In addition, the number of megakaryocytes increases under conditions that stimulate platelet production. Since recognizable megakaryocytes do not undergo cell division, the numerical increase results from additional DNA replication and cell division among megakaryocyte precursors. These added cells might arise by differentiation of resting cells in a pool of committed precursors, by cell division in a committed precursor compartment, or by recruitment from the pluripotential stem cell compartment that precedes the committed precursor compartment. In any case, additional DNA synthesis and cell division necessarily take place in some precursor compartment in order to maintain the cell line.

The foregoing results have indicated an increased synthesis of DNA in two stages of megakaryocytopoiesis in animals stimulated to greater production of platelets—among dividing precursors of megakaryocytes and among nondividing polyploid megakaryocytes of a later maturation stage (Table II). Whether thrombopoietin acts directly in one or both

TABLE II
CHANGES IN MEGAKARYOCYTOPOIESIS IN RESPONSE TO STIMULATION
OF PLATELET PRODUCTION

Parameter	Megakaryocytes	Platelets
Number	More DNA synthesis with cell division, more megakaryocytes	More platelets
Ploidy	More DNA synthesis without cell division, more cytoplasm synthesis	More and bigger platelets

of these cases to stimulate (or derepress) DNA synthesis or has a less direct mode of action remains unanswered. The molecular mechanism of action is, of course, completely unknown.

For comparison in another cell line, it has been proposed that a stimulus to produce more red blood cells instigates an additional cell cycle at the end of the dividing compartments, thereby amplifying the red cell population.

2. INCREASED RATE OF MATURATION

Labeling by [³H]TdR increased more rapidly in megakaryocytes of thrombocytopenic rats than in those of controls, suggesting that the maturation time of recognizable megakaryocytes is shortened in platelet-poor rats (Ebbe et al., 1968b; Odell et al., 1969). At the time the label was administered, however, more cells were entering the recognizable population in thrombocytopenic rats than in controls and so were subject to labeling. This large cohort of labeled cells, as it matured, would tend to dilute the preexisting unlabeled cells more rapidly than would a cohort of normal size in nonthrombocytopenic animals. This effect might be responsible at least in part for the more rapid labeling of the population of recognizable megakaryocytes, in contradistinction to a shortened maturation time. Moreover, the increase in ploidy level of megakaryocytes in stimulated individuals is presumably due to an increased number of generation cycles, and therefore the total maturation time should be longer rather than shorter. This apparent inconsistency remains to be explained.

V. REGULATION OF TERMINATION OF DNA SYNTHESIS

The population of mature, nondividing megakaryocytes is made up of cells of several ploidy levels. This raises another question about regulation of thrombocytopoiesis, namely the mechanism that terminates DNA synthesis in megakaryocytes. It has been suggested that the accumulation of thrombosthenin, a contractile protein peculiar to platelets, acts as a signal to stop DNA synthesis in megakaryocytes (Paulus, 1967), but this still does not explain how DNA synthesis can be stopped at a different ploidy level in one cell than in another. Perhaps the local environment of an individual megakaryocyte may influence its ultimate ploidy level. The importance of the microenvironment in the differentiation of hemopoietic cells has recently been emphasized (Trentin, 1970; Stohlman, 1970). It has also been postulated that the accumulation of a specific concentration of hemoglobin inhibits further DNA synthesis in developing erythrocytes (Stohlman, 1970).

VI. SUMMARY

Experimental results suggest that the number or mass of platelets in the peripheral circulation is monitored for changes from the normal level. When the monitoring mechanism detects a change from the normal species level, there is a corresponding change in the amount of a humoral agent that regulates megakaryocytopoiesis, and thus the production of new platelets. Changes in the amount of a humoral agent, called thrombopoietin, have been detected in the blood plasma, especially after thrombocytopenia. Changes in the rate of platelet production have likewise been documented in thrombocytopenic and thrombocytotic animals, and in animals injected with plasma or its products from thrombocytopenic donors. The nature of the monitoring mechanism is not known. Neither is it known where thrombopoietin is produced, although kidney and spleen have both been suggested as possible sites. Thrombopoietin has not been purified, and therefore its chemical nature is not known. Some results indicate that it may be a glycoprotein. It appears that thrombopoietin is produced and released into the bloodstream soon after the signal of thrombocytopenia, because its titer in the blood is increased in 2 hours. Moreover, the response time and other factors suggest that it may act on a committed diploid precursor cell, promoting differentiation and production of an increased number of megakaryocyte precursors as well as an increase in the average ploidy level of the recognizable megakaryocyte population. The promotion of DNA synthesis may be either direct or indirect. This megakaryocyte stimulation results in increased platelet production after 2–3 days.

REFERENCES

Abildgaard, C. F., and Simone, J. V. (1967). *Semin. Hematol.* **4**, 424.
Amorosi, E., Garg, S. K., and Karpatkin, S. (1971). *Brit. J. Haematol.* **21**, 227.
Ardaillou, N., Najean, Y., and Eberlin, A. (1971). *In* "Platelet Kinetics" (J. M. Paulus, ed.), p. 151. North-Holland Publ., Amsterdam.
Aster, R. H. (1967). *J. Lab. Clin. Med.* **70**, 736.
Batikyan, I. G. (1965). *Vop. Radiobiol., Sb. Tr.* **5**, 127.
Booyse, F. M., Hoveke, T. P., and Rafelson, M. E., Jr. (1968). *Biochim. Biophys. Acta* **157**, 660.
Chat, L. X., and Ebbe, S. (1962). *Med. Exp.* **7**, 317.

Cooney, D. P., Blatt, W. F., Louis-Ferdinand, R., and Smith, B. A. (1965). *Scand. J. Haematol.* **2**, 195.
Cooper, G. W. (1970). *In* "Regulation of Hematopoiesis" (A. S. Gordon, ed.), pp. 1611–1629. Appleton, New York.
Cooper, G. W., Cooper, B., and Chang, C-Y. (1970). *Proc. Soc. Exp. Biol. Med.* **134**, 1123.
Craddock, C. G., Adams, W. S., Perry, S., and Lawrence, J. S. (1955). *J. Lab. Clin. Med.* **45**, 906.
Cronkite, E. P., Bond, V. P., Fliedner, T. M., Paglia, D. A., and Adamik, E. R. (1961). *In* "Blood Platelets" (S. A. Johnson, R. W. Monto, J. W. Rebuck, and R. C. Horn, Jr., eds.), pp. 595–609. Little, Brown, Boston, Massachusetts.
de Gabriele, G., and Penington, D. G. (1967a). *Brit. J. Haematol.* **13**, 202.
de Gabriele, G., and Penington, D. G. (1967b). *Brit. J. Haematol.* **13**, 384.
Detwiler, T. C., Odell, T. T., Jr., and McDonald, T. P. (1962). *Amer. J. Physiol.* **203**, 107.
Ebbe, S. (1970). *In* "Regulation of Hematopoiesis" (A. S. Gordon, ed.), Vol. II, pp. 1587–1610. Appleton, New York.
Ebbe, S., and Phalen, E. (1970). *In* "Formation and Destruction of Blood Cells" (T. J. Greenwalt and G. A. Jamieson, eds.), pp. 128–142. Lippincott, Philadelphia, Pennsylvania.
Ebbe, S., and Stohlman, F., Jr. (1970). *Blood* **35**, 783.
Ebbe, S., Stohlman, F., Jr., Overcash, J., Donovan, J., and Howard, D. (1968a). *Blood* **32**, 383.
Ebbe, S., Stohlman, F., Jr., Donovan, J., Overcash, J. (1968b). *Blood* **32**, 787.
Evatt, B. L., and Levin, J. (1969). *J. Clin. Invest.* **48**, 1615.
Garcia, A. M. (1964). *J. Cell. Biol.* **20**, 342.
Garg, S. K., Amorosi, E. L., and Karpatkin, S. (1971). *In* "Platelet Kinetics" (J. M. Paulus, ed.), pp. 228–240. North-Holland Publ., Amsterdam.
Harker, L. (1968). *J. Clin. Invest.* **47**, 458.
Harker, L. A. (1970). *Amer. J. Physiol.* **218**, 1376.
Karpatkin, S., and Charmatz, A. (1970). *Brit. J. Haematol.* **19**, 135.
Kelemen, E., Lehoczy, D., Cserhati, I., Krizsa, F., and Rak, K. (1963). *Acta Haematol.* **29**, 16.
Krizsa, F. (1970). *In* "Abstract Volume. XIIIth Int. Congr. Hematol., Munich," p. 114. Lehmanns Verlag, Munich.
Krizsa, F., Gergely, G., and Rak, K. (1968). *Acta Haematol.* **39**, 112.
Levin, J. (1970). *In* "Formation and Destruction of Blood Cells" (T. J. Greenwalt and G. A. Jamieson, eds.), pp. 143–150. Lippincott, Philadelphia, Pennsylvania.
Linman, J. W., and Pierre, R. V. (1963). *J. Lab. Clin. Med.* **62**, 374.
Matter, M., Hartmann, J. R., Kautz, J., DeMarsh, Q. B., and Finch, C. A. (1960). *Blood* **15**, 174.
McClure, P. D., and Choi, S. I. (1968). *Brit. J. Haematol.* **15**, 351.
McDonald, T. P. (1973). *Blood* **41**, 219.
McDonald, T. P., Odell, T. T., Jr., and Gosslee, D. G. (1964). *Proc. Soc. Exp. Biol. Med.* **115**, 684.
Minter, F. M., and Ingram, M. (1971). *Brit. J. Haematol.* **20**, 55.
Najean, Y., and Ardaillou, N. (1969). *Scand. J. Haematol.* **6**, 395.
Odell, T. T., Jr. (1971). *In* "Manual on Radiation Haematology," pp. 109–116. Int. At. Energy Agency, Vienna.
Odell, T. T., Jr., and McDonald, T. P. (1964). *Amer. J. Physiol.* **206**, 580.

Odell, T. T., Jr., McDonald, T. P., and Detwiler, T. C. (1961). *Proc. Soc. Exp. Biol. Med.* 108, 428.
Odell, T. T., Jr., McDonald, T. P., and Asano, M. (1962). *Acta Haematol.* 27, 171.
Odell, T. T., Jr., McDonald, T. P., and Howsden, F. L. (1964). *J. Lab. Clin. Med.* 64, 418.
Odell, T. T., Jr., Jackson, C. W., and Reiter, R. S. (1967). *Acta Haematol.* 38, 34.
Odell, T. T., Jr., Jackson, C. W., and Reiter, R. S. (1968). *Exp. Cell Res.* 53, 321.
Odell, T. T., Jr., Jackson, C. W., Friday, T. J., and Charsha, D. E. (1969). *Brit. J. Haematol.* 17, 91.
Odell, T. T., Jr., Jackson, C. W., and Friday, T. J. (1971a). *Radiat. Res.* 48, 107.
Odell, T. T., Jr., Jackson, C. W., Friday, T. J., and Du, K. Y. (1971b). *Brit. J. Haematol.* 21, 233.
Paulus, J. M. (1967). *Blood* 29, 407.
Paulus, J. M. (ed.) (1971a). "Platelet Kinetics." North-Holland Publ., Amsterdam.
Paulus, J. M. (1971b). *In* "Platelet Kinetics" (J. M. Paulus, ed.), pp. 190–191. North-Holland Publ., Amsterdam.
Penington, D. G. (1969). *Brit. Med. J.* 4, 782.
Penington, D. G. (1970). *Brit. Med. J.* 1, 606.
Penington, D. G., and Olsen, T. E. (1970). *Brit. J. Haematol.* 18, 447.
Penner, J. A. (1970). *In* "Abstract Volume. XIII Int. Congr. Hematol, Munich," p. 112. Lehmanns Verlag, Munich.
Reed, R. E., and Penner, J. A. (1970). *In* "Program of Thirteenth Annual Meeting of the Amer. Soc. Hematol., San Juan," p. 129. Amer. Soc. Hematol.
Rolovic, Z., Baldini, M., and Dameshek, W. (1970). *Blood* 35, 173.
Schulman, I., Abildgaard, C. F., Cornet, J. A., Simone, J. V., and Currimbhoy, Z. (1965). *J. Pediat.* 66, 604.
Shreiner, D. P., and Levin, J. (1970). *J. Clin. Invest.* 49, 1709.
Spector, B. (1961). *Proc. Soc. Exp. Biol. Med.* 108, 146.
Spertzel, R. O., Bucci, T. J., and Ingram, M. (1963). *In* "Radiation Effects in Physics, Chemistry and Biology," *Proc. Int. Congr. Radiat. Res.*, 2nd (M. Ebert and A. Howard, eds.), pp. 342–344. YearBook Publ., Chicago, Illinois.
Steiner, M., and Baldini, M. (1969). *Blood* 33, 628.
Stohlman, F., Jr. (1970). *In* "Formation and Destruction of Blood Cells" (T. J. Greenwalt and G. A. Jamieson, eds.), pp. 65–84. Lippincott Co., Philadelphia, Pennsylvania.
Sullivan, L. W. (1969). *Clin. Res.* 17, 345 (abstract).
Trentin, J. J. (1970). *In* "Regulation of Hematopoiesis" (A. S. Gordon, ed.), pp. 161–186. Appleton, New York.
Witte, S. (1955). *Acta Haematol.* 14, 215.

HUMORAL ASPECTS OF BLOOD CELL DYSCRASIAS*

T. N. Fredrickson and P. F. Goetinck

I. INTRODUCTION

As other chapters in this book attest, humoral control of hematopoiesis has been the subject of intense investigations. During the last 2 years several books on various aspects of this subject have appeared (Stohlman, 1970; Gordon, 1970; Matoth, 1970; Clarke *et al.*, 1970; Krantz and Jacobson, 1970; Metcalf and Moore, 1971). This intensive study has been mainly directed to alleviating hematological diseases in humans. However, in the general area of biological growth and differentiation,

* Scientific Contribution No. 534 of the Agricultural Experiment Station, University of Connecticut, Storrs, Connecticut 06268. Preparation aided by Public Health Service Grant R01CA12815-01.
† Recipient of a Career Development Award of National Institute of General Medical Science (GM 16903).

hematological tissue is unique. Differentiation is directed by hormones, erythropoietin (EP), colony stimulating factor (CSF), and thrombopoietin and each stage of the process is reflected in morphological and functional changes. Humoral aspects of lymphopoiesis is reviewed by Havemann, Schmidt and Rubin in this monograph and therefore will not be considered here. Although the hematopoietic stem cell cannot be morphologically identified, it can be enumerated (Till and McCulloch, 1961) so that one may expect clarification of the level at which differentiating hormones act.

Constant loss of erythrocytes, granulocytes, and platelets in the peripheral circulation necessitates replacement with a continual influx of stem cells into precursor compartments. Thus, in the healthy animal, replication and differentiation for all cell lines take place simultaneously and at an even rate under normal conditions. Under conditions of excessive drain of differentiated cells these processes take place at an accelerated rate.

In dyscrasias associated with either under- or overproduction, the obvious abnormality would appear to be alterations in levels of hormones governing influx of stem cells and differentiation. It appears, however, that this is seldom the case in a wide range of erythroid dyscrasias, and levels of EP appear to respond in a highly efficient and very functional feedback system. Thus, except in rare instances, the etiology of the dyscrasia does not lie with humoral regulatory mechanisms but with a scarcity of raw materials required for hematopoiesis or with the inability of the precursor cells to synthesize required metabolites, allowing them to proceed down the maturation pathway, or with the accelerated destruction in the circulating blood. The stability of production and continued functionability of this humoral control mechanism indicates that it must represent the product of very successful evolutionary development and study of hematopoietic hormones has far-reaching implications.

Since other chapters deal with humoral mechanisms for control of each hematopoietic cell line, we shall attempt here to describe dyscrasias which have been studied most intensively from the standpoint of humoral control. The great potential for use of inbred mice with genetically directed errors in hematopoiesis and use of these animals as experimental models, are pointed out. Since it appears to us that unwarranted emphasis and reliance has been placed on clinical observations in humans, this contribution will not be a detailed review of hematopoiesis and its control, but will be limited to consideration of spontaneously occurring and experimentally induced hematopoietic dyscrasias in which humoral aspects have been clearly delineated.

II. HUMORAL ASPECTS OF ERYTHROPOIESIS

Within the erythroid compartments*, supplied by a constant stream of incoming stem cells, differentiation into erythroid precursors is mediated by the action of EP (Kubanek *et al.*, 1968; Fogh, 1971; Reissmann and Samorapoompichit, 1970). The precursor of EP, renal erythropoietic factor (REF) (Gordon and Zanjani, 1970) is synthesized in the kidney, possibly in the visceral epithelial cells of the glomerulus (Fisher *et al.*, 1970c) and reacts with a serum factor which is probably synthesized in the liver to form active EP (Zanjani *et al.*, 1967).

That pO_2 governs EP production was proved by perfusion of blood with different pO_2 through isolated kidneys and noting EP levels in effluent blood. There was a direct inverse relationship showing that anoxia promoted EP production (Pavlovic-Kentera *et al.*, 1965; Fisher and Langston, 1967). Hypoxic or anemic hypoxia induces elevation in EP levels as a result of reduction of pO_2 either due to reduced oxygen in the air or number of cells capable of oxygen transport. Reduction in red cell mass is a very effective stimulator of EP production and in most anemias levels of EP in plasma and urine are in an almost straight-line, inverse relationship with numbers of circulating cells. Direct evidence for effect of reduction of red cell mass, without complicating disease, was clearly shown in studies on effects of phlebotomy in healthy humans who reacted with distinct increases in EP (Pennington, 1966; Adamson, 1968). Similar results have been obtained in studies of animals subjected to phlebotomy and demonstration of a factor in plasma from bled animals, which induced erythrocytosis in recipients, led to identification of EP (Borsook *et al.*, 1954; Plzak *et al.*, 1955). Low pressure of O_2 in inspired air also induced an EP response in rats (Stohlman and Brecher, 1957), resulting in a secondary polycythemia. Once polycythemic animals are returned to normal ambient air, EP production ceases, hence the basis for the exhypoxic polycythemic mouse test for EP (Cotes and Bangham, 1961; DeGowin *et al.*, 1962). Direct measurements of EP plasma levels in mice under hypoxic conditions show that

---

* These are different for different animals according to their phylogenetic order; for example, insects lack erythroid cells, depending on circulation of hemolymph in their coelom, in teleost fishes the kidneys are the erythrogenic organs, in frogs the marrow only acts as an erythroid organ after hibernation or during metamorphosis (Siegel, 1970), the mouse has active erythropoiesis both in the marrow and normally in the spleen which acts as an organ of reserve in times of erythroid demand (Papayannopoulou and Finch, 1972), and in man erythropoiesis is normally restricted to the red marrow.

initial high levels of EP decrease after 15 hours exposure (Camiscoli and Gordon, 1970). Measurements of EP levels made soon after initiation of hypoxia show, however, a fairly clear-cut relationship between the degree of hypoxia and amount of the hormone in the blood (Gurney et al., 1965). Mice exposed to hyperbaric conditions, with increased O_2, also cease to have detectable levels of EP in their blood (Linman and Pierre, 1968) and reduction of oxygen from 10 to 8% in ambient air had considerable effects on iron uptake in polycythemic mice (Faura et al., 1968).

Thus, pO_2 has long been recognized as the primary mechanism for stimulation of erythropoiesis in a wide variety of animals. However, an additional control mechanism on erythropoiesis, independent of oxygen tension, has been proposed (Kilbridge et al., 1969; Nečas and Neuwirt, 1970; Whitcomb and Moore, 1968; Reynafarje et al., 1964; Krzymowski and Krzymowska, 1962).

Reynafarje et al. (1964) injected starved rats with sera from humans residing at sea level, at 4540 meters, and from recent arrivals at that altitude. Elevated levels of EP were noticed in sera from persons recently arrived at high altitudes, whereas plasma from natives of high altitudes brought down to sea level had a severe depressing effect on erythropoiesis in starved rats. This depressing effect was present for as long as 20 days of residence at sea level. Krzymowski and Krzymowska (1962) demonstrated presence of an inhibitory factor for erythropoiesis in the plasma of hypertransfused sheep, and Whitcomb and Moore (1968) showed similar inhibition, as tested in plethoric and in normal mice, after injection of plasma from polycythemic sheep. These results were analyzed on the basis of a fall of hemoglobin concentration and red cell count which were preceded by reduction in normoblasts of the marrow. These results indicated presence of a plasma inhibitory factor but more direct evidence was furnished by increasing red cell mass without increasing pO_2. Thus, Kilbridge et al. (1969) plethorized mice with two different types of erythrocytes, one obtained from a human with a hereditary methemoglobinemia (Met-RBC) and another from a normal human donor (Oxy-RBC). At hematocrits of 52% Oxy-RBC depressed EP levels to about 70% of control values, but Met-RBC did not reduce EP in test animals. With increasing levels of hematocrit, however, the difference in EP suppression of Oxy-RBC and Met-RBC decreased until there was no difference at a hematocrit level of 60%. At this level EP was suppressed to about 10% of control values. Inhibition of EP by red cell mass was clearly indicated by these experiments, particularly so if one considers that plethora was induced by cells unable to transport oxygen. The authors concluded that possible factors, other than pO_2, which may depress erythropoiesis at high hematocrits, might possibly include increase of blood volume, red cell mass or blood viscosity. Less convincing but supporting this view was the suppression of EP in hypoxic mice when injected with plasma from dehydrated mice.

Nečas and Neuwirt (1970) describe a series of experiments somewhat similar to those of Kilbridge *et al.* (1969) contrasting the effect of anemic hypoxia and hypoxic hypoxia on EP production in hypoxic rats. Anemia induced a much more severe EP response than hypoxic hypoxia although as the degree of hypoxic hypoxia increased, the difference between anemia and hypoxic hypoxia decreased when comparing 1 hour to 6 hours of hypoxia. Normovolemia was maintained in anemic rats and it appeared that EP production after hypoxic stimulus was affected by red cell mass. On the basis of their findings, the authors constructed a hypothesis on the nature of regulation of EP by two negative feedback loops, one based on oxygen pressure pO_2, the conventional EP regulating system, and the other on a humoral inhibitor associated with red cell mass which acted by decreasing EP production. These authors were careful to point out blood viscosity might be the mechanism for regulation rather than a humoral substance. By making animals plethoric, however, reduction of pO_2 may occur (Erslev and Thorling, 1968) and this must be ruled out before effect of plethora can be assessed with certainty. The general idea of control by red cell mass is consistent with a chalone concept as proposed by Rytomaa and Kiviniemi (1968).

A. Primary Alterations in EP Production

1. UNDERPRODUCTION OF EP

a. Uremia. Levels of EP below normal cannot be measured with certainty because of insensitive bioassay procedures. However, in most cases of anemia, EP levels are increased (Hammond *et al.*, 1968). A marked exception to this inverse relationship is uremic anemia in which EP levels are frequently undetectable (Naets and Heuse, 1962; Brown, 1966; Nathan *et al.*, 1968; Ward *et al.*, 1971). This points to EP underproduction as a likely cause for the anemia, but loss of mature cells is a complicating factor since hemolysis occurs in a degree proportional to the degree of uremia (Adamson *et al.*, 1968). The cause of hemolysis remains undetermined; Loge *et al.* (1958) found no increase in cell fragility although cytotoxic effects of uremic plasma have been described (Wardle, 1970). Deposition of platelets and fibrin, with resultant thrombosis of glomerular capillaries has also been suggested as a cause of lysis (Pennington and Kincaid-Smith, 1971). Reduced marrow stimulation, as indicated by low reticulocyte counts (Shaw and Schole, 1967), accounts for failure to compensate for the anemia and may be due only to decreased EP production. Restoration of plasma EP and elevation of hematocrit was clearly shown by measurements before and after renal transplantation (Nakao *et al.*, 1970) indicating the central role of the hormone in uremic anemia. Another possibility is presence in uremic anemia of a plasma inhibitor acting against EP. Such an inhibitor has been described in homogenates of kidneys from rabbits with nephrotoxic nephritis and, in addition, serum of these uremic rabbits also showed inhibitory action on EP which was ablated by bilateral nephrectomy (Moriyama and Shimotori, 1970). Other studies have also implicated an inhibitor of EP in uremic plasma (Fisher *et al.*, 1968b) and urine concentrates from anemic patients (Finne, 1968). Isolation of an inhibitor in urine from patients with chronic renal failure has been reported (Lindemann, 1970). Thus the inhibitors of EP have been associated with uremic conditions in several reports; however, further investigation will be required to assess their role in explaining the inactivity of the marrow in renal disease, because inhibitors, other than those associated with uremia, have also been reported, including a potent lipid inhibitor of EP isolated from renal homogenates of kidneys from normal rabbits (Erslev *et al.*, 1971), and a basic protein inhibitor from normal and anemic human urine (Lewis *et al.*, 1969). Uremic anemia is an obvious condition for therapy with exogenous EP and several attempts have been reported. Human anemic plasma restored erythropoiesis in a bilaterally nephrectomized dog (Naets, 1960), but large doses appeared to be necessary to alleviate uremic anemia in man (Van Dyke *et al.*, 1963). In a study of bilaterally nephrectomized rats, exogenous EP had little therapeutic effect (Bozzini *et al.*, 1966) and marrow may be less reactive to EP in uremic than in normal conditions and low EP levels may not be the sole or primary cause of anemia.

b. Chronic Inflammation. Moderately reduced erythropoiesis is associated with inflammatory conditions, particularly those of a chronic nature. In general, after a rapid decline, the hematocrit becomes stabilized at a lower than normal level, particularly when the infection is accompanied by severe systemic reactions (Cartwright, 1966). Reduction of erythrocyte survival time and hypoferremia associated with this condition complicate evaluation of the role of EP. A survey of patients with chronic disorders and anemia showed half had moderately elevated EP levels (Ward *et al.*, 1971). This inadequate level suggested inhibitors, which could not be demonstrated, or impaired synthesis or increased turnover of EP in chronic disorders. In one of the few experimental studies on turpentine induced abscesses, rats developed mild anemia. High doses of exogenous EP or exposure to hypoxic hypoxia increased red cell mass, above the levels of the rats without abscesses, and inadequate stimulus for erythrocytosis was considered as a possible cause of the anemia (Gutnisky and Van Dyke, 1963).

Bleeding reduces the number of marrow cells forming colonies *in vitro* (myeloid precursors) indicating a common stem cell for erythroid and myeloid cells (Metcalf, 1969). Such a relationship was also indicated by the observation that administration of endotoxin to mice increased number of colony forming cells (CFC) but reduced hematocrits (McNeill, 1970). Thus, one could speculate that the anemia of chronic inflammation may be associated with a drain of precursors which are diverted to the myeloid pathway of differentiation; however, this is a simplification of a complicated process since the spleen accumulates increased numbers of CFC (Rickard *et al.*, 1971).

c. Genetically Directed Underproduction. Genetic differences in the ability to produce EP in response to hypoxia have been reported by Shadduck *et al.* (1968). CAF_1 mice show a much lower increase in red cell mass in response to prolonged hypoxia than do CF_1 mice. This difference in response cannot be due to differences in sensitivity of EP sensitive cells to EP since both strains responded equally to the administration of exogenous EP. Both strains of mice were also similar in their oxygen dissociation curves, and responded equally to phenylhydrazine treatment and to posthemorrhagic anemia by increasing their EP production to the same extent. In the framework of the double negative feedback control, this could indicate that the strain differences affect the pO_2 control of EP production but not the red cell mass control mechanism (Kilbridge *et al.*, 1969; Nečas and Neuwirt, 1970).

Differences in EP production after hypoxic stimulus have been reported (Nohr, 1967) between other strains. Thus, BALB/c female mice do not respond to a 24-hour period of hypoxia caused by simulated altitude of 18,000 feet, while Swiss-Webster strain mice do by increasing EP production. A response at a simulated altitude of 25,000 feet was observed in the BALB/c mice but this was not as great as that observed in similarly treated Swiss-Webster mice. BALB/c males showed a measurable response to hypoxia which was comparable to that of BALB/c females but less than that measured in Swiss-Webster males. Response of BALB/c males to hypoxia was eliminated by castration. The observation is consistent with the known role of testosterone on erythropoiesis

(Fisher, 1969). The difference between the two strains in response to hypoxia is not due to a difference in ability to respond to EP, for female mice of both the BALB/c and the Swiss-Webster strains responded equally well to the exogenous administration of EP. BALB/c mice also responded to acute hemorrhage but not as well as Swiss-Webster mice. High blood pressure of BALB/c mice (Schlager, 1966) may have some relevance in its lack of response to hypoxia. However, this is not indicated since a strain of rats develops spontaneous hypertension resulting in increased red cell count and elevated EP production (Sen et al., 1972). This would be expected since, as noted below, hormones inducing high blood pressure are also associated with increased erythrocytosis, probably on the basis of increased EP production.

2. Overproduction of EP

   a. EP Producing Neoplasms. Certain neoplasms associated with production of EP and causing increased erythrocytosis include renal carcinoma (Hewlett et al., 1960), Wilm's tumor (Shalet et al., 1967; Thurman et al., 1966), cerebellar hemangioblastoma (Gallagher and Donati, 1968), pheochromocytoma (Bradley et al., 1961), adrenal adenoma (Gallagher and Donati, 1968), and hepatic carcinoma (Waldmann et al., 1968). Excision of some tumors has been associated with a reduction in plasma EP levels (Korst et al., 1959; Vertel et al., 1967). Not all cases of polycythemia have been associated with elevated serum levels of EP (Pennington, 1965) or presence of EP in saline extracts of tumor tissue (Waldmann et al., 1968). In an interesting approach to this problem, a transplantable adrenal cortical tumor of rats was used to study the pathogenesis of polycythemia associated with tumors (Waldmann et al., 1968). Implanted rats responded with a markedly increased red cell mass and there was a rapid decline in hematocrit following resection. Additional evidence of excessive EP was the observation that iron uptake in rats with small tumors and minimal increases in hematocrits, was not reduced after hypertransfusion as in normal mice. Although no EP activity could be measured in saline extracts of the tumors or sera of implanted polycythemic rats, the conclusion that erythrocytosis was caused by a factor other than EP is doubtful, since successful efforts to extract the hormone in this tumor have been reported (Meineke and Feleppa, 1968). Increased plasma levels of EP have also been found in mice with renal adenocarcinomas induced with dimethylnitrosamine (Mirand, 1968a). Increased levels of EP are not always observed in humans with polycythemia associated with tumors particularly uterine fibromas, and ovarian tumors and in some adrenal tumors secretion of other hormones may have been the cause for erythrocytosis (Fisher, 1969).

b. Hydronephrosis and Renal Cysts. Space-occupying, non-neoplastic lesions including hydronephrosis and renal cysts, have also been found to increase plasma levels of EP and EP activity could also be demonstrated when cystic fluid was inoculated into mice (Jones *et al.*, 1960; Rosse *et al.*, 1963; Pennington, 1965). Experimental elevation in plasma EP levels with polycythemia was induced by unilateral ligation of ureters of rabbits (Mitus *et al.*, 1968). About 60% of animals with ligated ureters had first a rise in EP levels followed shortly by an increase of blood cell volume and hemoglobin. Sixty days after ligation, EP levels returned to normal, although the polycythemic state was maintained. Secretion of EP in kidneys with ligated ureters was dependent on pressure, since overdistention of the renal pelvis caused cortical atrophy. Certain thresholds are required since spontaneous hydronephrosis in rats, with moderate uremia, did not cause anemia or elevated EP levels (Lozzio *et al.*, 1970).

B. Primary Alterations in Erythroid Cytopoiesis

1. STEM CELL DEFICIENCY

a. W Series Anemia. By far the most extensively investigated genetically determined anemia is caused by the action of the multiple allelic series at the dominant spotting or *W* locus in the mouse. These alleles have a pleiotropic effect in that they influence hematopoiesis, germ-cell production, and pigmentation. Mice of genotypes with two dominant mutant alleles have a severe macrocytic anemia; they are sterile and have black eyes and white hair. Viability and expressivity of the pleiotropic effect depend on the particular combination of alleles and on the genetic background in which these are maintained (Russell and Bernstein, 1966; Russell, 1970).

It seems quite certain that the blood defect in the *W* series is a cellular one and primarily of erythropoietic-sensitive cells (EPSC). This was shown by implantation of histocompatible normal cells from marrow, spleen, and fetal liver into anemic mice (Russell *et al.*, 1956, 1959; Russell, 1960; Bernstein, 1963; Russell and Bernstein, 1968) which cured the anemia of mutant mice. When *W* anemic marrow or spleen cells were implanted into lethally irradiated normal mice, however, erythropoietic and leukopoietic recovery were not observed. The leukopoietic

defect was unaffected by erythropoiesis since prolonged polycythemia
did not correct the defect (Bennett et al., 1968). W anemic mice produce EP normally and their plasma may contain slightly more EP than
normal mice (Russell, 1970). They can respond to exogenous EP
but in an extremely inefficient manner (Keighley et al., 1962). Polycythemic W/W^v mice require about 150 times as much EP as do $+/+$
polycythemic mice to achieve a similar degree of stimulation (Niece
et al., 1963). It seems, therefore, that EP may be regulated although in
an inefficient manner in W mice at the point of interaction of EP with
EPSC. That this may indeed be the case is suggested by the report of
Schooley et al. (1968) when they injected W/W^v mice with antierythro-
poietin serum. After 5 days these mice showed a decrease in marrow
erythroid cells, blood-iron incorporation, and reticulocyte levels. It is
important to point out that the 2 negative feedback regulatory steps,
referred to earlier, seem to be functional in the W mice. W anemic mice
responded to hypoxia with the same relative increase in erythrocyte
count as normal litter mates (Gruneberg, 1939; Keighley et al., 1966;
Bernstein et al., 1968), suggesting that the pO_2 control is functioning
normally. Furthermore, hypertransfusion suppressed EP completely
(Niece et al., 1963) suggesting that the production of EP is also responsive to the red cell mass control system. W anemic mice also showed a
hematopoietic response when both pO_2 and red cell mass were reduced
by phenylhydrazine treatment or phlebotomy (Russell, 1970).

b. Steel Anemia. The hereditary macrocytic anemia produced by the
combination of multiple alleles at the Steel locus (Sarvella and Russell,
1956) shows a strong phenotypic resemblance to those anemias produced
by the W locus. The phenotypic resemblance of the Sl anemias to the
W anemias involve in addition to the anemia also the sterility and pigmentation defect (Russell, 1970). Genetically, however, the two anemia
series are different in that they are in different linkage groups (Wolfe,
1963). The Sl and W anemias can also be shown to be different at
the cellular level as revealed by transplantation experiments.

Injection of Sl marrow cells into W hosts was able to permanently
cure the anemia in the latter (Bernstein et al., 1968) and parabiosis
of Sl anemic mice with W anemic mice resulted in normal blood values
in members of both genotypes. After separation of the parabionts, blood
cells returned to low levels in the Sl mice. The blood picture in the
W mice, however, remained normal after the separation of the
parabionts. This permanent cure was ascribed to the normal functioning
of the hematopoietic stem cells of the Sl partner which became implanted into the W marrow or spleen (Russell, 1970). Since Sl mice

are very unresponsive to exogenous EP administration, it seems that the response of the EPSC to EP depended on a particular microenvironment which appears defective in *Sl* mice. This explanation is further supported by the finding that suspensions of normal marrow or spleen cells when injected into *Sl* anemic mice were not able to overcome the anemia, although whole spleens were, suggesting that the structural integrity of this organ is critical (Russell, 1970).

As in the *W* series, there is evidence that the 2 negative feedback controls are also functional in *Sl* anemic mice. *Sl* mice have high levels of EP in their plasmas except under polycythemic conditions. This would indicate that EP production is not defective and that this production is sensitive to the red cell mass control mechanism. The fact that intermittent hypoxia elicits an erythropoietic response would indicate that the pO_2 control is functional.

2. IMBALANCES IN ERYTHROCYTOSIS

a. Substrate Deficiency. Studies on the effect of diet on EP production have shown that a drastic reduction in red cell mass occurs in the protein-deprived rat (Ito and Reissmann, 1966) which can be almost completely alleviated by exogenous EP therapy. In humans, however, normal or raised serum EP levels were observed in children with anemia associated with kwashiorkor. Although EP levels continued to rise with protein refeeding, a direct relationship between EP elevations and increase in hematocrit was difficult to accept in these children because, in addition to being deficient in protein, the diet was also deficient in vitamins and iron (McKenzie *et al.*, 1967). Serum EP levels in humans, on iron-deficient diets, were directly proportional to all levels of anemia including those of cases in which the degree was not severe. Urinary EP levels were also elevated but to a lesser extent than in serum (Movassaghi *et al.*, 1967). Caloric deprivation decreased EP levels in pigs with a concomitant reduction in red cell mass. Phlebotomy, however, elicited a prompt increase in EP and erythrocytosis indicating no deficiency of essential nutrients for erythropoiesis (Stekel and Smith, 1969).

b. Pancytopenia and Erythroleukemia. In humans with bone marrow failure, EP levels are inversely proportional to the degree of anemia (Alexanian and Alfrey, 1970). In another study it was found that in severe pancytopenia there were higher levels of EP than for other anemias of comparable severity. Cultured marrow cells varied in responsiveness to EP with more nonresponsive cells in patients with the severest anemias; despite responsiveness, however, anemia continued possibly due to an inhibitor of EP (Mizoguchi *et al.*, 1971). In an anemic patient with erythroleukemia, plasma levels of EP varied with the course of the disease and were initially high. In patients with DiGuglielmo's syndrome, EP levels were also high in keeping with the degree of anemia. Hypertransfusion depressed the EP levels (Adamson and Finch, 1970). These observations indicate that in human bone marrow failure or anemias associated with neoplastic disease of the erythropoietic system, the EP regulatory mechanism is operational.

c. *Hemoglobinopathies.* Certain hemoglobinopathies are associated with polycythemia in which the oxygen-dissociation curve is shifted to the left so that erythrocytes give up less oxygen to the tissues than normal (Weatherall, 1969). Levels of EP were not abnormally high, but phlebotomy caused extremely high values similar to the values found in humans with hypoxic polycythemia. Those hemoglobins which comprise a large proportion of the total, without causing hemolytic anemia, are associated with polycythemia (Charache, 1970). In thalassemia and sickle cell anemia, serum EP levels are increased (Gordon et al., 1964; Hammond et al., 1968).

d. *Polycythemia Vera.* With very rare exceptions, the red cell mass remains proportional to physiological needs. The exceptional cases of primary polycythemia or polycythemia vera are extremely interesting from the standpoint of humoral control. Levels of EP in plasma and urine are so uniformly depressed as to be unmeasurable (Stohlman, 1966; Adamson, 1970; Pennington, 1966). When subjected to phlebotomy to reduce red cell mass and to relieve the symptoms of their disease, patients with polycythemia vera respond with the production of measurable levels of EP (Adamson, 1970) but to a degree less than would be expected, considering the amount of blood withdrawn and usual degree of secondary polycythemia seen after phlebotomy (Adamson, 1968; Krantz, 1970). Polycythemia vera is associated with granulocytosis and thrombocytosis indicating a generalized proliferative disease of the marrow as do increased levels of serum muramidase, a granulocytic enzyme, in patients with the disease and the relatively high risk of acute myelogenous leukemia in long-term patients (Wasserman and Gilbert, 1966). Lack of response of cultured marrow cells from patients with polycythemia vera to EP stimulation was seen (Krantz, 1968). Along with red cell mass, the marrow increases in size (Van Dyke et al., 1970) and some investigators have proposed that the erythroid cells are of two types; one EP responsive which reacts to phlebotomy, and another EP nonresponsive which accounts for overproduction. This hypothesis based on a dual cell population presents some problems; erythroleukemia would seem to be a natural sequella to polycythemia, but this disease seldom if ever occurs despite widespread extramedullary hematopoiesis in advanced cases. It is also difficult to conceive of a large number of normal cells remaining in the erythroid precursor pool, ready to mature on signal from EP after phlebotomy. Lastly, the chronicity of the disease bespeaks an extremely benign "neoplastic" condition.

The inhibitor in plasma of polycythemics (Whitcomb *et al.,* 1969) would appear to act by inhibiting EP production and be analogous or identical to the inhibitor demonstrated in plasma from humans or animals with increased red cell mass, as mentioned above. Action of this inhibitor would seem to be evidenced by very low levels of EP in patients with polycythemia vera. The rapid rise in hematocrit after phlebotomy, followed by a much slower rise after normal levels are reached (Stohlman, 1966), could be due to increased amounts of inhibitor as the red cell mass increases. The central question as to why erythrocytosis continues in face of very low levels of EP remains to be answered, although chronic secondary polycythemia often is associated with unelevated EP levels.

e. Leukovirus Infection in Mice. Response of EP production in cases of infectious disease seem to have received very little attention even in those conditions primarily affecting the erythron. An exception is the interaction between the erythropoietic tissue of certain strains of mice and leukoviruses, particularly Friend leukemia virus (FLV) and Rauscher leukemia virus (RLV). FLV is a complex of different viral strains, some of which induce anemia, others polycythemia (Mirand, 1968b). However, RLV has been associated only with anemia. Polycythemia may be induced in Swiss mice which are plethoric at the time of infection (Mirand, 1968b) and have low EP levels. Thus, FLV appears to induce erythrocytosis in a manner similar to EP, and comparable increases in hematocrits occurred in mice given the virus as receiving large multiple doses of EP (Gurney *et al.,* 1961). Although EP levels are low in mice infected with FLV, hypoxic hypoxia induced EP production (Mirand, 1968b). When EP was injected into mice before and after infection with FLV, there was a much steeper rise in hematocrit compared to those which received only the virus (Mirand, 1966); however, anti-EP serum did not inhibit erythrocytosis induced by FLV (Sassa *et al.,* 1968). Comparison of the effects of EP and FLV indicates that the virus initiates hemoglobin synthesis and erythropoietic differentiation in EP-responsive cells (Tambourin and Wendling, 1971). In anemia induced with RLV, there was also no increase in plasma EP levels (Ebert *et al.,* 1972) indicating that in ineffective erythropoiesis, as well as in polycythemia, the hormone has no obvious role. Confirmation of low EP levels in anemic mice infected with RLV (Camiscoli et al., 1972) still does not account for many of the variables possibly interacting in causing anemia.

C. Factors Modulating EP Production

1. HORMONES

Androgens have been shown to have a stimulatory effect on EP in a variety of experimental animals (Gordon et al., 1966, 1968; Mirand et al., 1965; Fried and Gurney, 1968; Naets and Wittek, 1968) and in humans (Alexanian, 1966; Alexanian et al., 1967).

It has also been found that testosterone is effective in increasing erythropoiesis in patients with aplastic anemia (Gardner and Pringle, 1961; Rishpon-Meyerstein et al., 1968; Shahidi and Diamond, 1961). Since such patients generally have high levels of EP (Gurney et al., 1957; Naets and Heuse, 1962; Rishpon-Meyerstein et al., 1968; Van Dyke et al., 1957), the mechanism of action of testosterone is not understood. There may, however, be a connection with the observation of Fried and Gurney (1965) that testosterone alone has no effect unless its administration is preceded by a hypoxic stimulus. This hypoxic stimulus may render the kidney more sensitive to testosterone. That the action of androgens is to increase EP production is evident from several lines of investigation. No erythropoietic action was seen with anti-EP serum present (Schooley and Garcia, 1965; Fisher et al., 1967) or when mice are nephrectomized before exposure to hypoxia (Mirand et al., 1965). Investigations on the effect of testosterone on the isolated, perfused kidney (Malgor and Fisher, 1970) indicated that androgens are most effective in hypoxic kidneys and that they may be instrumental in releasing erythropoietic factor (REF) from the kidney (Fisher, 1969). Evidence is also available that androgens alone or in combination with EP may have an effect on erythroid cells in culture (Erslev, 1962; Reisner, 1966) although Malgor and Fisher (1970) were not able to demonstrate a direct stimulatory effect of testosterone on bone marrow cells of isolated, perfused hind limbs of dogs. This observation again suggests that the erythropoietic activity of testosterone is due to an increased elaboration of EP from the kidney. Other humoral factors which stimulate EP production by increasing the relative pO_2 deficit are the thyroid hormones and ACTH through its effect on the release of adrenal steroids. These hormones have been shown to increase the metabolic rate and to raise the red cell mass in hypophysectomized rats (Crafts and Meineke, 1959; Evans *et al.*, 1961; Fisher and Crook, 1962). The erythropoietic effect of these hormones was abolished by simultaneous administration of anti-EP serum (Fried and Gurney, 1968), indicating they act through the enhanced production of endogenous EP. Corticosteroids and thyroxine, alone or in combination, had no effect on restoring the reduced red cell mass, percentage of nucleated red cells in the bone marrow, or plasma EP levels in hypophysectomized monkeys. Thyroxine, in combination with human growth hormones, restored all the abnormal blood figures to normal levels (Vitale *et al.*, 1971).

2. DRUGS

Several drugs have been shown to affect erythropoiesis. The effect of these drugs is on the production of EP and it is probably exerted by the creation of a relative hypoxic state in the kidney. The most studied pharmacological agents are angiotensin, norepinephrine, and vasopressin. Infusion of angiotensin has been shown to elevate plasma levels of EP (Fisher and Crook, 1962; Fisher *et al.*, 1967; Bourgoignie *et al.*, 1968). Both angiotensin and norepinephrine have been shown to produce a reduced arterial blood flow and an increased blood pressure, as well as an increase in plasma EP in dogs. Mechanical renal vasoconstriction resulted in elevated levels of EP (Fisher *et al.*, 1968a). When hydralazine is given simultaneously with angiotensin, renal blood flow is restored to normal levels with a concomitant restoration of normal levels of plasma EP. These observations are taken to mean that the observed effect of the drug on EP levels is the result of a reduction of renal blood flow and therefore an activation of oxygen sensors (Malgor and Fisher, 1969).

3. COBALT

The administration of cobalt results in an increased rate of erythropoiesis and has been shown to correct the anemia caused by hypophysectomy (Garcia *et al.*, 1952; Crafts, 1952; Goldwasser *et al.*, 1958) and by protein starvation (Orten and Orten, 1945). The action of cobalt is thought to be directly on the kidney by affecting oxygen utilization. These conclusions were made from the observations that the erythropoietic action of cobalt was greatly reduced after bilateral nephrectomy (Jacobson *et al.*, 1957). Furthermore, perfusion of the isolated dog kidney with cobalt-treated blood resulted in increased release of EP in the perfusates (Fisher and Langston, 1968). These authors also showed that kidneys perfused with cobalt containing blood had a reduced rate of oxygen utilization, compared to controls, and must have been in a relative state of hypoxia.

Cobalt has been successfully used in elevating the red cell population in a number of human diseases. These diseased states include anemia secondary to rheumatoid arthritis, chronic inflammation, renal diseases, malignancies, and protein deprivation (Krantz and Jacobson, 1970). The mechanism here is also thought to be via an increase in EP production which results from the relative state of hypoxia in the cobalt-treated individuals. How the cobalt-induced EP works in humans, already in many instances producing high levels of EP due to the anemia, is not known. This cobalt effect is similar to the one observed after androgen treatments. In both the cobalt and androgen administration, elevation of EP may occur which corresponds to a pharmacological level of the hormone.

It is of particular interest that angiotensin (Bilsel et al., 1963; Mann et al., 1966; Fisher et al., 1968a), norepinephrine (Fisher et al., 1968a), testosterone (Gurney and Fried, 1965), cobalt (Fisher et al., 1968b), and triiodothyronine (Fisher et al., 1970a) do not have a stimulatory effect on the erythropoiesis of polycythemic animals. The failure of action is not understood, but two control factors must be taken into consideration. Either the polycythemia does not create an oxygen deficit even with the reduced blood flow or the negative feedback ascribed to red cell mass overpowers the pO_2 sensitive feedback system. That the regulation may very well act on production of EP and not the response of the EP-sensitive cell to EP is indicated by the erythropoietic response of these polycythemic animals to exogenous EP (Fisher, 1969).

III. HUMORAL ASPECTS OF THROMBOCYTOPOIESIS

It has been known for some time that depletion of thrombocytes by cross-transfusion with thrombocyte depleted blood, or use of antiplatelet antiserum, induced thrombocytosis (Odell et al., 1962) and, conversely, thrombocytosis, produced by infusion of platelets, induced a transient thrombocytopenia (de Gabriele and Pennington, 1967). Early studies were based on counts of circulating platelets, a procedure fraught with considerable difficulties particularly from the standpoint of reproducibility (Abildgaard and Simone, 1967). More precise procedures for measurement of thrombocytosis have been recently introduced based on incorporation of radioactive sulfur (Odell et al., 1964). Two such procedures, one using $Na_2[^{35}SO_4]$ as the source of isotope (Harker, 1970) and the other selenomethionine-[^{75}Se] (Levin, 1970; Evatt and Levin, 1969) employ animals made thrombocytotic by hypertransfusion. This reduced "endogenous" thrombocytosis provided a more sensitive bioassay system than previously, and allowed incorporation of isotope in circulating platelets as a precise measure of thrombocytopoiesis during the test period. Both techniques confirmed earlier conclusions that thrombocytopoiesis was governed by a humoral factor in a feedback system similar in concept to the EP-erythropoiesis system. Similarities with regard to stem cell recruitment are also striking (Ebbe *et al.*, 1971). The radiosulfate incorporation method in measuring thrombocytopoiesis was confirmed using rats made thrombocytopenic by anti-rat platelet antiserum (Cooper, 1970). Similar results were obtained in mice given muramidase (Gasic *et al.*, 1968). Complications for interpretation of platelet counts, i.e., possible destruction or splenic sequestration, have been largely obviated by these procedures and moderate responses of normal plasma (Harker, 1970) indicate that thrombopoietin acts as a regulator of normal thrombocytopoiesis. In an extensive review Abildgaard and Simone (1967) described studies which showed that thrombocytosis could be induced in mice, rats, and humans injected with either plasma, serum, or urine of humans with thrombocytosis, polycythemia vera, thrombocytopenic purpura, and aplastic anemia. They cautioned that in these studies thrombopoiesis in recipients of these materials was not dose dependent and in some studies negative results were obtained. An unequivocal case, studied for 10 years, of thrombocytopenia responsive to infusions of frozen serum has provided much information regarding response to thrombopoietin. Cycles of thrombocytosis were observed beginning 4 days after treatment, peaking at 9, and returning to low pretreatment levels at 22 days (Johnson *et al.*, 1971). The cycle was repeated with precision after every treatment. Thrombopoietic activity in plasma from patients with idiopathic thrombocytopenia has also been reported (Pennington, 1970). Earlier work in animals injected subcutaneously with ground glass or egg albumin showed a thrombocytosis (Odell *et al.*, 1964), indicating the large range of possible materials which may influence platelet production. Of particular importance in this study was the demonstration that plasma from these animals also induced thrombocytosis. Previously treatment with many substances had been shown to increase platelet counts (Abildgaard and Simone, 1967), but thrombopoietin content of serum of animals receiving these treatments was not assayed. Thus, most observations of thrombocytosis in experimental animals or human patients give little information on possible humoral mechanisms; as new bioassay techniques develop, definitive experiments will undoubtedly provide much exciting new information.

IV. HUMORAL ASPECTS OF LEUKOPOIESIS

Development of *in vitro* assay methods (Pluznik and Sachs, 1965; Bradley and Metcalf, 1966) has provided a technique for measurement of a humoral factor controlling myelopoiesis. This factor, using the nomenclature proposed by Metcalf, is the colony stimulating factor (CSF), which promotes formation of granulocytic and monocytic colonies in soft agar. Bioactivity has been demonstrated by induction of neutrophilia and monocytosis when CSF was injected into mice (Metcalf and Stanley, 1971). Further indications of bioactivity were elevations found during bacterial and viral infections in humans (Foster *et al.*, 1968; Wahren *et al.*, 1970). Low levels of CSF in germfree mice indicated that continuous stimulation was required for its presence in plasma; however, leukocyte counts of these animals did not reflect CSF levels (Metcalf *et al.*, 1967). Injection of Freund's complete adjuvant, or sheep red blood cells, in irradiated mice induced rises in levels of CSF (McNeill, 1970). Endotoxin of *Salmonella typhosa* has also been shown to increase CSF in plasma of mice with peak levels when injected animals were leukopenic (Chervenick, 1971). There was a direct relationship between plasma CSF levels and doses of endotoxin injected but endotoxin had, by itself, no colony-forming effect *in vitro*. It appears, however, that by mixing bacterial antigens minimal levels of CSF activity in plasma could be enhanced (Metcalf, 1971c). Plasma from mice made neutropenic by irradiation also enhanced colony formation *in vitro*, in proportion to the amount of marrow damage (Morley *et al.*, 1971). Time between irradiation and appearance of the plasma factor made it appear unlikely that it merely represented a product of cell destruction, and this corroborates other studies indicating that it is actively secreted by leukocytes (Otsuka *et al.*, 1972).

Several investigators have proposed a double feedback system for control of myelopoiesis, one fast loop acting on granulocytic storage pools in the marrow, and the other on precursor cells (Cronkite and Vincent, 1969; Sachs, 1970; Morley and Stohlman, 1970). Techniques demonstrating mobilization of marrow storage pools have been developed using perfused, isolated limbs of rats by measuring leukocyte counts in effluents after infusion with plasma from rats made leukopenic by leukophoresis (Gordon *et al.*, 1963). Also counts of peripheral cells in humans after injection of homologous plasma obtained during and after periods of leukopenia induced by chemotherapy, promoted leukocytosis (Marsh and Levitt, 1971). Plasma from humans made leukopenic by treatment with endotoxin caused mobilization of neutrophils (Boggs *et al.,* 1968) as did plasma from humans with myelofibrosis (Bierman and Hood, 1971), and dogs injected with plasma from those made neutropenic by vinblastine or nitrogen mustard also became neutrophilic (Boggs *et al.,* 1966). Any relationship between such leukocytosis inducing factors and CSF needs to be established. Many factors are known to induce leukocytosis (Fruhman, 1970) but few of these have been examined with regard to humoral mechanisms. Cortisone caused an acute fall in CSF levels in plasma of mice (Metcalf, 1969).

Of particular interest has been the role of CSF in leukemic conditions, either occurring spontaneously in humans or in mice induced with viruses or as transplantable tumors. Spontaneously leukemic and preleukemic AKR mice had elevated CSF levels compared to conventional or germfree Swiss mice (Metcalf *et al.,* 1967), but levels of mice infected by inoculation with leukemia viruses did not become elevated until leukemia was clearly established (Metcalf and Foster, 1967). In one study, humans with acute granulocytic leukemia had CSF levels above normal (Metcalf, 1971b), but in another study (Robinson and Pike, 1970), high levels were seen most strikingly in remission. Similar elevations in humans and animals treated with chemotherapeutic drugs have also been observed (Metcalf, 1971b). The role of inhibitors of CSF in leukemia is somewhat unclear, although it has been observed that blast cell levels in leukemic humans are higher when inhibitor levels were subnormal (Metcalf, 1971b). Stimulation of cultured human marrow cells by sera from patients with chronic myelogenous leukemia has been reported (Reisner, 1967).

Although *in vivo* activity of inhibitors of CSF appear equivocal (Metcalf, 1971a), they may act to depress myelopoiesis, and in mice there are correlations between development of monocytic as opposed to granulocytic colonies. Thus, colony-forming cells of BALB/c mice, with uniformly high inhibitor levels form mainly macrophage colonies and of C57Bl with low levels mainly granulocytic colonies (Chan, 1971). All normal human sera contain high levels of CSF inhibitors which are species-specific but not toxic to CFC *in vitro* (Chan *et al.,* 1971).

V. CONCLUSIONS

We have intended to point out in this article interrelationships between levels of control hormones and cell production in deranged hematopoiesis. It seems evident that a serious problem in all these studies has been limitations imposed by inadequacies and difficulties entailed

158   *T. N. Fredrickson and P. F. Goetinck*

in various assays for the hormones. Thus, even after 20 years of intensive study, the bioassay for EP cannot measure low levels of the hormone and the *in vitro* assay for colony forming cells remains a research tool. As mentioned, bioassays for thrombopoietin are being perfected but once more the prerequisite skill to perform them appears formidable. These problems are, of course, well recognized but advances in the area of endocrinological research, using such techniques as radioimmunoassay, portend great possibilities for investigation of humoral control of hematopoietic dyscrasias. However even in uremia, low levels of EP do not appear to be the sole cause of reduced hematopoiesis and proven cases of hematopoietic dyscrasias with absence of hormones or their etiology appear to be extremely rare. Such a case has been the patient with thrombocytopenia, described by Johnson *et al.* (1971), considered to be deficient in thrombopoietin or to possess a defective thrombopoietin. Similarly clear-cut cases of lack of EP can only be inferred by finding low levels in face of low hematocrits. Levels of CSF in such conditions as cyclic neutropenia would be of interest.

Certainly EP is a protein (glycoprotein) and in all likelihood the other hormones are probably also proteins. The primary structure of these proteins, therefore, reside in the hereditary material which must be subject to mutational events. The absence of clearly documented hereditary humoral disorders cannot be explained on the basis that these genes would be less mutable than others. Rather, assuming normal mutation rates, one can, in view of the essential roles of these hormones, predict that they would almost certainly result in lethal conditions. Their undetected presence is very likely the result of strong selection pressure against such mutations.

It seems, however, that the availability of humoral mutants would be extremely valuable in unraveling the entire regulatory process of hematopoiesis. This would be particularly valuable if they were used in conjunction with cellular mutants such as in Steel mice. It would seem that an effort should be made to detect, either spontaneous or induced mutants in genetically well-known experimental material such as the mouse, and to intensify investigation of hormonal control of hematopoiesis in experimental animals.

REFERENCES

Abildgaard, C. F., and Simone, J. V. (1967). *Semin. Hematol.* 4, 424.
Adamson, J. W. (1968). *Blood* 32, 597.
Adamson, J. W. (1970). *In* "Myeloproliferative Disorders of Animals and Man" (W. J. Clarke, E. B. Howard, and P. L. Hackett, eds.), pp. 440–452. USAEC, Oak Ridge, Tennessee.

Adamson, J. W., and Finch, C. A. (1970). *Blood* 36, 590.
Adamson, J. W., Eschbach, J., and Finch, C. A. (1968). *Amer. J. Med.* 44, 725.
Alexanian, R. (1966). *Blood* 28, 1007.
Alexanian, R., and Alfrey, C. P. (1970). *J. Clin. Invest.* 49, 1986.
Alexanian, R., Vaughn, W. K., and Ruchelman, M. W. (1967). *J. Lab. Clin. Med.* 70, 777.
Bennett, M., Cudkowicz, G., Foster, R. S., and Metcalf, D. (1968). *J. Cell. Comp. Physiol.* 71, 211.
Bernstein, S. E. (1963). *Radiat. Res.* 20, 695.
Bernstein, S. E., Russell, E. S., and Keighley, G. (1968). *Ann. N. Y. Acad. Sci.* 149, 475.
Bierman, H. R., and Hood, J. E. (1971). *Proc. Int. Symp. Comparat. Leukemia Res.*, 5th Karger, Basel (to be published).
Bilsel, Y., Wood, J. E., and Lange, R. D. (1963). *Proc. Soc. Exp. Biol. Med.* 114, 475.
Binder, R. A., and Gilbert, H. S. (1970). *Blood* 36, 228.
Boggs, D. R., Cartwright, G. E., and Wintrobe, M. M. (1966). *Amer. J. Physiol.* 211, 51.
Boggs, D. R., Chervenick, P. A., Marsh, J. C., Cartwright, G. E., and Wintrobe, M. M. (1968). *J. Lab. Clin. Med.* 72, 177.
Borsook, H., Graybiel, A., Keighley, G., and Windsor, E. (1954). *Blood* 9, 734.
Bourgoignie, J. J., Gallagher, N. I., Perry, H. M. Jr., Kurz, L., Warnecke, M. A., and Donati, R. M. (1968). *J. Lab. Clin. Med.* 71, 523.
Bozzini, C. E., Devoto, F. C. H., and Tomio, J. M. (1966). *J. Lab. Clin. Med.* 68, 411.
Bradley, J. E., Young, J. D. Jr., and Lentz, G. (1961). *J. Urol.* 86, 1.
Bradley, T. R., and Metcalf, D. (1966). *Aust. J. Exp. Biol. Med. Sci.* 44, 287.
Brown, R. (1966). *Lancet* 2, 319.
Camiscoli, J. F., and Gordon, A. S. (1970). *In* "Regulation of Hematopoiesis" (A. S. Gordon, ed.), pp. 369–393. Appleton, New York.
Camiscoli, J. F., LoBue, J., Alexander, P., Schultz, E. F., Hamburger, A., Gordon, A. S., and Fredrickson, T. N. (1972). *Cancer Res.* 32, 2843.
Carmena, A. O., Garcia de Testa, N., and Frias, F. L. (1967). *Proc. Soc. Exp. Biol. Med.* 125, 441.
Cartwright, G. E. (1966). *Semin. Hematol.* 3, 351.
Chan, S. H. (1971). *Aust. J. Exp. Biol. Med. Sci.* 49, 553.
Chan, S. H., Metcalf, D., and Stanley, E. R. (1971). *Brit. J. Haematol.* 20, 329.
Charache, S. (1970). *Mt. Sinai J. Med.* 37, 418.
Chervenick, P. A. (1971). *J. Lab. Clin. Med.* 79, 1014.
Clarke, W. J., Howard, E. B., and Hackett, P. L. (1970). "Myeloproliferative Disorders of Animals and Man." USAEC, Oak Ridge, Tennessee.
Cooper, G. W. (1970). *In* "Regulation of Hematopoiesis" (A. S. Gordon, ed.), pp. 1611–1629. Appleton, New York.
Cotes, P. M., and Bangham, D. R. (1961). *Nature (London)* 191, 1065.
Crafts, R. C. (1952). *Blood* 7, 863.
Crafts, R. C., and Meineke, H. A. (1959). *Ann. N. Y. Acad. Sci.* 77, 501.
Cronkite, E. P., and Vincent, P. C. (1969). *In* "Symposium on Hemopoietic Cellular Proliferation" (F. Stohlman, Jr., ed.), pp. 211–228. Grune and Stratton, New York.
de Gabriele, G., and Pennington, D. G. (1967). *Brit. J. Haematol.* 13, 202.

DeGowin, R. L., Hofstra, D., and Gurney, C. W. (1962). *Proc. Soc. Exp. Biol. Med.* **110**, 48.

Ebbe, S., Phalen, E., Overcash, J., Howard, D., and Stohlman, F. Jr. (1971). *J. Lab. Clin. Med.* **78**, 872.

Ebert, P. S., Maestri, N. E., and Chirigos, M. A. (1972). *Cancer Res.* **32**, 41.

Erslev, A. J. (1962). *Proc. Congr. Int. Soc. Haematol., 9th,* 143 (Abstract).

Erslev, A. J., and Thorling, E. B. (1968). *Ann. N. Y. Acad. Sci.* **149**, 173.

Erslev, A. J., Kazal, L. A., and Miller, O. P. (1971). *Proc. Soc. Exp. Biol. Med.* **138**, 1025.

Evans, E. S., Rosenberg, L. L., and Simpson, M. E. (1961). *Endocrinology* **68**, 517.

Evatt, B. L., and Levin, J. (1969). *J. Clin. Invest.* **48**, 1615.

Faura, J., Gurney, C. W., and Fried, W. (1968). *Ann. N. Y. Acad. Sci.* **149**, 456.

Finne, P. H. (1968). *Ann. N. Y. Acad. Sci.* **149**, 497.

Fisher, J. W. (1969). In "The Biological Basis of Medicine" (E. E. Bittar and N. Bittar, eds.), Vol. 3, pp. 41–79. Academic Press, New York.

Fisher, J. W., and Crook, J. J. (1962). *Blood* **19**, 557.

Fisher, J. W., and Langston, J. W. (1967). *Blood* **29**, 114.

Fisher, J. W., and Langston, J. W. (1968). *Ann. N.Y. Acad. Sci.* **149**, 75.

Fisher, J. W., Roh, B. L., and Halvorsen, S. (1967). *Proc. Soc. Exp. Biol. Med.* **126**, 97.

Fisher, J. W., Samuels, A., and Langston, J. (1968a). *Ann. N.Y. Acad. Sci.* **149**, 308.

Fisher, J. W., Hatch, F. E., Roh, B. L., Allen, R. C., and Kelley, B. J. (1968b). *Blood* **31**, 440.

Fisher, J. W., Samuels, A. I., and Malgor, L. A. (1970a). In "Erythropoiesis. Regulatory Mechanisms and Developmental Aspects" (Y. Matoth, ed.), pp. 70–78. Academic Press, New York.

Fisher, J. W., Roh, B. L., Malgor, L. A., and Noveck, R. J. (1970b). In "Erythropoiesis—Regulatory Mechanisms and Developmental Aspects" (Y. Matoth, ed.), pp. 155–162. Academic Press, New York.

Fisher, J. W., Busutill, R. L., and Roh, B. L. (1970c). In "Erythropoiesis. Regulatory Mechanisms and Developmental Aspects" (Y. Matoth, ed.), pp. 171–176. Academic Press, New York.

Fogh, J. (1971). *Radiat. Res.* **45**, 563.

Foster, R. S., Metcalf, D., Robinson, W. A., and Bradley, T. R. (1968). *Brit. J. Haematol.* **15**, 147.

Fried, W., and Gurney, C. W. (1965). *Proc. Soc. Exp. Biol. Med.* **120**, 519.

Fried, W., and Gurney, C. W. (1968). *Ann. N.Y. Acad. Sci.* **149**, 356.

Fruhman, G. J. (1970). In "Regulation of Hematopoiesis" (A. S. Gordon, ed.), pp. 873–915. Appleton, New York.

Gallagher, N. I., and Donati, R. M. (1968). *Ann. N.Y. Acad. Sci.* **149**, 528.

Garcia, J. F., Van Dyke, D. C., and Berlin, N. I. (1952). *Proc. Soc. Exp. Biol. Med.* **80**, 472.

Gardner, F. H., and Pringle, J. C., Jr. (1961). *Arch. Int. Med.* **107**, 112.

Gasic, G. J., Gasic, T. B., and Stewart, C. C. (1968). *Proc. Nat. Acad. Sci.* **61**, 46.

Goldwasser, E., Jacobson, L. O., Fried, W., and Plzak, L. (1958). *Blood* **13**, 55.

Gordon, A. S. (1970). "Regulation of Hematopoiesis." Appleton, New York.

Gordon, A. S., and Zanjani, E. D. (1970). In "Regulation of Hematopoiesis" (A. S. Gordon, ed.), pp. 413–458. Appleton, New York.

Gordon, A. S., Handler, E. S., Siegel, C. D., Dornfest, B. S., and LoBue, J. (1963). *Ann. N.Y. Acad. Sci.* 113, 766.

Gordon, A. S., Weintraub, A. H., Camiscoli, J. F., and Contrera, J. F. (1964). *Ann. N.Y. Acad. Sci.* 119, 561.

Gordon, A. S., Katz, R., Zanjani, E. D., and Mirand, E. A. (1966). *Proc. Soc. Exp. Biol. Med.* 123, 475.

Gordon, A. S., Mirand, E. A., Wenig, J., Katz, R., and Zanjani, E. D. (1968). *Ann. N.Y. Acad. Sci.* 149, 318.

Gruneberg, H. (1939). *Genetics* 24, 777.

Gurney, C. W., and Fried, W. (1965). *J. Lab. Clin. Med.* 65, 775.

Gurney, C. W., Goldwasser, E., and Pan, C. (1957). *J. Lab. Clin. Med.* 50, 534.

Gurney, C. W., Wackman, N., and Filmanowicz, E. (1961). *Blood* 17, 531.

Gurney, C. W., Munt, P., Brazell, I., and Hofstra, D. (1965). *Acta Haematol.* 33, 246.

Gutnisky, A., and Van Dyke, D. (1963). *Proc. Soc. Exp. Biol. Med.* 112, 75.

Hammond, D. G., Shore, N., and Movassaghi, N. (1968). *Ann. N.Y. Acad. Sci.* 149, 516.

Harker, L. A. (1970). *Amer. J. Physiol.* 218, 1376.

Hewlett, J. S., Hoffman, G. C., Senhauser, D. A., and Battle, J. D. (1960). *New England J. Med.* 262, 1058.

Ito, K., and Reissmann, K. R. (1966). *Blood* 27, 343.

Jacobson, L. O., Goldwasser, E., Fried, W., and Plzak, L. (1957). *Trans. Ass. Amer. Phys.* 70, 305.

Johnson, C. A., Abildgaard, C. F., and Schulman, I. (1971). *Blood* 37, 163.

Jones, N. F., Payne, R. W., Hyde, R. D., and Price, T. M. L. (1960). *Lancet* 1, 299.

Keighley, G., Russell, E. S., and Lowy, P. H. (1962). *Brit. J. Haematol.* 8, 429.

Keighley, G. H., Lowy, P., Russell, E. S., and Thompson, M. W. (1966). *Brit. J. Haematol.* 12, 461.

Kilbridge, T. M., Fried, W., and Heller, P. (1969). *Blood* 33, 104.

Korst, D. R., Whalley, B. E., and Bethell, F. H. (1959). *J. Lab. Clin. Med.* 54, 916.

Krantz, S. B. (1968). *J. Lab. Clin. Med.* 71, 999.

Krantz, S. B. (1970). *Med. Clin. North Amer.* 54, 173.

Krantz, S. B., and Jacobson, L. O. (1970). "Erythropoietin and the Regulation of Erythropoiesis." Univ. of Chicago Press, Chicago, Illinois.

Krzymowski, T., and Krzymowska, H. (1962). *Blood* 19, 38.

Kubanek, B., Tyler, W. S., Ferrari, L., Porcellini, A., Howard, D., and Stohlman, F., Jr. (1968). *Proc. Soc. Exp. Biol. Med.* 127, 770.

Levin, J. (1970). In "Formation and Destruction of Blood Cells" (T. J. Greenwalt and G. A. Jamieson, eds.), pp. 143–149. Lippincott, Philadelphia, Pennsylvania.

Lewis, J. P., Neal, W. A., Morres, R. R., Gardner, E. Jr., Alford, D. A., Smith, L. L., Wright, C.-S., and Welch, E. T. (1969). *J. Lab. Clin. Med.* 74, 608.

Lindemann, R. (1970). In "Erythropoiesis. Regulatory Mechanisms and Developmental Aspects" (Y. Matoth, ed.), pp. 185–189. Academic Press, New York.

Linman, J. W., and Pierre, R. V. (1968). *Ann. N.Y. Acad. Sci.* 149, 25.

Loge, J. P., Lange, R. D., and Moore, C. V. (1958). *Amer. J. Med.* 24, 4.

Lozzio, B. B., McDonald, T. P., and Lange, R. D. (1970). In "Erythropoiesis.

Regulatory Mechanisms and Developmental Aspects" (Y. Matoth, ed.), pp. 179–184. Academic Press, New York.

Malgor, L. A., and Fisher, J. W. (1969). Amer J. Physiol. 216, 563.

Malgor, L. A., and Fisher, J. W. (1970). Amer. J. Physiol. 218, 1732.

Mann, D. L., Donati, R. M., and Gallagher, N. I. (1966). Proc. Soc. Exp. Biol. Med. 121, 1152.

Marsh, J. C., and Levitt, M. (1971). Blood 37, 647.

Matoth, Y. (1970). "Erythropoiesis. Regulatory Mechanisms and Developmental Aspects." Academic Press, New York.

McKenzie, D., Friedman, R., Katz, S., and Lanzkowsky, P. (1967). So. Afr. Med. J. 41, 1044.

McNeill, T. A. (1970). Immunology 18, 61.

Meineke, H. A., and Feleppa, A. E. Jr. (1968). Proc. Soc. Exp. Biol. Med. 128, 944.

Metcalf, D. (1969). Proc. Soc. Exp. Biol. Med. 132, 191.

Metcalf, D. (1971a). Immunology 21, 427.

Metcalf, D. (1971b). Aust. J. Exp. Biol. Med. Sci. 49, 351.

Metcalf, D. (1971c). Med. J. Aust. 2, 739.

Metcalf, D., and Foster, R. S. (1967). J. Nat. Cancer Inst. 39, 1235.

Metcalf, D., and Moore, M. A. S. (1971). "Haematopoietic Cells." Frontiers of Biology, Vol. 24. North-Holland Publ., Amsterdam and American Elsevier, New York.

Metcalf, D., and Stanley, E. R. (1971). Brit. J. Haematol. 21, 481.

Metcalf, D., Foster, R. D., and Pollard, M. (1967). J. Cell. Comp. Physiol. 70, 131.

Mirand, E. A. (1966). In "National Cancer Institute Monograph 22" (M. A. Rich and J. B. Moloney, eds.), pp. 483–503. Nat. Cancer Inst., Bethesda, Maryland.

Mirand, E. A. (1968a). Ann N.Y. Acad. Sci. 149, 94.

Mirand, E. A. (1968b). Ann. N.Y. Acad. Sci. 149, 486.

Mirand, E. A., Gordon, A. S., and Wenig, J. (1965). Nature (London) 206, 270.

Mitus, W. J., Toyama, K., and Brauer, M. J. (1968). Ann. N.Y. Acad. Sci. 149, 107.

Mizoguchi, H., Miura, Y., Takaku, F., Sassa, S., Chiba, S., Nakao, U. (1971). Blood 37, 624.

Moriyama, Y., and Shimotori, T. (1970). Acta Hematol. 44, 321.

Morley, A., and Stohlman, F. Jr. (1970). New England J. Med. 282, 643.

Morley, A., Rickard, K. A., Howard, D., and Stohlman, F. Jr. (1971). Blood 37, 14.

Movassaghi, N., Shore, N. A., and Hammond, D. (1967). Proc. Soc. Exp. Biol. Med. 126, 615.

Naets, J. P. (1960). J. Clin. Invest. 39, 102.

Naets, J. P., and Huese, A. (1962). Lab. Clin. Med. 60, 365.

Naets, J. P., and Wittek, M. (1968). Ann. N.Y. Acad. Sci. 149, 366.

Nakao, K., Takaku, F., and Chiba, S. (1970). In "Erythropoiesis. Regulatory Mechanisms and Developmental Aspects" (Y. Matoth, ed.), pp. 164–168. Academic Press, New York.

Nathan, D. G., Beck, L. H., Hampers, C. L., and Merrill, J. P. (1968). Ann. N.Y. Acad. Sci. 149, 539.

Nečas, E., and Neuwirt, J. (1970). Blood 36, 754.

Niece, R. L., McFarland, E. C., and Russell, E. S. (1963). Science 142, 1468.

Nohr, M. L. (1967). Amer. J. Physiol. 213, 1285.

Odell, T. T. Jr., McDonald, T. P., and Asano, M. (1962). *Acta Haematol.* **27,** 171.

Odell, T. T., Jr., McDonald, T. P., and Howsden, F. L. (1964). *J. Lab. Clin. Med.* **64,** 418.

Orten, J. M., and Orten, A. U. (1945). *Amer J. Physiol.* **144,** 464.

Otsuka, A., Robinson, W. A., and Entringer, M. A. (1972). *Proc. Leucocyte Culture Conf., 6th* (M. R. Schwartz, ed.), pp. 37–53. Academic Press, New York.

Papayannopoulou, T., and Finch, C. A. (1972). *J. Clin. Invest.* **51,** 1179.

Pavlovic-Kentera, V., Hall, D. P., Bragassa, C., and Lange, R. D. (1965). *J. Lab. Clin. Med.* **65,** 577.

Pennington, D. G. (1965). *Proc. Roy. Soc. Med.* **58,** 488.

Pennington, D. G. (1966). *Proc. Roy. Soc. Med.* **59,** 1091.

Pennington, D. G. (1970). *Brit. Med. J.* **1,** 606.

Pennington, D. V., and Kincaid-Smith, P. (1971). *Brit. Med. Bull.* **27,** 136.

Pluznik, D. H., and Sachs, L. (1965). *J. Cell. Comp. Physiol.* **66,** 319.

Plzak, L., Fried, W., Jacobson, L. O., and Berthard, W. F. (1955). *J. Lab. Clin. Med.* **46,** 671.

Reisner, E. H. (1966). *Blood* **27,** 460.

Reisner, E. H. (1967). *Cancer* **20,** 1679.

Reissmann, K. R., and Samorapoompichit, S. (1970). *Blood* **36,** 287.

Reynafarje, C., Ramos, J., Faura, J., and Villavicencio, D. (1964). *Proc. Soc. Exp. Biol. Med.* **116,** 649.

Rickard, K. A., Morley, A., Howard, D., and Stohlman, F., Jr. (1971). *Blood* **37,** 6.

Rishpon-Meyerstein, N., Kilbridge, T., Simone, J., and Fried, W. (1968). *Blood* **31,** 453.

Robinson, W. A., and Pike, B. L. (1970). *New England J. Med.* **282,** 1291.

Rosse, W. F., Waldmann, T. A., and Cohen, P. (1963). *Amer. J. Med.* **34,** 76.

Russell, E. S. (1960). *Fed. Proc.* **19,** 573.

Russell, E. S. (1970). *In* "Regulation of Hematopoiesis" (A. S. Gordon, ed.), pp. 649–675. Appleton, New York.

Russell, E. S., and Bernstein, S. E. (1966). *In* "Biology of the Laboratory Mouse" (E. L. Green, ed.), 2nd ed., pp. 351–372. McGraw-Hill, New York.

Russell, E. S., and Bernstein, S. E. (1968). *Arch. Biochem. Biophys.* **125,** 594.

Russell, E. S., Smith, L. J., and Lawson, F. A. (1956). *Science* **124,** 1076.

Russell, E. S., Bernstein, S. E., Lawson, F. A., and Smith, L. J. (1959). *J. Nat. Cancer Inst.* **23,** 557.

Rytomaa, T., and Kiviniemi, K. (1968). *Cell Tissue Kinet.* **1,** 329, 341.

Sachs, L. (1970). *In* "Regulation of Hematopoiesis" (A. S. Gordon, ed.), pp. 217–233. Appleton, New York.

Sassa, S., Takaku, F., and Nakao, K. (1968). *Blood* **31,** 758.

Sarvella, P. A., and Russell, L. B. (1956). *J. Hered.* **47,** 123.

Schlager, G. (1966). *Nature (London)* **212,** 519.

Schooley, J. C., and Garcia, J. F. (1965). *Blood* **25,** 204.

Schooley, J. C., Garcia, J. F., Cantor, L. N., and Havens, V. N. (1968). *Ann. N.Y. Acad. Sci.* **149,** 266.

Sen, S., Hoffman, G. C., Stowe, N. T., Smeby, R. R., and Bumpus, F. M. (1972). *J. Clin. Invest.* **51,** 710.

Shadduck, R. D., Howard, D., and Stohlman, F. Jr. (1968). *Proc. Soc. Exp. Biol. Med.* **128,** 132.

Shahidi, N. T., and Diamond, L. K. (1961). *New England J. Med.* **264,** 953.

164   T. N. Fredrickson and P. F. Goetinck

Shalet, M. F., Holder, T. M., and Walters, T. R. (1967). *J. Pediat.* **70**, 615.

Shaw, A. B., and Schole, M. C. (1967). *Lancet* **1**, 799.

Shreiner, D. P., and Levin, J. (1970). *J. Clin. Invest.* **49**, 1709.

Siegel, C. D. (1970). In "Regulation of Hematopoiesis" (A. S. Gordon, ed.), pp. 67–76. Appleton, New York.

Stekel, A., and Smith, N. J. (1969). *Pediat. Res.* **3**, 320.

Stohlman, F. Jr. (1966). *Semin. Hematol.* **3**, 181.

Stohlman, F. Jr. (1970). "Symposium on Hemopoietic Cellular Proliferation." Grune and Stratton, New York.

Stohlman, F., and Brecher, G. (1957). *J. Lab. Clin. Med.* **49**, 890.

Tambourin, P., and Wendling, F. (1971). *Nature New Biol.* **234**, 230.

Till, J. E., and McCulloch, E. A. (1961). *Radiat. Res.* **14**, 213.

Thurman, W. G., Grabstald, H., and Lieberman, P. H. (1966). *Arch. Int. Med.* **117**, 280.

Van Dyke, D. C., Garcia, J. F., and Lawrence, J. H. (1957). *Proc. Soc. Exp. Biol. Med.* **96**, 541.

Van Dyke, D. C., Keighley, G., and Lawrence, J. H. (1963). *Blood* **22**, 838.

Van Dyke, D., Lawrence, J. H., and Anger, H. O. (1970). In "Myeloproliferative Disorders of Animals and Man" (W. J. Clarke, E. B. Howard, and P. L. Hackett, eds.), pp. 721–733, USAEC, Oak Ridge, Tennessee.

Vertel, R. M., Morse, B. S., and Prince, J. E. (1967). *Arch. Int. Med.* **120**, 54.

Vitale, L., Shahidi, N. T., Kerr, G. E., Wolf, R. C., Korst, D. R., and Meyer, R. K. (1971). *Proc. Soc. Exp. Biol. Med.* **138**, 418.

Wahren, B. K., Lantorp, K., Sternes, G., and Espmark, A. (1970). *Proc. Soc. Exp. Biol. Med.* **133**, 934.

Waldmann, T. A., Rosse, W. F., and Swarm, R. L. (1968). *Ann. N.Y. Acad. Sci.* **149**, 509.

Ward, H. P., Kurnick, J. E., and Pisarczyk, M. J. (1971). *J. Clin. Invest.* **50**, 332.

Wardle, E. N. (1970). *Acta Haematol.* **43**, 129.

Wasserman, L. R., and Gilbert, H. S. (1966). *Semin. Hematol.* **3**, 199.

Weatherall, D. J. (1969). *New England J. Med.* **28**, 604.

Whitcomb, W. H., and Moore, M. (1968). *Ann. N.Y. Acad. Sci.* **149**, 462.

Whitcomb, W. H., Moore, M., and Rhoads, J. P. (1969). *J. Lab. Clin. Med.* **73**, 584.

Wolfe, H. G. (1963). *Mouse News Lett.* **29**, 40.

Zanjani, E. D., Contrera, J. F., Gordon, A. S., Cooper, G. W., Wong, K. K., and Katz, R. (1967). *Proc. Soc. Exp. Biol. Med.* **125**, 505.

REGULATION OF ERYTHROPOIESIS IN LOWER VERTEBRATES*

Esmail D. Zanjani, Albert S. Gordon, Anthony S. Gidari, and
Robert A. Kuna†

I. INTRODUCTION

Our knowledge of mechanisms involved in the complex series of events that lead to the formation of red blood cells in mammalian species has expanded greatly during the last 20 years. One of the most significant observations responsible for progress in this field was the discovery that production of red blood cells in adult mammals is subject to humoral

* Original work reported in this Chapter was supported by U.S. Public Health Service Research Grants 5 R01 HE03357-15 from the National Heart and Lung Institute and 1 R01 AM15525-02 from the National Institute of Arthritis and Metabolic Diseases.
† Predoctoral U.S. Public Health Service Trainee (Grant 5 T01 HE05645-07).

control (Erslev, 1953; Stohlman *et al.*, 1954; Gordon, 1959). Thus the principle, erythropoietin (Ep), has been found in increased quantities in the body fluids of animals subjected to a variety of hypoxic stimuli which lead to increased production of erythrocytes (Gordon, 1959). In addition, the suppressive effects of starvation and of transfusion-induced elevation of circulating red cell mass on erythropoiesis is probably the result of a decreased production of Ep (Reissmann, 1964; Gordon *et al.*, 1966). Moreover, the presence of Ep in plasma and urine of normal humans (Weintraub *et al.*, 1963; Adamson *et al.*, 1966) and the fact that mice treated with anti-Ep show markedly suppressed erythropoiesis (Schooley *et al.*, 1968), provide evidence for the physiological role of Ep in erythropoiesis. Studies utilizing Ep have revealed the existence in the body of undifferentiated cells which upon stimulation by Ep give rise to hemoglobin-synthesizing cells (Stohlman *et al.*, 1968). Thus the Ep-responsive cell (ERC) is believed to originate from a more primitive cell (pluripotential stem cell) which in addition to the ERC can give rise to elements destined to form the other cellular elements of the blood in the presence of appropriate stimuli, probably humoral in nature (McCulloch, 1970). The transition from stem cells to ERC (or other "sensitive" cells) may be dependent upon the type of microenvironment to which the stem cell becomes exposed (Trentin, 1970).

Chemically, Ep is a glycoprotein with a molecular weight of 48,500. The carbohydrate moiety essential for its biological activity in the organism is sialic acid (Goldwasser and Kung, 1968). Thus treatment of the hormone with neuraminidase renders it biologically inactive. In adult mammals the kidney represents the major source of Ep formation (Jacobson *et al.*, 1957). The mechanism by which the kidney participates in this process is not yet fully understood. One of the proposed schemes suggests the formation by the kidney of a protein substance (erythrogenin) which upon reacting with a component of serum (probably also proteinaceous) is capable of generating Ep (Gordon and Zanjani, 1970). Like Ep, the erythrogenin-serum system does not exhibit any measurable degree of species-specificity among mammals (Zanjani *et al.*, 1969a). Thus erythrogenin extracted from rat kidneys produces quantitatively similar amounts of Ep when incubated with sera obtained from humans, sheep, dogs, pigs, rabbits, and rats (Zanjani *et al.*, 1969a). By comparison, our knowledge of the mechanisms controlling red cell production in submammalian species is at best sketchy. However, data are available suggesting that a regulatory system similar to that found in adult mammals may be operative in lower vertebrates. The bulk of evidence in support of this contention is derived from experiments involving frogs

(Rosse *et al.*, 1963), birds (Rosse and Waldmann, 1966; Cohen, 1969a, b), and fish (Zanjani *et al.*, 1969b).

II. STUDIES IN FROGS

Although hypoxia is the prime stimulus of erythropoiesis in mammals (Gordon, 1959), chronic hypoxia failed to augment production of red blood cells in frogs (Sokolov, 1941). Similarly, no erythropoietic effect was seen when frogs were treated with cobalt (Rosse *et al.*, 1963). Bleeding, on the other hand, effectively stimulated erythropoiesis in this animal. Utilizing the latter procedure, Rosse *et al.* (1963) observed that following induction of anemia, an erythrostimulatory factor could be demonstrated in the circulation of these animals. Thus sera collected from anemic frogs augmented erythropoiesis when administered to normal frogs. Serum obtained from anemic human subjects, however, failed to influence red cell production in normal frogs. In this connection, administration of anemic frog serum to polycythemic mice also failed to affect erythropoiesis (Rosse *et al.*, 1963). It should be noted that polycythemic mice are presently the most widely used animals for the bioassay of erythropoietin from mammalian sources (Camiscoli and Gordon, 1970). The fact that mammalian Ep was ineffective in the frog and that anemic frog serum did not influence mammalian erythropoiesis, may be indicative of the existence of class specificity with regards to the biological action of Ep. However, as will become apparent later in this chapter, such specificity is probably relative rather than absolute.

III. STUDIES IN BIRDS

While subjection to lowered atmospheric pressure was found to be ineffective in frogs, birds responded to this type of oxygen deficiency (hypoxic hypoxia) with an enhanced rate of erythropoiesis. Thus, in at least three species of birds (Japanese quail, domestic chicken, and pintail ducks), hypoxic hypoxia resulted in a significant increase in the rate of red cell production (Rosse and Waldmann, 1966; Cohen, 1969a). In this regard, exposure of quails to 3 weeks of intermittent hypoxia resulted in the development of polycythemia in these animals. Upon return to room pressure, a significant decrease to below normal levels in the production rate of erythrocytes was noted in these animals (Rosse and Waldmann, 1966). In pintail ducks exposure to reduced atmospheric pressure resulted in the elevation of circulating red cell mass and total blood volume (Cohen, 1969a). Upon return to ambient pressure these parameters decreased rapidly, reaching normal levels after approximately 20 days. The decrease in erythropoiesis following induction of poly-cythemia also occurs in mammalian species following the elevation of the circulating red cell mass (induced by transfusion with homologous red cells or by chronic exposure to hypoxia) (Camiscoli and Gordon, 1970). In addition to hypoxic hypoxia, birds also responded to anemic hypoxia (induced by bleeding).

That the erythrostimulatory influences of hypoxic hypoxia and bleed-ing in birds is humorally mediated, was shown by the finding that sera from both anemic chickens and hypoxic quails caused enhancement of red cell production in the polycythemic quail (Rosse and Waldmann, 1966). Once again these sera failed to stimulate erythrocyte formation in polycythemic mice. In addition, Ep-rich human serum was without effect in the polycythemic quail (Rosse and Waldmann, 1966). More-over, antibody prepared in rabbits against human Ep failed to influence the activity of hypoxic quail serum when assayed in the polycythemic quail. On the other hand, hypoxic quail serum was rendered biologically inactive with trypsin, while no change in activity was seen after treat-ment with neuraminidase (Rosse and Waldmann, 1966). It appears, therefore, that although avain erythropoiesis-stimulating factor requires a protein structure for its biological integrity, as does mammalian Ep, it differs from the latter in that it does not require sialic acid for activity. While these results imply, once again, the possible existence of a class specificity, it should be noted that administration of mammalian Ep to a urodele (*Desmognathus phoca*) (Gordon, 1960) and, as is discussed below, to two species of fish resulted in the augmentation of erythropoiesis.

IV. STUDIES IN FISH

Studies in the blue gourami (*Trichogaster trichopterus*) revealed that in the fish, as in mammals, amphibians and birds erythropoiesis is influ-enced by a circulating humoral agent. Moreover, the data showed that erythropoiesis in the fish, as in the birds, can be quantitated by the same investigative procedures as are employed to assess RBC formation in mammalian species. Of interest was the finding that administration of very large doses of mammalian Ep to this fish (Zanjani et al., 1969b) as well as another species of fish, namely the carp (*Cyprinus carpio* L.) (Krzymowska et al., 1960), resulted in a significant increase in erythrocyte formation. A detailed description of the studies in the blue gourami follows.

Adult male and female blue gourami, weighing 4–6 gm, were used in all experiments. During the study period the animals were kept in disposable plastic aquaria maintained at temperatures ranging from 24°–28°C. All groups of fish (except those used in starvation studies) were fed equal amounts of dried food and a mixture of cooked beef liver, and oatmeal.

A. Effect of Starvation and Refeeding on Erythropoiesis

Four groups of fish were established. Group I was deprived of food for 10 days and then sacrificed. Group II was deprived of food for 10 days, but in addition received an intracardial injection of 0.05 μCi [^{59}Fe]Cl$_3$ one day before sacrifice. The animals in Groups III and IV were starved for the same period as were Groups I and II, and then placed on a diet of dry food for 10 days before exsanguination. The fish in Group IV also received an injection of 0.05 μCi radioiron 24 hours before sacrifice. At the time of sacrifice, each fish in the group was bled via the heart and the data presented in Table I were obtained from these samples.

Table I shows that lack of food caused a significant decrease in the rate of erythropoiesis in the gourami. This is indicated by the lower values of the RBC counts, hemoglobin concentrations, hematocrits, and percent incorporation of radioiron into the circulating RBC in these fish. Although the total volume of blood was not altered significantly, starvation resulted in a significant decrease in the total circulating red

TABLE I

EFFECT OF STARVATION ON ERYTHROPOIESIS IN BLUE GOURAMI[a]

	Normal gourami	Starved gourami[b]
RBC ($\times 10^{-5}$)/mm^3	30.64 ± 0.73[c] (20)[d]	14.84 ± 0.43 (40)
WBC ($\times 10^{-3}$)/mm^3	18.00 ± 1.00 (8)	19.40 ± 1.45 (7)
Hematocrit (%)	27.30 ± 1.47 (15)	13.14 ± 0.42 (36)
Hb (gm %)	10.42 ± 0.36 (20)	5.41 ± 0.21 (40)
Total blood volume (% body weight)	4.46 ± 0.83 (15)	3.97 ± 0.54 (15)
Total RBC volume (% body weight)	1.22 ± 0.14 (15)	0.52 ± 0.08 (15)
Percent [^{59}Fe] RBC-incorporation	19.08 ± 2.16 (14)	3.94 ± 0.32 (32)

[a] From Zanjani *et al.*, 1969b.
[b] Fish were starved for 10 days.
[c] Standard error of the mean.
[d] Number of animals.

TABLE II

INFLUENCE OF REFEEDING ON ERYTHROPOIESIS IN STARVED BLUE GOURAMI[a]

	Starved gourami[b]	Starved–limited food[c]	Starved–unlimited food[d]
RBC ($\times 10^{-5}$)/mm^3	17.92 ± 0.78[e] (8[f])	21.06 ± 1.31 (5)	41.12 ± 1.82 (5)
Hematocrit (%)	16.80 ± 1.98 (8)	20.35 ± 2.10 (5)	29.10 ± 0.82 (6)
Hb (gm %)	5.90 ± 0.90 (8)	7.99 ± 0.76 (8)	11.09 ± 1.92 (5)
Total blood volums (% body weight)	4.11 ± 1.00 (7)	4.91 ± 1.10 (5)	4.76 ± 1.70 (4)
Total RBC volume (% body weight)	0.69 ± 0.16 (7)	1.00 ± 0.18 (5)	1.39 ± 0.21 (4)
Percent [^{59}Fe] RBC-incorporation	6.81 ± 1.11 (8)	21.10 ± 2.32 (7)	32.86 ± 2.92 (8)

[a] From Zanjani et al., 1969b.
[b] Fish were starved for 10 days.
[c] These fish were fed for 10 days one-third the amount of food given to unlimited group.
[d] Full diet for 10 days.
[e] Standard error of the mean.
[f] Number of animals.

cell volume (Table I). These results are similar to those reported to occur in mammals following food deprivation (Reissmann, 1964). Also, as in mammalian species, it is apparent from the data in Table II that the depressive effect of starvation on erythropoiesis in these fish can be overcome by refeeding. In this regard, it is of interest that when the amount of food given these previously starved animals was restricted to one-third of their normal intake, the rate of erythropoiesis was only partially increased (Table II). On the other hand, full replacement of diet completely reversed the inhibitory effects of food deprivation on erythropoiesis in these animals (Table II, unlimited food). In fact, in this group, the percent RBC-radioiron incorporation values were considerably above normal (Table I). Such phenomena also occur in mice recovering from the suppressive influences of irradiation or chemotherapy (Lajtha, 1970). The similarity in response between the fish and mammalian species is also evident when the effect of bleeding on erythropoiesis in these fish is considered.

B. Effect of Bleeding on Erythropoiesis

Groups of fish were bled via heart puncture (see Zanjani et al., 1969b, for details of method). Adequate amounts of blood can be obtained from a single bleeding of one fish. Repeated bleeding of these fish at 72-hour intervals can be performed with little mortality. In this study, however, normal and 10-day starved fish were bled only once. The volume of blood drawn from each fish ranged from 0.05–0.08 ml (approximately 25–30% total blood volume). The animals were then maintained as before (i.e., no food was given to the starved group) for 5 additional days. At the end of this period they were exsanguinated and the various indices of erythropoiesis determined. Table III reveals that RBC regeneration after bleeding in the gourami, occurred with great rapidity. Thus, by as early as the fifth day following bleeding in both fed and starved fish, considerably greater values than those in non-bled fish were obtained for the various erythropoietic indices including radioiron incorporation into the circulating erythrocytes (Table III). It is evident from the foregoing that in the fish the factors controlling red cell production respond to conditions such as protein-deprivation and anemia in a manner quite similar to those described for the mammalian system. Whether a humoral factor participates at the level of the fish was examined in the following study.

C. Effect of Anemic Gourami Plasma on Erythropoiesis in Starved Fish

In order to determine whether an erythrostimulatory factor(s) is present in the circulation of anemic fish, it was necessary to employ an assay procedure that permitted the detection and evaluation of stimulatory effects with reasonable speed and accuracy. For the mammal, experience had indicated the best procedure as one utilizing animals in which the production of red blood cells was markedly depressed. Both protein-deprived rats and mice rendered polycythemic either by exposure to lowered atmospheric pressures or by hypertransfusion with homologous red cells have been used. Of these, the polycythemic mouse assay has proved to be superior (Camiscoli and Gordon, 1970). However, attempts to render the blue gourami polycythemic were not successful. Administration of washed gourami RBC to these fish resulted in a high mortality, thus rendering the procedure not feasible. On the other hand, as is apparent from the data in Table I, endogenous erythropoiesis in this fish was suppressed significantly when the animal was deprived of food. For this reason, the starved fish was used for the assays.

Fish were made anemic by a single bleeding via cardiac puncture. Heparin (4 mg/ml) served as the anticoagulant. The plasma separated from this bleeding was designated as normal plasma. Forty-eight hours after the initial bleeding, the animals were sacrificed by exsanguination and the plasma obtained was indicated as anemic plasma. The erythropoietic activity of these plasma samples was determined in starved blue

TABLE III

EFFECT OF BLEEDING ON ERYTHROPOIESIS IN NORMAL AND STARVED GOURAMI[a]

	Normal gourami		Starved gourami[b]	
	Before bleeding	After bleeding[c]	Before bleeding	After bleeding[c]
RBC ($\times 10^{-5}$)/mm³	30.60 ± 1.73[d] (6[e])	41.22 ± 2.95 (5)	17.12 ± 1.95 (8)	39.80 ± 1.42 (5)
Hematocrit (%)	28.30 ± 2.58 (8)	33.75 ± 1.44 (5)	12.88 ± 0.90 (8)	34.20 ± 2.60 (5)
Hb (gm %)	9.30 ± 0.42 (8)	13.63 ± 1.30 (5)	4.16 ± 0.42 (8)	14.46 ± 2.10 (5)
Total blood volume (% body weight)	5.01 ± 1.16 (5)	4.73 ± 1.76 (4)	Not determined	Not determined
Total RBC volume (% body weight)	1.42 ± 0.19 (5)	1.60 ± 0.31 (4)	Not determined	Not determined
Percent [59Fe] RBC-incorporation	22.81 ± 3.10 (5)	42.90 ± 4.68 (5)	8.14 ± 1.63 (8)	50.11 ± 6.92 (5)

[a] From Zanjani et al., 1969b.
[b] Starved for 10 days.
[c] Five days after bleeding.
[d] Standard error of the mean.
[e] Number of animals

TABLE IV

INFLUENCE OF ANEMIC GOURAMI PLASMA ON ERYTHROPOIESIS
IN STARVED FISH[a]

Material assayed	Number of assay fish	Percent [59Fe] RBC-incorporation
0.15 ml Saline	20	4.08 ± 0.58[b]
0.15 ml Normal gourami plasma	15	6.76 ± 0.90
0.15 ml Anemic gourami plasma	30	24.16 ± 1.96

[a] Fish were starved for 4 days prior to use and were kept starved for
the duration of the assay.
[b] Standard error of the mean.

gourami according to the following schedule. The fish were starved
for 4 days prior to receiving a single intraperitoneal injection of 0.15
ml of either normal or anemic plasma. Forty-eight hours after the injection, each fish was administered 0.05 µCi of radioiron via the heart.
The animals were killed by exsanguination 24 hours later and the percent
RBC-radioiron incorporation values determined. Blood volume of the
starved fish was calculated to be 3.8% of body weight. Food was with-
held from the fish for the duration of the assay.

The data in Table IV support the contention that an erythropoiesis-
stimulating principle was evoked in the blue gourami by a single bleeding. Thus administration of 0.15 ml of anemic plasma to fish with suppressed erythropoiesis due to starvation resulted in a highly significant
increase in the rate of radioiron incorporation into the red cells of these
animals. Plasma obtained from normal fish produced only borderline
activity when compared to saline-treated controls (Table IV). This indicates that the significant erythropoietic response of the starved blue
gourami to anemic serum cannot be attributed to the generalized effect
of nonspecific plasma proteins in protein-deprived animals (Gordon and
Weintraub, 1962), since the degree of response to anemic serum, when
compared to normal plasma was of considerably greater magnitude.

V. SPECIES AND CLASS SPECIFICITY OF MAMMALIAN AND SUBMAMMALIAN EP

A lack of specificity for the Ep among members of the mammalian
class is shown from observations that the Ep of human urinary or plasma
origin can stimulate erythropoiesis in mice, rats, rabbits (Gordon, 1959),
and man (Van Dyke *et al.*, 1963). Experiments involving the study
of the effect of mammalian Ep on erythropoiesis in frogs (Rosse *et*

al., 1963) and birds (Rosse and Waldmann, 1966), however, revealed the existence of class specificity for this hormone. On the other hand, our studies as well as others (Krzymowska et al., 1960) indicate that, at least in the fish, one may speak of class specificity only in relative terms.

The studies in blue gourami involved administration of single graded doses of human urinary Ep to different groups of starved gouramis and determining the effect on the rate of incorporation of radioiron into red blood cells. The results shown in Table V revealed that erythropoietic effects did not occur until large doses of human Ep were used. In fact, the first clear response was obtained only when 0.8 units of the Ep was administered. This dose of Ep corresponds to about 16 units of Ep/100 gm body weight in the fish. It should be noted that the polycythemic mouse is sensitive to approximately 0.2 units/100 gm body weight (Camiscoli and Gordon, 1970). It appears, therefore, that mice are approximately 80 times more sensitive than the fish in responding to mammalian Ep.

Even greater response was noted with higher doses of the hormone. Thus the greatest erythropoietic response was noted when the starved gourami received 4 units of the hormone (Table V). That the stimulatory

TABLE V

EFFECT OF HUMAN URINARY[a] AND SHEEP PLASMA[b]
ERYTHROPOIETIN ON ERYTHROPOIESIS IN STARVED GOURMAI[c]

Material assayed (IRP units)[d]	Number of assay fish[e]	Percent [59Fe] RBC-incorporation
Saline	10	6.30 ± 1.10[f]
1 unit Sheep plasma Ep	10	22.12 ± 1.98
0.01 unit Human Urinary Ep	8	5.81 ± 0.90
0.05 unit Human Urinary Ep	8	8.76 ± 0.74
0.20 unit Human Urinary Ep	8	10.21 ± 0.98
0.50 unit Human Urinary Ep	8	10.67 ± 0.60
0.80 unit Human Urinary Ep	8	21.85 ± 2.16
1.00 unit Human Urinary Ep	8	28.63 ± 1.48
4.00 unit Human Urinary Ep	8	36.84 ± 2.91

[a] Prepared from urine of a child with severe hypoplastic anemia by dialysis against Carbowax (Adamson et al., 1966).

[b] Supplied by the Erythropoietin Distribution Committee of the National Heart Institute (Step IIIA Lot #147-192 A).

[c] From Zanjani et al., 1969b.

[d] International Reference Preparation of Ep.

[e] Fish were starved for 4 days prior to use and were kept starved for the duration of the assay.

[f] Standard error of the mean.

effect of mammalian Ep in the fish is not limited to human Ep is shown by the fact that a significant erythropoietic response was evoked with 1 unit of sheep plasma Ep (Table V).

The possibility that the observed stimulatory effect of mammalian Ep (also that of anemic gourami plasma) on RBC-radioiron incorporation in the gourami was due in part to direct effect on the circulating erythrocytes was examined *in vitro*. Such a study was considered necessary since it is known that circulating erythrocytes of submammalian species are capable of synthesizing hemoglobin (Harris, 1967). Suspensions consisting of whole blood (0.5 ml) or washed RBC from 0.5 ml of whole blood resuspended in saline (0.5 ml) from normal and starved gourami, together with 0.1 μCi radioiron (in 0.1 ml saline) and either anemic fish plasma (0.1 ml) or human Ep (4 units in 0.1 ml), were incubated for 1 hour at 28°C in a water bath with shaking. At the end of the incubation period, and radioactivity of the content of each vessel was determined. The cells were then washed with saline, until no further significant activity appeared in the wash. The cells were then brought to their original volume with saline and counted. Radioiron uptake was determined, using the following formula:

$$\text{Percent RBC-radioiron uptake} = \frac{\text{CPM after wash} \times 100}{\text{CPM before wash}}$$

The results shown in Table VI indicate that while the gourami RBC take up radioiron *in vitro*, no stimulatory effect on this process occurred upon addition of anemic plasma or mammalian Ep to the incubation mixture. It is of interest that the RBC from starved gourami accumulated radioiron *in vitro* to a smaller extent than cells obtained from normally fed fish (Table VI). Thus, while small quantities of mammalian Ep were ineffective in the fish, very large doses of this hormone stimulated red cell formation in this form. Therefore, the reported failure of mammalian Ep to stimulate erythropoiesis in birds and frogs may have been due to the use of insufficient amounts of this factor. It is conceivable that all vertebrate erythropoietins possess a common biologically active core but that slight differences in attached chemical groupings are responsible for the class specificity reported. Thus giving very large doses of one type might produce an effect in another class, despite the existence of physiological class specificity. Utilizing a different approach, we have obtained data indicating that, at least in birds, a plasma principle exists which when activated gives rise to factor(s) capable of stimulating red cell production in the mouse (Zanjani *et al.*, 1969a). A brief description of these studies follows.

TABLE VI

Effect of Anemic Gourami Plasma and Human Erythropoietin on the Uptake of Radioiron by Peripheral Erythrocytes in Vitro[a]

Donor fish[b]	State of blood	Material added to blood in addition to radioiron	Percent-[^{59}Fe] RBC incorporation
Normal (40)	Whole	None	7.57 ± 1.51[c]
Normal (38)	Washed RBC	None	7.36 ± 1.70
Starved (46)	Whole	None	2.10 ± 0.50
Starved (46)	Washed RBC	None	3.27 ± 0.68
Normal (38)	Whole	Anemic gourami plasma[d]	6.80 ± 0.80
Normal (35)	Washed RBC	Anemic gourami plasma	8.51 ± 1.49
Starved (49)	Whole	Anemic gourami plasma	4.51 ± 2.30
Starved (39)	Washed RBC	Anemic gourami plasma	3.28 ± 0.26
Normal (35)	Whole	Human Ep[e]	6.06 ± 1.27
Normal (38)	Washed RBC	Human Ep	10.09 ± 2.35
Starved (41)	Whole	Humen Ep	6.37 ± 0.72
Starved (43)	Washed RBC	Human Ep	4.38 ± 1.19

[a] From Zanjani et al., 1969b.
[b] The values for normal blood have been corrected for the differences in hematocrit levels in the normal and starved groups. "Starved" means 10-day starved fish.
[c] Standard error of the mean.
[d] 0.10 ml Plasma/0.5 ml blood.
[e] 4 IRP units of erythropoietin/0.5 ml blood.

It has been demonstrated that a hypotonic extract of the "light-mito-chondrial" fraction of mammalian kidneys contains a principle (erythro-genin, Eg) which, when incubated with normal serum, engenders the production of Ep *in vitro* (Gordon *et al.*, 1967). The Eg-serum system does not exhibit species-specificity within the class Mammalia (Zanjani *et al.*, 1969a). This system has been used to examine the question of class specificity among some representative animals of four classes of vertebrates, namely the mammal (rat), bird (duck), amphibian (frog) and fish (carp). Light-mitochondrial extracts were prepared from kidneys of adult rats (Long-Evans strain), ducks (Long Island), frogs (*Rana pipiens*), and carps (*Cyprinus carpio* L.). Using stan-dard procedures, sera from these animals were dialyzed against disodium ethylenediamine tetraacetate (EDTA) as previously reported (Zanjani *et al.*, 1967). Among mammals this dialysis procedure prevents the operation of an Ep-inactivating agent present in Eg-serum mixtures. To generate Ep, incubation systems, each containing 6 ml of dialyzed serum and 6 ml of the light-mitochondrial extract (equivalent to 3 gm of original kidney tissue), were established. The mixtures were incubated for 60 minutes at 37°C in a water bath with constant shaking. All incuba-tions were open to air. Two-ml aliquots of each mixture were assayed for activity in the hypoxia-induced polycythemic mouse (Camiscoli *et al.*, 1968).

The data presented in Table VII show that incubation of renal extracts from fish, frogs, and ducks with sera of these species did not result in the generation of erythropoiesis-stimulating activity when tested in polycythemic mice. As reported earlier (Zanjani *et al.*, 1969a), the incubation of a similarly prepared renal extract from rats with equal volumes of rat serum, however, was accompanied by the appearance of considerable quantities of Ep. The lack of Ep-generating capacity of the renal extracts from the 3 nonmammalian species was also shown from the failure of these extracts to produce Ep when incubated with rat serum (Table VII). To determine whether sera from these species contained the factor for interaction with rat Eg, incubations consisting of hypoxic rat Eg and dialyzed sera from carp, frog, and duck were performed. Table VII shows that, while incubation of rat Eg with sera from the carp and frog did not produce detectable activity, the incuba-tion mixtures containing rat Eg and serum from normal ducks evoked considerable erythropoietic activity in the test polycythemic mice. The observation that incubation of mammalian Eg with duck serum resulted in the generation of erythropoietic activity detectable in the poly-cythemic mouse indicates that mammalian type substrate for Eg is present in the blood of the bird. This finding is indicative of a closer

TABLE VII

In Vitro ERYTHROPOIETIN-GENERATING CAPACITY
OF "LIGHT-MITOCHONDRIAL" EXTRACTS OBTAINED
FROM KIDNEYS OF RAT, CARP, FROG, AND DUCK[a]

Composition of incubation mixture	Percent [^{59}Fe] RBC-incorporation for 2 ml of incubate/mouse
6 ml NRS-EDTA[b] + 6 ml rat LME[c]	9.05 ± 1.14[d]
6 ml NFS-EDTA[e] + 6 ml frog LME	2.98 ± 0.42
6 ml NFS-EDTA + 6 ml rat LME	1.38 ± 0.32
6 ml NRS-EDTA + 6 ml frog LME	1.49 ± 0.35
6 ml NCS-EDTA[f] + 6 ml carp LME	2.21 ± 0.53
6 ml NCS-EDTA + 6 ml rat LME	2.49 ± 0.33
6 ml NRS-EDTA + 6 ml carp LME	1.29 ± 0.31
6 ml NDS-EDTA[g] + 6 ml duck LME	3.07 ± 1.35
6 ml NDS-EDTA + 6 ml rat LME	9.80 ± 1.53
6 ml NRS-EDTA + 6 ml duck LME	4.03 ± 1.17
Controls	
Saline	1.14 ± 0.29
0.05 IRP units Ep	7.51 ± 1.38
0.20 IRP units EP	15.15 ± 2.65

[a] From Zanjani et al., 1969a.
[b] EDTA-dialyzed normal rat serum.
[c] "Light-mitochondrial" extract.
[d] Standard error of the mean.
[e] EDTA-dialyzed normal frog serum.
[f] EDTA-dialyzed normal carp serum.
[g] EDTA-dialyzed normal duck serum.

similarity in the serum factor of birds and mammals than exists between the latter and the other submammalian species studied. This is also evident from the observation that of the several species examined only birds respond to hypoxia in a manner similar to mammalian species (Rosse and Waldmann, 1966; Cohen, 1969a). In addition, as is shown in Table VIII and discussed below, erythropoiesis in the starved gourami was stimulated by the administration of anemic duck plasma. No effect in the gourami was seen with plasmas obtained from anemic frogs and turtles. These studies were undertaken to examine the feasibility of employing the starved gourami as an assay animal in the search for erythropoietins in lower vertebrates.

In another series of experiments frogs were given several injections of phenylhydrazine hydrochloride over a 2-week period. Turtles were rendered anemic according to the method outlined by Hirschfeld and Gordon (1965) and exsanguinated after the anemia was developed. Adult ducks were bled twice at weekly intervals. At each bleeding,

TABLE VIII

Effect of Normal and Anemic Frog, Turtle, and Duck
Plasma on Erythropoiesis in Starved Gourami

Material assayed	Number of assay fish[a]	Percent [59Fe] RBC-in-corporation
Saline	14	5.93 ± 0.89[b]
Human urinary Ep	15	24.09 ± 3.82
Anemic gourami plasma	18	21.86 ± 1.72
Normal frog plasma	18	7.13 ± 1.08
Anemic frog plasma	18	6.69 ± 0.97
Normal turtle plasma	24	6.72 ± 0.90
Anemic turtle plasma	22	9.42 ± 2.34
Normal duck plasma	24	8.18 ± 1.23
Anemic duck plasma	19	18.73 ± 1.67

[a] Fish were starved for 4 days prior to use and were kept starved
for the duration of the assay.
[b] Standard error of the mean.

a volume of blood corresponding to 1.2% of body weight was removed. The ducks were sacrificed by exsanguination 12 days after the second bleeding. Plasmas from all animals were assayed in the starved gourami as described above. Control groups received either saline, anemic gourami plasma, human Ep or plasmas from normal frogs, turtles and ducks. Results are presented in Table VIII.

It is evident that, in addition to anemic gourami plasma and human Ep, only plasma from anemic ducks was effective in the starved gourami. Administration of plasma from nonanemic ducks, however, to these fish failed to influence red cell production. Similarly no significant effect was demonstrable when plasmas from normal and anemic turtles were assayed (Table VIII).

VI. COMMENTS AND CONCLUSIONS

Available evidence indicates that in lower vertebrates (fish, amphibians, and birds), as in mammals, erythropoiesis is regulated by a circulating principle. Unfortunately little attention has been devoted to the further elucidation of mechanisms controlling red blood cell production in lower vertebrates. Results obtained from studies already concluded point to a strong similarity between the system in mammals and submammalian species. Thus the main stimulus of erythropoiesis in all these classes appears to be a reduction in the supply of oxygen to the body tissues (Gordon, 1934; Cohen, 1969a, b; Jalavisto *et al.*, 1965; Weiss *et al.*, 1965; Rosse *et al.*, 1963; Rosse and Waldmann, 1966; Hirshfeld and Gordon 1965; Zanjani *et al.*, 1969b). In air breathing species a reduction in delivery of oxygen to the tissues can be caused with much less effort than in nonair breathers. Thus subjection of mammals and birds to reduced atmospheric pressures for a relatively short period is sufficient to enhance the production of red blood cells (Gordon, 1959; Rosse and Waldmann, 1966).

However, experience indicates that more drastic measures are required to create a hypoxic condition in those species where in addition to breathing air, oxygen can be obtained by other means. In these animals, the most effective approach involves a physical reduction in the numbers of oxygen carriers (i.e., red blood cells). In this regard, subjection of frogs and turtles to lowered atmospheric pressures did not influence erythropoiesis, while bleeding caused the stimulation of red cell production at a level significantly above normal (Altland and Parker, 1955; Rosse *et al.*, 1963; Hirshfeld and Gordon, 1965). On the other hand, Gordon (1934) has shown that erythropoiesis in *Necturus maculosus* was significantly enhanced following exposure to hypoxic hypoxia. It appears, therefore, that while hypoxia remains the prime stimulus of erythropoiesis in lower vertebrates, various species respond differently to the same stimulus. This may be a manifestation of the differences in the metabolic rate of these animals. Thus birds are known to possess a more efficient system for transport of oxygen in correlation with their greater metabolic activity (Simons, 1960). Perhaps by employing a more severe hypoxic environment, it will be possible to demonstrate an effect in species such as frogs and other poikilothermic vertebrates.

That erythropoiesis in lower vertebrates is regulated by mechanisms similar to that found in adult mammals, is further supported by the effect on red cell production exhibited by starvation and increased circulating red cell mass. In the fish, a near complete suppression of erythropoiesis occurred following food deprivation (Zanjani *et al.*, 1969b). Moreover, while no change in peripheral hemogram was observed in turtles following starvation (probably due to the long life span of the turtle erythrocyte), these animals were incapable of red cell production even when stimulated by bleeding (Hirschfeld and Gordon, 1965). In addition, erythropoiesis in birds (Japanese quail) was significantly suppressed by induction of polycythemia (Rosse and Waldmann, 1966); also, a reduced hematocrit was observed in chickens exposed to hyperoxic environment (Weiss *et al.*, 1965). In this regard, starvation and increased circulating red cell mass, also potent inhibitors of mammalian erythropoiesis, probably suppress red cell production by inhibiting the production of erythropoietin in these animals. This is evident from the increased production of red blood cells in both the starved gourami and polycythemic quail following administration of anemic gourami plasma (Zanjani *et al.*, 1969b) and hypoxic quail plasma (Rosse and Waldmann, 1966), respectively.

It appears that humoral regulation of erythropoiesis may be a prominent feature of not only the mammalian species but also of many of the lower vertebrate species. The inability to demonstrate the existence of "erythropoietin" in some of the lower vertebrate species may be a manifestation of the differences in the metabolic rate and possible sensitivity of the erythrocyte-producing tissue in these animals. In the frog, for example, a shift in the site of formation of blood cells accompanies the change from the larval organization to adult form. This occurs when the higher rate of metabolism is evident in the animal (Jordan and Speidel, 1923). Adult birds and mammals which possess even greater overall metabolic rate show predominantly marrow erythropoiesis. It is of interest that the latter group of animals also exhibit a more sensitive system concerned with the regulation of erythropoiesis.

REFERENCES

Adamson, J. W., Alexanian, R., Martinez, C., and Finch, C. A. (1966). *Blood* **28**, 354.
Altland, P. D., and Parker, M. (1955). *Amer. J. Physiol.* **180**, 421.
Camiscoli, J. F., and Gordon, A. S. (1970). In "Regulation of Hematopoiesis" (A. S. Gordon, ed.), pp. 369–393. Appleton, New York.
Camiscoli, J. F., Weintraub, A. H., and Gordon, A. S. (1968). *Ann. N.Y. Acad. Sci.* **149**, 40.
Cohen, R. R. (1969a). *Physiol. Zool.* **42**, 108.
Cohen, R. R. (1969b). *Physiol. Zool.* **42**, 120.
Erslev, A. J. (1953). *Blood* **8**, 349.
Goldwasser, E., and Kung, C. K. H. (1968). *Ann. N.Y. Acad. Sci.* **149**, 49.
Gordon, A. S. (1934). *Proc. Soc. Exp. Biol. Med.* **32**, 820.
Gordon, A. S. (1959). *Physiol. Rev.* **31**, 1.
Gordon, A. S. (1960). In "Haemopoiesis" (G. E. W. Wolstenholme and M. O'Connor, eds.), pp. 325–368. Churchill, London.
Gordon, A. S., and Weintraub, A. H. (1962). In "Erythropoiesis" (L. O. Jacobson and M. Doyle, eds.), pp. 1–15. Grune and Stratton, New York.
Gordon, A. S., and Zanjani, E. D. (1970). In "Regulation of Hematopoiesis" (A. S. Gordon, ed.), pp. 413–457. Appleton, New York.
Gordon, A. S., Katz, R., Zanjani, E. D., and Mirand, E. A. (1966). *Proc. Soc. Exp. Biol. Med.* **123**, 475.
Gordon, A. S., Cooper, G. W., and Zanjani, E. D. (1967). *Semin. Hematol.* **4**, 337.
Harris, H. (1967). *J. Cell Biol.* **2**, 23.

Hirschfeld, W. J., and Gordon, A. S. (1965). *Anat. Rec.* **153**, 317.

Jacobson, L. O., Goldwasser, E., Fried, W., and Plzak, L. (1957). *Nature (London)* **179**, 633.

Jalavisto, H. E., Kuotinka, I., and Kallastinen, M. (1965). *Acta Physiol. Scand.* **63**, 486.

Jordan, H. E., and Speidel, C. C. (1923). *J. Exp. Med.* **38**, 529.

Krzymowska, H., Iwanska, I., and Winnicki, A. (1960). *Acta Physiol. Pol.* **11**, 425.

Lajtha, L. G. (1970). *In* "Regulation of Hematopoiesis" (A. S. Gordon, ed.), pp. 133–159. Appleton, New York.

McCulloch, E. A. (1970). *In* "Regulation of Hematopoiesis" (A. S. Gordon, ed.), pp. 133–159. Appleton, New York.

Reissmann, K. R. (1964). *Blood* **23**, 137.

Rosse, W. F., and Waldmann, T. (1966). *Blood* **27**, 654.

Rosse, W. F., Waldmann. T., and Hull, F. (1963). *Blood* **22**, 66.

Schooley, J. C., Garcia, J. F., Cantor, L. N., and Havens, V. W. (1968). *Ann. N.Y. Acad. Sci.* **149**, 266.

Simons, J. R. (1960). *In* "Biology and Comparative Physiology of Birds" (A. J. Marshall, ed.), pp. 345–362. Academic Press, New York.

Sokolov, O. M. (1941). *Medii Zh. Akad. Nauk. USSR* **11**, 145.

Stohlman, F., Jr., Rath, C. E., and Rose, J. C. (1954). *Blood* **9**, 721.

Stohlman, F., Jr., Ebbe, S., Morse, B., Howard, D., and Donovan, J. (1968). *Ann. N.Y. Acad. Sci.* **149**, 156.

Trentin, J. J. (1970). *In* "Regulation of Hematopoiesis" (A. S. Gordon, ed.). pp. 161–186. Appleton, New York.

Van Dyke, D., Lawrence, J. H., Pollycove, M., and Lowy, P. (1963). *In* "Hormones and the Kidney" (P. C. Williams, ed.), pp. 221–230. Academic Press, New York.

Weintraub, A. H., Gordon, A. S., and Camiscoli, J. F. (1963). *J. Lab. Clin. Med.* **62**, 743.

Weiss, H. S., Wright, R. A., and Hiatt, E. P. (1965). *J. Appl. Physiol.* **20**, 1227.

Zanjani, E. D., Contrera, J. F., Cooper, G. W., Gordon. A. S., and Wong, K. K. (1967). *Science* **156**, 1367.

Zanjani, E. D., Gordon, A. S., Wong, K. K., and McLaurin, W. D. (1969a). *Proc. Soc. Exp. Biol. Med.* **131**, 1095.

Zanjani, E. D., Yu, M-L., Perlmutter, A., and Gordon, A. S. (1969b). *Blood* **33**, 573.

9

HUMORAL REGULATION OF LYMPHOCYTE GROWTH in Vitro

*Klaus Havemann, Manfred Schmidt, and Arnold D. Rubin**

* This work was supported in part by the Deutsche Forschungsgemeinschaft;
The Max Kade Foundation; the U.S. Public Health Service Grant CA 10478 from
the National Cancer Institute and Contract AT(11-1) 3363 from the Atomic Energy
Commission.

184   K. Havemann, M. Schmidt, and A. Rubin

I. INTRODUCTION

The rapid accumulation of evidence during the past 10 years has established the key role of the small lymphocyte in the immune response. It is the lymphocyte which recognizes an antigen as such and mounts a reaction resulting in the phenomena known as humoral and cellular immunity. Lymphocyte proliferation appears to be the means whereby a single cell antigen encounter can be scaled up to the level of the whole organism. Only a small percentage of lymphocytes are capable primarily of reacting to a specific antigen. However, the actual site of cellular infiltration in response to antigenic exposure contains mostly nonsensitized cells (McCluskey et al., 1963). This type of evidence suggests an amplification of the response through a recruitment of nonsensitized cells to proliferate at the site of invasion (Dumonde, 1970; Riggio et al., 1969). These experiments point to a humoral factor(s) as the means whereby recruitment is effected. In addition, lymphocyte proliferation is controlled in such a way that the given immune response can be limited, still leaving a finite number of memory cells to await a subsequent invasion by that antigen. This growth-inhibiting phenomenon also seems to be influenced by the feedback of soluble factors elaborated by stimulated lymphocytes (Garcia-Giralt et al., 1970; Lasalvia et al., 1970; Moorhead et al., 1967; Panayi, 1970).

The experimental reproduction of lymphocyte proliferation in vitro has afforded great insight into the possible mechanism regulating lymphocyte growth. This technique has greatly facilitated the study of soluble factors. It is well known that specific antigens to which small lymphocytes have been sensitized as well as a variety of nonspecific mitogens will stimulate growth and proliferation in an in vitro system. Data obtained from the in vitro system have established that the resting small lymphocyte is incapable of mitosis. Only under stimulation, will this cell enlarge into a blast, replicate its DNA, and proceed on to mitosis. The following discussion will review the evidence that these in vitro phenomena can be influenced by a variety of humoral factors derived from the stimulated lymphocytes themselves or from interactions between lymphocytes, other leukocytes, and their culture medium. It will then attempt to establish the significance of these factors in the regulation of lymphocyte growth. The actual factors to be discussed are listed in Tables I and II and summarized in Table III.

TABLE I
Soluble Activities Stimulating or Potentiating
Lymphocyte Growth *in Vitro*

Derived from leukocytes
Blastogenic factor
Mitogenic factor
Lymphocyte transforming and potentiating factor
Mitogenic factor released by nonspecific mitogen and in continuous cultures
Transfer factor
Reconstituting factor
Immune and informational RNA?
Present in culture medium
Albumin rich serum fraction
Serum-antiproteases
Hormones (Growth-, Neuro-, Parathyroid-H, Prolactin)
Bradykinin, Kallikrein?, Kallidin?
Peptides
Leucogenenol?

TABLE II
Soluble Substances Inhibiting Lymphocyte
Growth *in Vitro* Derived from Leukocytes

Inhibiting activities in cultures stimulated
with: antigen
nonspecific mitogen
in: continuous cultures
Lymphocytic chalone?
Present in culture environment
α_2-Glycoprotein
Blocking antibodies?
Serum factors from patients with different diseases

II. SOLUBLE ACTIVITIES DERIVED FROM LEUKOCYTES STIMULATING OR POTENTIATING LYMPHOCYTE GROWTH IN VITRO

A. Blastogenic Factor

In 1964, Bain *et al.* reported that if peripheral leukocytes of two individuals are cultured together, some of the cells enlarge and undergo mitosis. Further studies by these and other authors (Bain *et al.*, 1964; Bach and Hirschhorn, 1964) demonstrated that the mitogenic response in mixed leukocyte cultures is due to differences in histocompatibility antigens between lymphocyte populations present in culture. Kasakura and Lowenstein (1965) and Gordon (1965) first reported a blastogenic factor in the cell-free supernatants of mixed leukocyte cultures. These

TABLE III
BIOLOGICAL EFFECTS OF GROWTH MODIFYING FACTORS DERIVED FROM LEUKOCYTES

			Production or release in culture						
	Effect *in vitro*	Origin	Spontaneous	Plus antigen	Nonspecifiic plus mitogen	In continuous culture	Effect dependent on antigen	Specific for lymphocytes	Species specific
Blastogenic factor	Mitogenic for allogeneic and xenogeneic ly.	Lymphocytes	+	++	n.d.	(++)	probably is antigen	+	0
Mitogenic factor	Mitogenic for allogeneic, autologous and xenogeneic, sensitized and nonsensitized ly.	Lymphocytes	(+)	++	(+)	(+)	0	n.d.	0
Lymphocyte transforming factor	Sensitizes nonsensitive ly	Lymphocytes	0	++	n.d.	n.d.	+	+	n.d.
Transfer factor	Sensitizes small number of nonsensitive ly.	Lymphocytes	0	+	n.d.	n.d.	+	+	(+)
Reconstituting factor	Reconstitutes response of pure ly. to antigen	Adherent cells	+	n.d.	n.d.	n.d.	+	+	0
Inhibiting factor?	Inhibits proliferation of autologous and allogeneic ly.	Lymphocytes	(+)	++	(+)	(+)	n.d.	n.d.	n.d
Lymphocytic chalone	Inhibits proliferation of allogeneic and exeogeneic ly.	Lymphocytes	n.d.	n.d.	n.d.	n.d.	n.d.	+	0

Legend:
0 No production of factor.
(+) Factor production questionable.
\+ Factor production demonstrated.
++ Factor production demonstrated to marked degree.
n.d. Not done.

authors demonstrated blastogenic activity, generated in unstimulated lymphocyte cultures from single donors, which stimulated allogenic lymphocytes but which showed only slight or no effect on autologous cells. As gleaned from studies of identical twins (Kasakura, 1970a), this blastogenic factor seems to arise from soluble histocompatibility antigens released into the medium. It has been shown that purified HL-A antigen induces blastogenesis when added to cultures of allogeneic lymphocytes from single donors (Viza *et al.*, 1968) and that cell-free media recovered from cultures of dog spleen can induce homograft sensitivity (Mannick *et al.*, 1964). The active material which stimulates RNA and DNA synthesis after 5–7 days can, in part, be sedimented, yielding a molecular weight of 100,000 (Kasakura and Lowenstein, 1965, 1967). As supernatants of mixed leukocyte cultures already contained blastogenic activity after 18–24 hours, the generation of this activity precedes the appearance of blastoid cells (Kasakura and Lowenstein, 1967; Kasakura, 1971). Maximal production is obtained after 3–5 days (Kasakura and Lowenstein, 1967). Sufficient irradiation of lymphocytes did not prevent formation of blastogenic factor. Furthermore, studies on factor production in mixed lymphocyte culture, in which one population was X-irradiated, show enhanced factor release by the X-irradiated cells. Whereas most of the activity released stimulated only allogenic lymphocytes, some of it seems to be related to nonspecific mitogenicity (Kasakura, 1971).

These studies suggest that the blastogenic factor which activates allogeneic cells is antigenically specific and is, in its effect, comparable to the reaction of lymphocytes with specific antigen. It is also possible that this factor may play a role in regulating lymphocyte growth *in vitro* as well as *in vivo*. For instance, surface antigens of a subpopulation of lymphocytes might be changed by environmental factors, as in virus transformed cells (Junge *et al.*, 1970).

In contrast to the original reports, careful studies on the effect of cell-free media from unmixed leukocyte cultures on allogeneic and autologous cells showed that autologous cells also are stimulated by these supernatants but to a small extent and inconsistently (Kasakura, 1971, 1970b; Havemann *et al.*, 1971a). The relation of this activity to the potentiating factor described by Janis and Bach (1970) and to the mitogenic factor described by Dumonde *et al.* (1969) and Maini *et al.* (1969) will be discussed below.

B. Mitogenic Factor

Dumonde *et al.*, (1969) reported that sensitized guinea pig lymphocytes responding to specific antigen in culture release a soluble factor capable of activating guinea pig lymphocytes otherwise unresponsive to the antigen. This so-called mitogenic factor was induced under conditions of immunological specificity. Thus, only cells capable of responding to the antigen released mitogenic factor. The factor activated lymphocytes from an immunized animal as well as lymphocytes derived from other animals not sensitized to the specific antigen. This mitogenic activity could easily be distinguished from the previous blastogenic factor by its effect on autologous cells, by the lack of activity of supernatants from unstimulated cultures on allogeneic cells, and by the absence of activity in a 100,000 sediment (Wolstencroft, 1971). In its specificity of induction and its nonspecificity of action, mitogenic factor is similar to other biological activities released in the supernatants of antigen stimulated lymphocyte cultures such as migration inhibiting factor (MIF) (David, 1971; Bloom and Bennett, 1970), cytotoxic factor (Granger et al., 1969; Williams and Granger, 1969), and intradermal inflammation factor (Dumonde et al., 1969).

The activity of mitogenic factors was measured by the increase of [3H]thymidine incorporation induced in secondary cultures of both sensitized and nonsensitized autologous and allogeneic cells. No data, so far, are available concerning the effect on RNA and protein synthesis of lymphocytes. The results of Dumonde et al. (1969) have been confirmed by other workers using guinea pig lymph node cells of outbred and inbred strains (Spitler and Fudenberg, 1970; Wolstencroft, 1971), unpurified lymphoid populations of spleen cells (Wolstencroft, 1971), thymus cells (Spitler and Fudenberg, 1970), peripheral blood leukocytes (Flad and Hochapfel, 1971; Spitler and Fudenberg, 1970; Spitler and Lawrence, 1969; Maini et al., 1969; Havemann et al., 1971a), and peritoneal exudate cells (Wolstencroft, 1971). The release of mitogenic factor after exposure to antigen has also been shown in humans (Kasakura, 1970b; Maini et al., 1969; Flad and Hochapfel, 1971; Havemann et al., 1971a) in rat (Wolstencroft, 1971), in mouse, rabbit, and chicken (Wolstencroft, 1971). In none of these studies have purified lymphocyte suspensions been used. Therefore, the possibility remains that the presence of macrophages or granulocytes in this system may influence the rate of production of mitogenic factor. It is also possible to observe the mitogenic activity induced in one species by testing it on the lymphoid cells of another, although the ability to cross the species barrier varies in different combinations (Wolstencroft, 1971). Production of mitogenic factor by sensitized lymphocytes is dependent on the dose of antigen employed. That is, an increase in factor production has been found by increasing the dose of antigen (Spitler and Fudenberg, 1970; Spitler and Lawrence, 1969; Wolstencroft, 1971; Havemann et al., 1971a,

Wolstencroft and Dumonde, 1970). At higher antigen doses a plateau of mitogenic factor production, or sometimes, an inhibition of production in respect to lower doses was observed (Wolstencroft, 1971; Wolstencroft *et al.*, 1971). Kinetic studies indicated a maximal production after 24–36 hours in guinea pig and in man (Wolstencroft, 1971; Havemann *et al.*, 1971a). After longer incubation periods with antigen, decreasing factor production has been shown to fall below the control level at 72 hours (Havemann *et al.*, 1971a). This low or absent response on longer incubation periods might have been due to medium consumption or to an inhibitor produced after longer incubation. The latter possibility (vide infra) might point to the existence of a control mechanism regulating lymphocyte activation nonspecifically, or perhaps, as analogous to tissue specific chalones. It is not clear whether DNA synthesis is required for cultures to generate mitogenic activity. Maximal production at 24 hours of incubation with antigen occurs before the onset of DNA synthesis. It has been shown by Valentine (1971) that, in a hyperbaric oxygen atmosphere, the DNA synthesis of PPD stimulated lymphocytes can be inhibited whereas the production of a mitogenic activity is unchanged. Similar studies with vincristine and "hot [^{3}H]thymidine kill" (Bloom, 1971b) also indicated that cells with their nuclear chromatin damaged by irradiation, continue factor production in the absence of DNA synthesis.

From the data accumulated so far, it is apparent that the thymus dependent (T) lymphocyte elaborates these factors, and, it has been claimed, that the B lymphocyte is the cell upon which the factor acts (Dumonde, 1970; Wolstencroft *et al.*, 1971; Hartmann, 1971). These points remain unsettled. Equally uncertain is whether the mild stimulating activity present in the supernatants of unstimulated, as well as in the supernatants of antigen-reconstituted unstimulated cultures, is due to a small amount of mitogenic factor released by dormant lymphocytes or to soluble histocompatibility antigens as discussed above. The decrease in this activity in supernatants after sedimenting out at factors heavier than 30,000 (Havemann and Burger, 1971) and late appearance of this activity, maximal at 6–7 days (Havemann and Burger, 1971), would support the latter possibility. However, the small amount of activity in the supernatants of unstimulated or antigen-reconstituted cultures on autologous cells would imply that the mitogenic or growth promoting activity also may be released by incubation of unstimulated cells (Kasakura, 1970a,b; Havemann and Burger, 1971; Janis and Bach, 1970).

It has been shown in guinea pigs by Dumonde *et al.* (1969), Wolstencroft and Dumonde (1970), Spitler and Fudenberg (1970), Spitler and Lawrence (1969), and by Kasakura (1970b) and Havemann and Burger

(1971) in man, that, with all probability, mitogenic factor acts on lymphocytes in secondary cultures independent of the presence of antigen. Supernatants of primary cultures of antigen-activated guinea pig lymphocytes were still capable of stimulation if the antigen was removed by salt precipitation or by column chromatography (Dumonde et al., 1969; Wolstencroft, 1971; Wolstencroft et al., 1971). Media from guinea pig or human lymphocyte cultures supplemented with additional antigen (Havemann and Burger, 1971; Kasakura, 1970b) exhibited no increased activity in secondary culture. Furthermore, mitogenic activity could be produced after incubation of human lymphocytes with PPD for a short period followed by washing and culturing in antigen-free media (Kasakura, 1970b).

Gel chromatography of mitogenic factor induced by antigen revealed a molecular weight between 25 and 60,000 with peak activity at 40,000, implying a certain heterogeneity in molecular weight (Dumonde, et al., 1969; Wolstencroft, 1971; Horvat, et al., 1972). Mitogenic factor can be precipitated at ammonium sulfate concentrations higher than 50% (Dumonde et al., 1969; Wolstencroft, 1971), and is recovered after acrylamide electrophoresis in the albumin rich fraction (Wolstencroft, 1971; Havemann, unpublished results). Whereas mitogenic factor can be induced in guinea pigs in a serum-free medium (Wolstencroft, 1971), it is quite difficult to obtain sufficient activity in serum-free cultures stimulated with PPD in man (Horvat et al., 1972). The necessity of heat-labile serum components for the induction, as well as for the effect, of MF in man is intriguing (Dosch et al., 1971). Heat inactivation of the serum used in primary and in secondary cultures considerably reduced the effect of mitogenic factor. As far as we know, this effect seems to be independent of complement components. The mitogenic factor-rich culture supernatants show only a slight decrease of activity after heat inactivation (Spitler and Fudenberg, 1970; Havemann, unpublished results) but are unstable after storage at −70°, a loss of activity is observed (Bloom and Glade, 1971). The effect of mitogenic factor is dependent on the presence of bivalent cations (Havemann, unpublished) as is also the case for the effect of other mitogens (Kay, 1971).

The biochemical nature of this factor cannot be predicted on the basis of present data. So far, it is not known whether other biological activities released from lymphocytes such as migration inhibitory factor (David, 1971) and chemotactic factor for mononuclear cells (David, 1971) can be attributed to a single molecule or to multiple molecular entities. Slight differences in buoyant density and, perhaps, in molecular weight may indicate more than one molecule (David, 1971; Remold and David, 1971). This view has been confirmed by the findings that

MIF activity can be readily abolished by certain antiproteases (Havemann *et al.*, 1972; Havemann *et al.*, 1971c) which do not affect activity of mitogenic factor (Havemann *et al.*, 1971). Furthermore, similarities with physiochemical properties of blood clotting, fibrinolytic, kinin releasing or complement factors have yet to be clarified. Whereas relationships to immunoglobulin and antigen-antibody complexes can be almost excluded on the basis of molecular weight, present evidence does not exclude a possible relationship to fragments of immunoglobulins. While apparently not a protease, mitogenic factor may still bear a relationship to other lysosomal enzymes.

C. Lymphocyte Transforming and Potentiating Factor

Valentine and Lawrence (1968, 1969) described a soluble material which was generated in sensitized human lymphocyte cultures in the presence of antigen and which induced nonsensitive lymphocytes to enlarge and proliferate in response to antigen *in vitro*. The conditions for production are comparable to those described for mitogenic factor. In contrast to mitogenic factor, the effect of transforming factor depends on the presence of antigen in the secondary culture. These results led the authors to believe that the factor mediates sensitization to the corresponding antigen. The activity could not be sedimented at 100,000 and was destroyed by heating at 56°. The nature of the antigen requirement cannot yet be resolved in terms of Dumonde and co-workers (1969, 1970). Perhaps technical problems as washing cultures after addition of antigen, the dose of the antigen, or the quality and the purity of cells employed, may be involved in this phenomenon. Although the studies of mitogenic factor in man would indicate an independence from antigen (Spitler and Fudenberg, 1970; Kasakura, 1970b; Havemann and Burger, 1971), it is apparent that the discrepancies can be overcome only when the mitogenic activity in humans becomes separated from the antigen. The preliminary results described by Janis and Bach (1970) point to the existence of a potentiating factor, which, after induction in primary cultures of autologous or allogenic, sensitized or nonsensitized cells (with or without antigen), potentiates the response to an antigen in secondary culture. A stimulation of secondary cultures by antigen was also obtained in an autologous system of nonsensitized cells. These rather intriguing observations can only be explained, in part, by the findings of Dumonde *et al.* (1969), Maini *et al.* (1969), and Valentine and Lawrence (1969), and should be repeated in using a larger number of experiments to allow statistical analysis.

D. Mitogenic Activity Generated by Nonspecific Mitogen and
by Continuous Cultures of Lymphoid Cells

Nonspecific mitogens would be of special interest as inducing agents
for factor generation as, under these circumstances, a large proportion
of lymphocytes are stimulated, antigen and antibody play no role, and
activation is much easier to obtain in a serum-free system. However,
because of the lack of specificity, accurate controls are hard to obtain.
It is therefore important to aspire to the complete removal of the mitogens from the culture supernatants. This cannot be done only by simple
washing because of binding and subsequent release of these mitogens
during culture. It was shown in experiments using lymphocyte cultures
preincubated and reconstituted with PHA that an additional mitogenic
activity is released during the early course of the culture (Dosch *et
al.*, 1971). A comparable mitogen has been found in the supernatants
after washing the cells free of PHA and culturing in fresh medium
for an additional 20 hours (Wolstencroft, 1971). Gel chromatography
to separate the active material from the PHA present in culture was
only partially successful. Using Concanavalin A (Con A) as the inducing
agent, it was possible to remove this mitogen completely owing to its
ability to bind on cross-linked dextrans (Horvat *et al.*, 1972). The
molecular weight of the mitogenic activity in the Con A free supernatants, however, was found to be larger than 100,000, quite different
from the molecular weight of the mitogenic factor induced by specific
antigen in a serum containing system. It remains to be shown whether
the difference in molecular weight is due to a different substance or
to a polymerization of mitogenic factor. The latter activates allogeneic
as well as autologous cells and cannot be compared with the autostimu-
lating factor reported by Powles *et al.* (1971).

E. Transfer Factor

Although most of the information on transfer factor concerns *in vivo*
experiments, some recent results indicate the possibility that lymphocytes
can be sensitized by transfer factor *in vitro*. It has been shown that,
after incubation of nonsensitized lymphocytes with low molecular
weight transfer factor obtained from a sensitized donor, stimulated a
weak proliferative response developed in the presence of antigen (Fireman *et al.*, 1967; Lawrence and Valentine, 1970).

Some other authors failed to show such activity with the low
molecular weight product (Baram and Condoulis, 1970; Paque *et al.*,
1969), whereas *in vitro* transfer of sensitivity with the nondialyzable

fraction was obtained. The liberation of transfer factor from sensitized lymphocytes into the supernatants some hours after incubation with antigen (Lawrence and Pappenheimer, 1957) would imply a possible role in lymphocyte reaction *in vitro*. The minute response (only 3% of cells became transformed) and the very late reaction to transfer factor (after 7–9 days of incubation) (Baram and Condoulis, 1970), make it improbable that a sensitization by transfer factor is involved in the *in vitro* recruitment of proliferation of lymphocytes.

The other mitogenic activities described above (Dumonde *et al.*, 1969; Wolstencroft, 1971; Horvat *et al.*, 1972) can be demarcated from both transfer factors by their molecular weight.

III. FACTORS INVOLVED IN LYMPHOCYTE–MACROPHAGE INTERACTION

It has been postulated that soluble mediators released by sensitized lymphocytes represent the means whereby these cells influence macrophages as a cooperative venture in the cellular immune response (Carpenter, 1963; Kronman *et al.*, 1969; Rocklin *et al.*, 1970; Mackler, 1971). The rapid accumulation of macrophages at sites of antigen invasion (McCluskey *et al.*, 1963) has stimulated much of this discussion. Several mediators produced by lymphocytes after incubation with specific antigens or unspecific plant mitogens affect macrophages *in vitro* (David, 1971; Bloom, 1971a; Pick and Turk, 1972).

A. Migration Inhibitory Factor (MIF)

This is the most extensively studied factor. It inhibits the migration of macrophages from different sources [peritoneal cavity, lung, lymph node, spleen, peripheral blood (in suspension out of capillary tubes); Rocklin *et al.*, 1970; Bloom, 1971a; David *et al.*, 1964; Bloom and Bennett, 1970], in agar or the outgrowth from tissue explants is released (Salvin and Nishio, 1971; Svejcar *et al.*, 1971). Evidence has been presented recently that MIF by lymphoid and fibroblast cell lines also depends upon the activation of cells to enter the mitotic cycle (Tubergen *et al.*, 1972).

B. Chemotactic Factor for Macrophages and Monocytes

This activity influences the migration of macrophages through a Millipore filter (Ward *et al.*, 1969, 1971) and can be separated from MIF by electrophoresis on acrylamide gels (Ward et al., 1970; Remold et al., 1971). Components of the complement system or their cleavage products seem to be involved (Ward et al., 1971; Snyderman et al., 1971).

C. Macrophage Aggregation Factor (MAF)

This causes an aggregation of peritoneal exudate cells of nonsensitive animals (Gothoff et al., 1971; Lolekha et al., 1970), but its correlation to delayed type hypersensitivity or transplantation immunity remains controversial (Douglas and Goldberg, 1972).

D. Macrophage or Monocyte Activating Factor

The activation of peripheral monocytes or macrophages from spleen lymph node or peritoneal cavity by lymphocyte culture supernatants or separated components has been reported to affect these cells by an increase in the rate of attachment to culture vessels (Rocklin et al., 1971; Mooney and Waksman, 1970; Adams et al., 1970; Nathan et al., 1971; Schmidt et al., 1972). Morphological changes such as increased cell size, increased numbers of lysosomes, appearance of vacuoles, clumping (Mooney and Waksman, 1970; Adams et al., 1970), increase in the rate of phagocytosis of inert material (Nathan et al., 1971), bacteria (Nathan et al., 1971), or sheep red blood (Barnet et al., 1968; Jaffe et al., 1971), enhanced hexose monophosphate oxidation and elevated RNA-precursor uptake (Nathan et al., 1971; Schmidt et al., 1972), improved resistance against bacteria (enhanced killing, prolongation of survival time) (Simon and Sheagren, 1971; Godal et al., 1971; Patterson and Youmans, 1971; Krahenbuhl and Remington, 1971), release of biological material (skin reactive factor = SRF) (Pick et al., 1970).

Furthermore, lymphocyte factors can cause an inhibition of macrophage spreading in vitro (Fauve and Dekaris, 1971) and the disappearance of macrophages from the peritoneal cavity (Sonozaki and Cohen, 1971).

E. Macrophage Factors Influencing Lymphocytes

No conclusive data are available regarding factors derived from monocytes or macrophages which influence or affect lymphocyte growth in vitro or in vivo. A soluble reconstituting factor has been detected by Bach et al. (1970) and Alter et al. (1970) who prepared a cell-free supernatant of adherent cells (macrophages) and were able to stimulate antigen-induced growth in a pure suspension of lymphocytes. Presumably this macrophage derived substance substituted for intact macrophages, the presence of which is required for optimal growth in antigen-stimulated (Hersh and Harris, 1968; Oppenheim *et al.*, 1968) and mixed lymphocyte cultures (Gordon, 1968).

Reports on the restoration of the immune response in purified lymphocytes or macrophage-deprived spleen cells by macrophage culture supernatants (Bach *et al.*, 1970; Hoffmann and Dutton, 1971) need further confirmation. In both studies, surface attached cells have been employed as source of the supernatants so that the participation of other "sticky" cells cannot be excluded.

There is still a considerable controversy concerning the role and biological function of the so-called immunogenic RNA synthesized or released by peritoneal cells (Bishop and Gottlieb, 1970). It has been shown in previous studies that antibody synthesis in lymphoid cells of unimmunized animals can be induced by RNA extracted from peritoneal cells of immunized individuals. This immunogenicity has been attributed to a unique informational RNA by one group (Fishman and Adler, 1963a,b, 1967; Adler *et al.*, 1966), or to the presence of antigen or fragments complexed to macrophage RNA by others (Friedman, 1966; Askonas and Rhodes, 1965; Gottlieb and Glisin, 1967). This RNA was different from messenger RNA. A light density ribonucleoprotein associated with small antigen portions has been described as unique for macrophages (Gottlieb, 1969; Gottlieb and Straus, 1969). This was not confirmed by other investigators who found a merely adjuvant effect of RNA in priming animals for immune responses (Roelants and Goodman, 1969).

In conclusion, there is, at the present time, no proof that lymphocyte growth is influenced *in vitro* by macrophage-derived factors. But macrophages are affected *in vitro* by a variety of lymphokines which may all, with the exception of the chemotactic factor, identify with MIF. The interesting concept that the macrophage-lymphocyte interaction in the cell-mediated immune response *in vivo* may, at least, partially be mediated by soluble effector molecules derived from both cell types remains to be confirmed.

IV. SOLUBLE ACTIVITIES PRESENT IN THE CULTURE ENVIRONMENT STIMULATING OR POTENTIATING LYMPHOCYTE GROWTH *IN VITRO*

Growth activating factors derived from leukocytes are specifically or nonspecifically released during a certain time period of culture. In contrast, nutritional requirements in the culture medium, as well as in the serum used in culture, are invariably present from the beginning. Their function as a reservoir of essential nutrients lies beyond the scope of this discussion. Still, serum contains, or may contain, a number of substances which potentiate or even stimulate cell growth in culture. The effect of these substances are nonspecific as they exert their activity on various growing tissues including lymphocytes. When enumerating humoral factors which may be involved in lymphocyte regulation, however, these environmental factors should be discussed because of the existence of a possible interrelationship with the factors derived from the cells.

Because of its different composition when obtained from different sources, serum is an important source of variability in analyzing short-term lymphocyte cultures. To obtain comparable results, to avoid antigenic stimulation by foreign serum proteins, and to maintain a balance between growth promoting and inhibiting serum proteins, it is essential to employ fresh frozen or lyophylized serum from allogeneic donors used as a pool for one series of experiments. Despite these technical problems, the macromolecular and low molecular substances present in the serum appear to directly or indirectly regulate lymphocyte growth.

Macromolecular Serum Proteins Promoting Lymphocyte Growth

It has been shown by Puck *et al.* (1968) that fetuin, a component of fetal calf serum consisting of two proteins comparable to human α_2-macroglobulin and α_1-antitrypsin, exhibits strong nonspecific growth promoting activity on many cultivated tissues. Both proteins are antiproteases which inactivate the proteolytic activity of trypsin and some other proteolytic enzymes. It has been claimed by Landureau and Steinbruch (1969) and Wallis *et al.* (1969), that their effect is due to inactivation of lyosomal enzymes released in culture. Studies were performed in which PHA-stimulated lymphocyte cultures were supplemented with different amounts of 19 S, 7 S and 4 S serum fraction obtained by gel chromatography (Havemann *et al.*, 1970b). Whereas the activity of the 19 S fraction could be attributed to α_2-macroglobulin, part of the growth enhancing activity in the 4 S fraction was presumably due to α_1-antitrypsin (Havemann *et al.*, 1970b). The release of esterases in lymphocyte cultures, the binding of labeled lymphocyte proteins to α_2-macroglobulin and α_1-antitrypsin and the promoting effect of other antiproteases of plant or animal origin on PHA and antigen stimulated lymphocytes (Havemann *et al.*, 1971b) constitute further evidence for the binding of proteolytic enzymes by these proteins which seems to be essential for optimal growth. Whether these substances exert their effect by simply inactivating proteolytic enzymes released by the cells, which might be cytotoxic, or whether they are necessary to stabilize growth promoting proteases is at present unknown. The latter possibility, although still speculative, would be of some interest in relation to the known ability of proteolytic enzymes to induce cell growth. Thus, nerve growth factor, epidermal growth factor and other growth inducing agents (Greene *et al.*, 1971) contain proteolytic enzymes as a subunit of their molecules. Furthermore, there is some indication that, under certain conditions, trypsin induces or promotes the proliferation of lymphocytes (Mazzei *et al.*, 1966). Future studies will show if this has some relevance in the actual regulation of lymphocyte growth.

Beside the effect of serum antiproteases, it has been shown that haptoglobin, orosomucoid, and albumin (Auger *et al.*, 1971) stimulate cell growth of continuous cultures of lymphoid origin. Recently, it has been demonstrated that C reactive protein stimulates lymphocytes *in vitro* at low concentration and is cytotoxic for lymphocytes at high concentration (Hornung and Fritsch, 1971). The mechanism of action of these serum proteins is unknown. Other factors such as hormones (Whitfield *et al.*, 1968) the bradykinin-kallikrein system (Mandel *et al.*, 1966; Perris and Whitfield, 1969; Schachter, 1968), peptides (Amborski and Moskowitz, 1968; Ito and Moore, 1969), and leucogenenol (Rice and Ciavarra, 1971) may be important regulators of lymphocyte growth.

V. BIOLOGICAL SIGNIFICANCE OF DIFFERENT STIMULATING OR POTENTIATING ACTIVITIES

The major event in the induction of the proliferative response *in vitro* is the contact of antigen with a receptor site of the sensitized lymphocyte. Prior sensitization of the lymphocytes to the appropriate antigen is a prerequisite for antigen-induced lymphocyte response *in vitro* (Hirschhorn *et al.*, 1963). In general, the proliferation upon addition of antigen *in vitro* has been found to correlate with the presence of delayed-type hypersensitivity in the cell donor (Valentine, 1971). Although still open to discussion, the majority of antigen binding cells present in the lymphocyte populations *in vitro* seem to be thymus dependent (Bach, 1971). It is, furthermore, known from experiments *in vivo*, as well as *in vitro*, that the stimulated T-cells are able to help the induction or clone formation of nonsensitized or sensitized antibody producing B-cells (Mitchison, 1969; Hartmann, 1971), although the role of this helper function in lymphocytes held *in vitro* is still unknown. This background knowledge has to be taken into consideration if discussing the biological effects of certain factors derived from leukocytes which may, in part, be responsible for humoral regulation of lymphocyte growth *in vitro*. The main question which arises is whether the factors which exhibit mitogenic activity are necessary for recruitment of lymphocytes not initially responsive to antigen. Although the number of dividing cells certainly indicates that lymphocyte stimulation occurred which may be proportional to the number of cells initially responding to antigen, it is not equal to the number of cells initially responding to that antigen because of the proliferation that has taken place. Therefore, the number of 20–30% of blast cells observed after 5–7 days of culture must be the outcome of only a few cells proliferating at the start of the culture. The number of responsive cells has been estimated by different techniques and in different lymphatic tissues including peripheral blood. In peripheral blood, for instance, the number of reacting lymphocytes for one antigen varies from 0.3 to 2% (Jimenez *et al.*, 1971; Coulson and Chalmers, 1967; Marshall *et al.*, 1969; Donald and Beck, 1971). These results, however, were partly obtained with techniques which do not exclude a recruitment of lymphocytes during the early course of stimulation. Marshall *et al.* (1969) cultured lymphocytes with different antigens for 2–4 days in tubes and then transferred them to micro chambers. The growing cells in the chambers were followed up to 6 days by time lapse cinematography. Clonal proliferation of single cells could be observed through several generations. From the number of blast cells present at 2 days and the final number of dividing cells at the end of the culture, he concluded that only clonal proliferation but no recruitment occurred during the later period of lymphocyte cultures, and, that if there was recruitment at all, it took place during the early period of culture (1–2 days). The clonal proliferation assumed by these authors would be in keeping with the well-known exponential increase of thymidine incorporation during an antigen stimulated lymphocyte culture (Jimenez *et al.*, 1971; Marshall *et al.*, 1969; Havemann and Burger, 1971; Valentine, 1971). However, because of the start of the single cell culture after 2–4 days, it could not be proven that the cells which showed clonal proliferation were sensitized cells originating from cells initially responding to the antigen. There are several indirect and direct indications for the occurrence of recruitment during the early culture period: (1) The early production of mitogenic factor maximal after 24 hours points to a recruitment present at this time period when significant proliferative activity cannot be observed by the methods employed (Wolstencroft, 1971; Havemann and Burger, 1971). If supernatants containing mitogenic factor and antigen and control supernatants containing the same amount of antigen were tested on autologous cells from a sensitized donor, an exponential incorporation of [^{3}H]thymidine could be demonstrated in both culture containing mitogenic factor. A slight increase above control levels were already visible after 48–60 hours, whereas the maximal difference appeared after the longest time of cultivation (Havemann and Burger, 1971).

(2) Furthermore, the effect of addition of nonsensitized lymphocytes from HL-A indentical sibs to sensitized cells from the other sib has been studied (Schellekens and Eijsvoogel, 1971). It has shown that 1×10^6 sensitized cells mixed with 2×10^6 lymphocytes from the nonsensitized HL-A identical sibs gave a response similar to that obtained using 3×10^6 sensitized lymphocytes alone. However, if the metabolism of the nonsensitized cells was inhibited by previous incubation with iodoacetic acid, no such restoration occurred. Indirect evidence for a presence of a recruitment was obtained by further experiments in which the addition of two weak antigens resulted in a summation of response, whereas no additive effect is achieved by strong antigens. The latter experiments may be explained· by maximal reactivity being limited by the size of the pool of recruitable cells.

(3) In this respect, the findings of Jimenez *et al.* (1971) are of special interest. They showed, by a virus plaque assay, that only a linear, vinblastin resistant, increase of antigen-sensitive cells could be observed in culture, whereas the increase of thymidine incorporation was exponential in time.

In vivo studies, as for instance the enormous accumulation of cells not primarily sensitive after intradermal injection of antigen (McCluskey *et al.*, 1963), as well as the striking proliferation seen in the lymph nodes after injection of mitogenic factor into the afferent lymph (Dumonde, 1970), would also point to the recruitment of lymphocytes.

On the other hand, *in vitro* experiments of Wilson and Nowell (1970) and Wilson *et al.* (1967) using parental F_1 mixed leukocytes cultures with a sex-chromatin marker were not able to detect the response of F_1 cells. A-strain cells from a male tolerant to B and normal A-strain female cells were mixed with an AB F_1. The vast majority of mitotic figures were from normal A-strain female cells.

Despite these contradictory results which have been obtained in a rather complicated animal system, the vast majority of the present evidence is directed to the existence of a recruitment, which amplifies the response in short-term lymphocyte cultures. The assumption that a soluble mitogenic factor is responsible for this recruitment does not eliminate the possibility that a close cell-to-cell contact in lymphocyte cultures can also lead to an amplification of response. It has been shown that lymphocytes do not readily survive and grow at low cell concentrations (Moorhead *et al.*, 1967; Valentine, 1971). In general, the best growth conditions are those with the highest cell concentrations the medium used will support. On the other hand, continuous agitation markedly inhibits the mitogen induced transformation.

Leventhal and Oppenheim (1969) described the effect of cell density on the response of lymphocyte cultures and concluded that lymphocyte-lymphocyte contact may be required for *in vitro* proliferation if no macrophages are present. After close contact between cells, a formation of bridges has been shown (Borek *et al.*, 1969), which allow the exchange of ions and larger molecules. It has been found by Schellekens and Eisjvoogel (1971) and Valentine (1971) that a close cell to cell contact is only necessary during the first 24 hours of culture. Afterwards, it plays no role if the surface of the culture vessels is increased. This effect has only been demonstrated with antigen stimulated cultures, whereas the response to PHA seems to be independent of the surface area (Schellekens and Eisjvoogel, 1971). Taking these results into account, it may be that, under the influence of the appropriate antigen, the few sensitized lymphocytes become a focus from which other non-sensitized cells are recruited. The initially sensitized cells produce a factor which is either secreted, by which in a dense cell sediment high concentrations are achieved, or it is transferred by direct cytoplasmic contact which also is facilitated by cell proximity.

The results of Valentine and Lawrence (1969) with lymphocyte transforming factor, the results with low and macromolecular transfer factor (Havemann *et al.*, 1971; Baram and Coudoulis, 1970; Paque *et al.*, 1969) and the results with immune or informational RNA (Fishman and Adler, 1963a,b; Jacherts, 1970) would imply that recruitment is the expression of a cell sensitization leading to a higher number of sensitized cells in culture. In this case, a purified factor would only be able to exert its effect in the presence of antigen. However, this supposition is highly questionable from a number of experiments in animal and man which show the independence in effect of mitogenic factor from the presence of antigen (Dumonde *et al.*, 1969; Kasakura, 1970b; Spitler and Fudenberg, 1970; Spitler and Lawrence, 1969; Havemann and Burger, 1971). In this respect, the findings of Jimenez *et al.* (1971) are of great interest. They showed that the number of antigen-sensitive cells was only slightly increased during culture and that they did proliferate, whereas the proliferating cells increased exponentially. The authors speculated that the antigen-sensitive cell, which probably was the T lymphocyte itself, did not proliferate, whereas an innocent bystander cell, perhaps the B lymphocyte, was the only cell involved in the proliferative response. Comparable conclusions have been observed by Dumonde (1970), Wolstencroft *et al.* (1971), and Panayi (1970). On a more speculative basis, these authors assumed that mitogenic activity is of "fundamental importance in maintaining antigen-sensitive cells of both lymphocyte compartments in a physiological sensitive state" which may "lead to local activation of immunocytes of different antigenic specificities" (Wolstencroft et al., 1971). The production of nonspecific immunoglobulins, as well as desensitization or even high dose tolerance, have been claimed by these authors to the result of a sterile activation due to the nonspecific action of mitogenic factor.

The possible importance of this factor for adoptive immunity is apparent although there are no data at this time available to suggest a mechanism in short-term lymphocyte cultures. The biological significance of the other factors derived from leukocytes is even less conclusive. It can be assumed that blastogenic factor seems to operate like an antigen (Kasakura and Lowenstein, 1967). There are still indications that sensitization may occur in culture to a certain degree. This has been shown by Valentine and Lawrence (1969) with transforming factor and by Janis and Bach (1970) with potentiating factor. It is also suggested by the experiments with transfer factor (Fireman et al., 1967; Valentine, 1971; Baram and Condoulis, 1970). However, in these experiments, a low reactivity to the antigen has not been entirely excluded. Some of these results may be explained by a factor which is derived from adherent cells and which reconstitute the response to antigen in vitro (Bach et al., 1970; Hoffmann and Dutton, 1971). It seems possible that the immunogenicity of an antigen is enhanced by this factor similar to the theoretical increase of immunogenicity produced inside macrophages.

The role of the other soluble nonlymphocyte derived physiological substances in nonspecifically initiating or potentiating lymphocyte response is unknown. Their enumeration may be of value, however, for at this time there appear some relationships to the soluble substances derived from the cells.

VI. SOLUBLE SUBSTANCES DERIVED FROM LEUKOCYTES INHIBITING LYMPHOCYTE GROWTH IN VITRO

The difficulties in maintaining proliferation in lymphocyte cultures beyond a certain time interval may simply be a function of depleted nutrients. However, there are some indications which strengthen the assumption of a physiological inhibitor also operating in culture. In testing inhibitors for lymphocyte growth, several problems have to be considered:

1. In the isolated, artificial system of lymphocyte culture, the inducing agent (specific antigen, nonspecific mitogen) is, in most instances, present during the entire culture period. The physiological feedback of proliferation, i.e., the removal of the antigen, is, therefore, not operational. In vivo, and only under special circumstances in vitro (Ginsburg et al., 1971), the enlarged blast cells will transform back to small lymphocytes.

2. The lymphocytes have to be stimulated by the antigens, nonspecific mitogens, in mixed lymphocyte cultures or in continuous cultures before an inhibitor can be tested. Conceivably, an inhibiting substance does not interfere with the stimulating agent simply by inactivating it. Thus, the existence of an inhibitor might be demonstrated in a system involving different antigens and mitogens. Nonspecific mitogens such as PHA and Concanavalin A exhibit binding reactions with glycoproteins (as α_2-macroglobulin, IgM) and lipoproteins (Havemann et al., 1970a). The complexes between PHA and these proteins are, in themselves, stimulatory, but the dose-response of the mitogen is changed by addition of these proteins (Havemann et al., 1970a).

3. In most of the techniques for short-term lymphocyte culture, the medium is not changed. Therefore, particularly after longer incubation, an exhaustion of the accumulation of toxic products may appear. These toxic byproducts would be unrelated to physiological inhibitors. In general, the problem of medium consumption or accumulation of toxic products is hard to ignore.

4. In general, proliferation of cells is tested by [3H]- or [14C]thymidine incorporation. Inhibition by release of cold nucleotides from the cells must be considered.

In contrast to the reports on mitogenic factors, very little is known about inhibiting factors. Indirect evidence for such an inhibitor(s) in short-term lymphocyte culture has been offered by experiments in which sensitized lymphocytes were grown in the presence of PPD (Havemann and Burger, 1971; Wolstencroft et al., 1971). After a culture period of 48–72 hours, the supernatants, when tested on secondary cultures, often contained inhibitory activity when compared to zero time controls. This inhibiting activity was also operative if supernatants were added in small amounts excluding, in part, the consumption of nutritional products. The inhibiting activity seemed to be dialyzable in part (Wolstencroft 1971; Wolstencroft, et al., 1971), whereas some of it remained in a macromolecular fraction (Dosch et al., 1971). Beside this inhibiting fraction, mitogenic activity still was present in the supernatants (Dosch et al., 1971). Inhibiting substances were much more easily obtained if heat inactivated serum was used in culture, suggesting the removal of heat labile growth promoting serum protein. Results which may be comparable to these have been obtained in cultures of mouse fibroblasts (Salas and Green, 1971) where a synthesis of growth promoting and inhibiting proteins could be demonstrated. The production of the inhibiting protein could only be achieved at low serum concentrations. The growth inhibiting protein, partly purified, was bound specifically to DNA.

Further indications for the existence of an inhibitor present in certain lymphocyte cultures were offered by the experiments of Smith *et al.* (1970) and Havemann (unpublished results) and Dosch *et al.* (1971). Culture fluids removed from PHA stimulated cells, depending on the time of removal, had the capacity to inhibit or to activate fresh cultures (Dosch *et al.*, 1971; Smith *et al.*, 1970; Havemann, unpublished results).

Although at this time no evidence exists as to the production in lymphocyte cultures, a specific inhibitor of lymphocyte DNA synthesis has been extracted from different sources of lymphatic tissue by Moorhead *et al.*, (1967), Bullough and Lawrence (1970), Garcia-Giralt *et al.* (1970), Lasalvia *et al.* (1970), Houck *et al.* (1971), Kiger (1971), and Carpenter *et al.* (1971, 1970, 1971b). This substance inhibits DNA synthesis and mitosis of human lymphocytes stimulated with antigen, PHA, or allogeneic cells *in vitro* (Moorhead *et al.*, 1967; Garcia-Giralt *et al.*, 1970; Houck *et al.*, 1971; Carpenter *et al.*, 1970), the DNA synthesis of established cell lines derived from normal and acute lymphocytic leukemia human lymphocytes (Lasalvia *et al.*, 1970; Houck *et al.*, 1971; Kiger, 1971), the DNA synthesis in the lymphatic tissues of mice (Garcia-Giralt *et al.*, 1970; Kiger, 1971), the production of plaque forming cells (Garcia-Giralt *et al.*, 1970; Kiger, 1971), and the graft-versus-host reaction (Garcia-Giralt *et al.*, 1970). Whereas the effect of this inhibitor derived from the lymphatic tissues of different species could be demonstrated on all mammalian lymphatic tissues investigated so far, the inhibitor seems to show no effect on cells other than lymphocytes (Houck *et al.*, 1971; Garcia-Giralt *et al.*, 1970). The activity seems to be a specific inhibitor of DNA synthesis without effecting protein and RNA synthesis. In contrast to the original findings of Cooperband *et al.* (1969) an inhibition of the production of migration inhibitory factor could not be confirmed (Carpenter *et al.*, 1970, 1971). This substance therefore, has been called a "lymphocytic chalone" (Bullough and Laurence, 1970; Garcia-Giralt *et al.*, 1970; Kiger, 1971). It is concluded that this partially purified natural globulin exerts its modulating effect on immune responses by preventing new DNA synthesis in the stimulated cell and not by blocking the early recognition of response to antigen (Carpenter *et al.*, 1970, 1971).

Beside this lymphoid protein, several tissue extracts from lymphoid and other tissues reveal inhibitory activity on lymphoid cells as well as on established cell lines derived from various tissues (Nilsson and Philipson, 1968; Rounds, 1970). This inhibiting effect is assumed to be nonspecific. Whether these inhibiting activities have any relevance to the control of growth *in vitro* is unknown. Because of considerable cell death which occurs in lymphocyte culture, the specific or nonspecific inhibiting effect of lymphocyte degradation products on the growing cells should be kept in mind.

VII. SOLUBLE SUBSTANCES PRESENT IN THE CULTURE ENVIRONMENT INHIBITING LYMPHOCYTE GROWTH *IN VITRO*

Despite extensive studies attempting to employ only artificial media in lymphocyte cultures, the addition of serum is still required to obtain optimal growth conditions. As it has been stated, serum contains a promotor of lymphocyte growth in the albumin-rich fraction. There are also several reports concerning inhibiting serum proteins. The prolongation of survival of skin allografts by a crude serum α-globulin fraction was first described by Kamrin (1969). A decrease in production of γ-globulins and diminished antibody responses has also been shown (Kamrin, 1969). These results later were confirmed by employing a partial purified bovine α-glycoprotein, which was associated with ribonuclease activity (Mowbray, 1963a, b). Contradictory results obtained by Cooperband *et al.* (1968, 1969) with a human preparation using PHA stimulated lymphocytes, makes it questionable whether ribonuclease activity associated with this α-globulin is necessary for its inhibiting activity (Cooperband *et al.*, 1969). Further studies by Mannick and Schmid (1967), Mowbray (1963a,b), and Milton (1971) showed an inhibiting effect of this serum α_2-glycoprotein on the *in vitro* proliferation of human lymphocytes stimulated by PHA, PPD, or allogeneic cells. A linear dose-response curve was obtained for the inhibition of maximally stimulated PHA cultures but no parallel phenomenon was observed in antigenically stimulated cultures (Milton, 1971). Preincubation with α_2-globulin had no effect on the subsequent stimulation with PHA, but did inhibit subsequent stimulation with PPD or allogeneic cells (Milton, 1971). The effect of α_2-glycoprotein was not species specific (Mowbray, 1963a,b), and it has not been shown that it exerts its effect only on lymphocytes. It may be related to a 7 S protein strongly inhibiting the PHA response in culture (Havemann *et al.*, 1970b). Recent studies by Carpenter *et*

al. (1971b; 1970) indicated a close similarity in effect between serum
α-globulin and a thymus fraction which seemed to be identical with
the lymphocytic chalone in its biological effect does not show a complete
biochemical identity with serum α-globulin. The mechanism of action
is still unknown. As it is discussed by Kamrin (1969), the effect could
be related to the ability of glycoprotein rich α-globulins to interfere
nonspecifically with the outer membranes of the immune competent
cells, perhaps by occlusion of the antigen binding sites. However, if
the view of Carpenter (1963) is correct that the serum fraction and
the thymus fraction which does not influence the production of MIF
and does not interfere with antigen reception sites, Kamrin's hypothesis
of blocking the early recognition of antigen seems to be questionable.
Its importance for regulation of lymphocyte growth *in vitro* is unknown.
The obvious existence of balance between growth promoting and inhibiting serum proteins in a given serum specimen, together with the apparent instability of the promoting protein, may point to the possibility
that, after a certain time of culture growth, the balance between promoting and inhibiting agents is changed in favor of the inhibiting one.
This possibility would be reinforced by results showing an increase of
inhibiting serum capacity and a decrease of production of mitogenic
factor after heat inactivation of serum which markedly reduces the
growth promoting capacity of the serum (Dosch *et al.*, 1971). There
may be a relationship to a toxic serum protein present in heat inactivated
sera which was present in the prealbumin region and which was cytotoxic for continuous cell lines of malignant and nonmalignant origin
(Beno *et al.*, 1969).

Another inhibitory event which should be taken into consideration is
the effect of blocking antibodies possibly produced by the lymphocytes
in culture or the effect of such antibodies in forming complexes with
the antigen present. Although the production of antibodies in cultures
seems to be very small, this possibility should be kept in mind in discussing the role of inhibiting substances on cultures of lymphocytes *in vitro*
(Smith *et al.*, 1971). Antibodies directed to lymphocytes have been demonstrated in the sera of patients with different diseases showing lymphocyte toxicity or inhibition of lymphocyte response *in vitro*. There is
also evidence which indicates that anti-HL-A antibodies specific for
either stimulating or responding cells may inhibit the response in one-
way allogeneic mixed leukocyte cultures (Grumet and Leventhal, 1970).
Although many of the inhibitors present in lymphatic tissues and in
serum have been tested on stimulated lymphocytes, they may play no
role as inhibitors of humoral regulation *in vitro*. Mainly for technical
reasons (e.g., the disadvantage of a closed system), the cell death due

to exhaustion of nutrients and the accumulation of toxic products which are released from the cells or which are generated by interference with substances present in the environment, it is questionable whether the system of short-term lymphocyte cultures can be applied for final clarification of the effect of inhibitors on lymphocyte growth *in vitro*. However, the use of other systems, as, for instance, the growth of lymphocytes on feeder layers or in *in vivo* cultures of Millipore chambers, would open a variety of other problems. Therefore, studies should focus on the isolation and biochemical characterization of inhibiting substances released during lymphocyte culture.

VIII. STUDIES ON GROWTH MODIFYING AGENTS IN DIFFERENT DISEASES

Most of our present knowledge regarding these regulatory factors concerns the situation present under physiological circumstances. However, a blastogenic response has been reported by Junge *et al.* (1970) when the transformed cells in peripheral blood from patients with infectious mononucleosis were kept deeply frozen and were later cultivated in mixed lymphocyte culture with normal autologous lymphocytes. The supernatant of these responding cultures stimulated autologous cells. Since the activity was heat labile, nondialyzable, and could be sedimented at 200,000, it seems to be comparable to blastogenic factor. There are some preliminary findings on the production of mitogenic factor in immediate hypersensitivity and in patients with lymphoproliferative disorders (Miami *et al.*, 1969; Flad and Hochapfel, 1971; Horvat and Havemann, 1972). In grass pollen allergy, 90% of the patients generated a mitogenic factor in response to the antigen, whereas only 30% were able to produce migration inhibitory factor. Retrospective inquiry revealed that the latter group had either received desensitization therapy or showed only minor forms of clinical allergy (Maini *et al.*, 1969). Lymphocytes from patients with chronic lymphocytic leukemia (CLL), Hodgkin's disease, and sarcoidosis exhibited a response to mitogenic factor (Flad and Hochapfel, 1971; Horvat and Havemann, 1972) which was comparable to normal nonsensitized lymphocytes in patients with sarcoidosis (Horvat and Havemann, 1972), but which was suppressed in high cell count CLL and in Hodgkin's disease (Flad and Hochapfel, 1971; Horvat and Havemann, 1972). In low count CLL, however, there was no significant difference in response to the normal controls (Horvat and Havemann, 1972). It would seem, therefore, that the decreased response to mitogenic factor in these patients runs in parallel to the impaired reactivity seen with nonspecific mitogens. Preliminary results on the production of mitogenic activity released is dependent on the sensitivity to the inducing antigen (Horvat and Havemann, 1972).

Indications for the elevation of levels of lymphocyte suppressive α-globulin in sera of patients undergoing renal allograft rejection has been given by Riggio *et al.* (1969).

Cytotoxic or inhibiting serum factors exerting their effect on autologous or allogeneic lymphocytes *in vitro* have been described in various diseases. These factors might be related either to cytotoxic or blocking antibodies specific for HL-A antigen or to serum factors still to be described. They have been described in the serum or plasma of patients with various diseases such as systemic lupus erythematosus, rheumatoid arthritis, rheumatic heart disease (Terasaki *et al.*, 1970), after vaccination (Kreisler *et al.*, 1970), in tuberculosis and multiple sclerosis (Knowles *et al.*, 1968; Stjernholm *et al.*, 1970), in hepatitis (Paronetto and Popper, 1970), in chronic mucocutaneous candidiasis (Canales *et al.*, 1969), in ataxia telangiectasia (McFarlin and Oppenheim, 1969), in syphilis (Levene *et al.*, 1969), and in patients with various malignancies of lymphoid or nonlymphoid origin (Gatti, 1971; Langner *et al.*, 1971; Scheurlen *et al.*, 1968; Silk, 1967; Trubowitz *et al.*, 1966; Whittaker *et al.*, 1971). As shown in some instances, these inhibitors appear or disappear with time and clinical status of the same patient (Gatti, 1971; Stjernholm *et al.*, 1970). The impaired response of lymphocyte cultures from these patients performed with autologous plasma or serum can be related to these inhibitors if the response in homologous serum is normal. However, it is not known if comparable inhibitors are involved in regulating lymphocyte growth *in vitro* under physiological conditions. The effect of antilymphocytic antibodies in controlling immune response *in vivo* has been discussed in detail by Uhr and Moller (1968).

IX. CONCLUSION

In vitro studies of lymphocyte proliferation has added considerably to our knowledge of the immune response. The demonstration of soluble growth promoting or inhibiting factors elaborated by cultivated lymphocytes is intriguing. It carries great potential significance for understanding the regulation of lymphocyte proliferation *in vivo* and in particular, it offers a plausible mechanism to explain how a limited number of antigen sensitive cells can recruit large numbers of active cells into an area of inflammation. However, the *in vitro* data do not establish an actual role for soluble mediators in immunity or, for that matter, even in the proliferative reaction *in vitro*. Soluble factors influencing growth were identified from a variety of sources, some of these bearing no direct relation to cultivated lymphocytes. Therefore, more precise conclusions must await precise chemical characterization and comparison of the different factors isolated from *in vivo* and *in vitro* systems. Ultimately one would like to have access to a group of soluble factors derived from human sources capable of influencing the immune response for therapeutic purposes.

REFERENCES

Adams, D. O., Bieseker, J. L., and Koss, L. G. (1970). *Fed. Proc.* **29**, 359 (Abstr.).
Adler, F. L., Fishman, M., and Dray, S. (1966). *J. Immunol.* **97**, 554.
Alter, B., Zoschke, D., Solliday, S., and Bach, F. H. (1970). Int. Congr. Transplant. Brussel, Abstract, p. 99.
Amborski, R. L., and Moskowitz, M. (1968). *Exp. Cell. Res.* **53**, 117.
Askonas, B. A., and Rhodes, J. M. (1965). *Nature (London)* **205**, 470.
Auger, M. A., Tiollais, P., and Jayle, M. F. (1971). *Rev. Eur. Etudes Clin. Biol.* **16**, 130.
August, C. S., Rosen, F. S., Filler, R. M., Janeway, C. A., Markowski, B., and Ray, H. E. M. (1968). *Lancet* **2**, 1210.
Bach, F., and Hirschhorn, K. (1964). *Science* **143**, 814.
Bach, F. H., Alter, B. J., Solliday, S., Zoschke, D. C., and Janis, M. (1970). *Cell. Immunol.* **1**, 219.
Bach, J. F. (1971). In "Cell-mediated Immunity; In Vitro Correlates" (J. T. Revillard, ed.), p. 51. Karger, Basel.
Bain, B., Vas, M. R., and Lowenstein, L. (1964). *Blood* **23**, 118.
Baram, P., and Condoulis, W. (1970). *J. Immunol.* **104**, 769.
Barnet, K., Pekarek, J., and Johannovsky, J. (1968). *Experientia* **24**, 948.
Beno, D. W., Lytle, R. I., and Edwards, E. A. (1969). *J. Nat. Cancer Inst.* **42**, 899.
Bishop, D. A., and Gottlieb, A. A. (1970). *Current Topics Microbiol.* **51**, 1.
Bloom, B. R. (1971a). *Advan. Immunol.* **14**, 101.
Bloom, B. R. (1971b). As summarized in "Progress in Immunology" (B. Amos, editor), p. 1527. Academic Press, N.Y.
Bloom, B. R., and Bennett, B. (1970). *Ann. N.Y. Acad. Sci.* **169**, 258.
Bloom, B. R., and Glade, P. R. (1971). In "In Vitro Methods in Cell-mediated Immunity" (B. R. Bloom and P. R. Glade, eds.), p. 571. Academic Press, New York.
Borek, C., Higashino, S., and Lowenstein, W. R. (1969). *J. Membrane Biol.* **1**, 274.
Bullough, W. S., and Laurence, E. B. (1970). *J. Cancer* **6**, 525.
Canales, L., Middlemas, R. T., Louro, J. M., and South, M. A. (1969). *Lancet* **ii**, 567.
Carpenter, R. R. (1963). *J. Immunol.* **91**, 803.
Carpenter, C. B., Philipps, S. M., Boylston, A. W., and Merrill, J. P. (1970). Int. Congr. Transplant. Brussel, Abstract, p. 247.

Carpenter, C. B., Boylston, II., A. W., and Merrill, J. P. (1971a). *Cell. Immunol.* **2**, 435.

Carpenter, C. B., Boylston, A. W., and Merrill, J. P. (1971b). *Cell. Immunol.* **2**, 425.

Cooperband, S. R., Davis, R. C., Schmid, K., and Mannick, J. A. (1968). *Science* **159**, 1243.

Cooperband, S. R., Davis, R. C., Schmid, K., and Mannick, J. A. (1969). *Trans. Proc.* **I**, 516.

Coulson, A. S., and Chalmers, D. G. (1967). *Immunology* **12**, 417.

David, J. R. (1971). *In* "Progress in Immunology" (B. Amos, ed.), p. 399. Academic Press, New York.

David, J. R., Al-Askari, D., Lawrence, H. S., and Thomas, L. (1964). *J. Immunol.* **93**, 264.

Donald, D., and Beck, J. S. (1971). *Nature* **231**, 183.

Dosch, H. M., Havemann, K., Malchow, H., Sodomann, C. P., and Schmidt, M. (1971). *In* "The Role of Lymphocytes and Macrophages in the Immunological Response" (D. C. Dumonde, ed.), p. 62. Springer-Verlag, Heidelberg.

Douglas, S. D., and Goldberg, M. (1972). *Vox Sang.* **23**, 214.

Dumonde, D. C. (1970). *Proc. Roy. Soc. Med.* **63**, 27.

Dumonde, D. C., Wolstencroft, R. A., Panayi, G. S., Matthew, M., Morley, J., and Howson, W. T. (1969). *Nature (London)* **224**, 38.

Fauve, R. M., and Dekaris, D. (1971). *In* "In Vitro Methods of Cell-Mediated Immunity" (B. R. Bloom and P. R. Glade, eds.), p. 313. Academic Press, New York.

Fireman, P., Boesman, M., Haddad, Z. H., and Gitlin, D. (1967). *Science* **155**, 337.

Fishman, M., and Adler, F. L. (1963a). *J. Exp. Med.* **117**, 595.

Fishman, M., and Adler, F. L. (1963b). *In* "Immunopathology" (P. Grabar and P. Miescher, eds.), 3rd Int. Symposium, p. 799. Schwabe, Basel.

Fishman, M., and Adler, F. L. (1967). *Quant. Biol.* **32**, 343.

Flad, H. D., and Hochapfel, G. (1971). *In* "The Role of Lymphocytes and Macrophages in the Immunological Response" (D. C. Dumonde, ed.), p. 58. Springer-Verlag, Heidelberg.

Friedman, H. M. (1966). *Bact. Proc.* 38.

Garcia-Giralt, E., Lasalvia, E., Florentin, I., and Mathe, G. (1970). *Eur. J. Clin. Biol. Res.* **15**, 1012.

Gatti, R. A. (1971). *Lancet* **i**, 1351.

Ginsburg, H., Hollander, N., and Feldman, M. (1971). *J. Exp. Med.* **134**, 1062.

Godal, T., Rees, R. J. W., and Lamvik, J. O. (1971). *Clin. Exp. Immunol.* **8**, 625.

Gordon, J. (1968). *Proc. Soc. Exp. Biol.* (N.Y.) **127**, 30.

Gordon, J., and MacLean, L. D. (1965). *Nature (London)* **208**, 795.

Gothoff, S. D., Lolekha, S., and Dray, S. (1971). *In* "In Vitro Methods in Cell-Mediated Immunity" (B. R. Bloom and P. R. Glade, eds.), p. 327. Academic Press, New York.

Gottlieb, A. A. (1969). *Biochemistry* **8**, 2111.

Gottlieb, A. A., and Glisin, V. (1967). *Fed. Proc.* **26**, 527.

Gottlieb, A. A., and Straus, D. S. (1969). *J. Biol. Chem.* **244**, 3224.

Granger, G. A., Shacks, S. J., Williams, T. W., and Kolb, W. P. (1969). *Nature (London)* **221**, 1155.

Greene, L. A., Tomita, J. T., and Varon, S. (1971). *Exp. Cell. Res.* **64**, 387.

Grumet, F. C., and Leventhal, B. G. (1970). *Transplantation* **9**, 405.

Hartmann, K. U. (1971). *In* "In Vitro Studies of Cellular Cooperation During Immune Induction" (D. C. Dumonde, ed.), p. 22. Springer-Verlag, Heidelberg.

Havemann, K., and Burger, S. (1971). *Eur. J. Immunol.* 1, 285.
Havemann, K., Burger, S., and Dosch, H. M. (1970a). *Z. Ges. Exp. Med.* 153, 308.
Havemann, K., Dosch, H. M., and Burger, S. (1970b). *Z. Ges. Exp. Med.* 153, 297.
Havemann, K., Burger, S., and Dosch, H. M. (1971a). *Acta Haematolica* 46, 282.
Havemann, K., Horvat, M., and Sodomann, C. P. (1971b). *Int. Arch. Allergy* 30, 58.
Havemann, K., Sodomann, C. P., and Horvat, M. (1971c). *Lancet* ii, 980.
Havemann, K., Horvat, M., Sodomann, C. P., and Burger, S. (1972). *Eur. J. Immunol.* 2, 97.
Hersh, E. M., and Harris, J. E. (1968). *J. Immunol.* 101, 1184.
Hirschhorn, K., Bach, F., Kolodny, R. L., and Firschlin, I. L. (1963). *Science* 142, 1185.
Hoffmann, M., and Dutton, R. W. (1971). *Science* 172, 1047.
Hornung, M., and Fritsch, S. (1971). *Nature New Biol.* 230, 84.
Horvat, M., and Havemann, K. (1972). *In* "Leukemia" (R. Gross, Ed.) p. 443. Springer-Verlag, Heidelberg.
Horvat, M., Havemann, K., and Sodomann, C. P. (1972). *Int. Arch. Allergy Appl. Immunol.* 43, 446.
Houck, J. C., Irausquin, H., and Leikin, S. (1971). *Science* 173, 1139.
Ito, T., and Moore, G. E. (1969). *Exp. Cell Res.* 56, 10.
Jacherts, D. (1970). *J. Immunol.* 104, 746.
Jaffe, C. J., Pegram, C. N., and Vazquez, J. J. (1971). *Fed. Proc.* 30, 592 (Abstr.).
Janis, M., and Bach, F. H. (1970). *Nature* (*London*) 225, 238.
Jimenez, L., Bloom, B. R., Blume, M. R., and Oettgen, H. F. (1971). *J. Exp. Med.* 134, 740.
Junge, U., Hoekstra, J., and Deinhardt, F. (1970). *Lancet* ii, 217.
Kamrin, B. (1969). *Trans. Proc.* 1, 506.
Kasakura, S. (1970a). *Nature* (*London*) 227, 507.
Kasakura, S. (1970b). *J. Immunol.* 105, 1162.
Kasakura, S. (1971). *Transplantation* 11, 117.
Kasakura, S., and Lowenstein, L. (1965). *Nature* (*London*) 208, 794.
Kasakura, S., and Lowenstein, L. (1967). *Nature* (*London*) 215, 80.
Kay, J. E. (1971). *Exp. Cell Res.* 68, 11.
Kiger, N. (1971). *Rev. Eur. Etudes Clin. Biol.* 16, 566.
Knowles, M., Hughes, D., Caspary, E. A., and Field, E. J. (1968). *Lancet* ii, 1207.
Krahenbuhl, J. L., and Remington, J. S. (1971). *Infect. Immunity* 4, 337.
Kreisler, M. J., Hirata, A. A., and Terasaki, P. I. (1970). *Transplantation* 10, 411.
Kronman, B. S., Wepsic, H. T., Churchill, W. H., Zbar, B., Borsos, T., and Rapp, H. J. (1969). *Science* 165, 296.
Landureau, J. C., and Steinbruch, M. (1969). *Z. Naturforsch.* 25, 231.
Langner, A., Pawinska-Proniewska, M., Glinski, W., and Maj., S. (1971). *Brit. J. Dermatol.* 85, 7.
Lasalvia, E., Garcia-Giralt, E., and Maciera-Coelho, A. (1970). *Rev. Eur. Etudes Clin. Biol.* 15, 789.
Lawrence, H. S., and Pappenheimer, A. M. (1957). *J. Clin. Invest.* 36, 908.
Lawrence, H. S., and Valentine, F. T. (1970). *Ann. N.Y. Acad. Sci.* 169, 269.
Levene, G. M., Turk, J. L., Wright, D. J. M., and Grimble, A. G. S. (1969). *Lancet* ii, 246.

Leventhal, B. G., and Oppenheim, J. J. (1969). *Proc. Ann. Leukocyte Culture Conf.,*
    3rd (W. D. Rieke, Ed.). Appleton, New York.
Lolekha, S., Dray, S., and Gotsoff, S. P. (1970). *J. Immunol.* **104,** 296.
Mackler, B. F. (1971). *Lancet* ii, 297.
Maini, R. N., Bryceson, A. D. M., Wolstencroft, R. A., and Dumonde, D. C. (1969).
    Nature (London) **224,** 43.
Mandel, P., Mantz, J. M., Deleman, M., Michaelidis, P., Rodesch, J., and Massart,
    A. (1966). *Communaute Eur. Energ. At. Euratom. Eur.* 2477.
Mannick, J. A., and Schmid, K. (1967). *Transplantation* **5,** 1231.
Mannick, J. A., Graxiani, J. T., and Egdahl, R. H. (1964). *Transplantation* **2,** 321.
Marshall, W. H., Valentine, F. T., and Lawrence, H. S. (1969). *J. Exp. Med.*
    130, 327.
Mazzei, D., Novi, C., and Bazzi, C. (1966). *Lancet* ii, 232.
McCluskey, R. T., Benacerraf, B., and McCluskey, J. W. (1963). *J. Immunol.*
    90, 466.
McFarlin, D. E., and Oppenheim, J. J. (1969). *J. Immunol.* **103,** 1212.
Milton, J. D. (1971). *Immunology* **20,** 205.
Mitchison, N. A. (1969). *In* "Mediators of Cellular Immunity" (H. S. Lawrence
    and M. Landy, eds.), p. 73. Academic Press, New York.
Mooney, J. J., and Waksman, B. H. (1970). *J. Immunol.* **105,** 1138.
Moorhead, J. F., Connòlly, J. J., and McFarland, W. (1967). *J. Immunol.* **99,** 413.
Mowbray, J. F. (1963a). *Immunology* **6,** 217.
Mowbray, J. F. (1963b). *Fed. Proc.* **22,** 441.
Nathan, C. F., Karnovsky, M. L., and David, J. R. (1971). *J. Exp. Med.* **133,**
    1356.
Nilsson, G., and Philipson, L. (1968). *Exp. Cell Res.* **41,** 275.
Oppenheim, J. J., Leventhal, B. G., and Hersh, E. M. (1968). *J. Immunol.* **101,**
    262.
Panayi, G. S. (1970). *Brit. Med. J.* ii, 656.
Paque, R. E., Kniskern, P., Dray, S., and Baram, P. (1969). *J. Immunol.* **103,**
    1014.
Paronetto, F., and Popper, H. (1970). *N. England Med. J.* **283,** 277.
Patterson, R. J., and Youmans, G. P. (1971). *Infect. Immunity* **1,** 600.
Perris, A. D., and Whitfield, J. F. (1969). *Proc. Soc. Exp. Biol. Med.* **130,** 1198.
Pick, E., and Turk, J. L. (1972). *Clin. Exp. Immunol.* **10,** 1.
Pick, E., Krejci, J., and Turk, J. L. (1970). *Nature (London)* **225,** 236.
Powles, R., Balchin, L., Currie, G. A., and Alexander, P. (1971). *Nature (London)*
    231, 161.
Puck, T. T., Waldren, C. A., and Jonas, C. (1968). *Biochemistry* **59,** 192.
Remold, H. G., and David, J. R. (1971). *J. Immunol.* **197,** 1090.
Remold, H. G., Ward, P. A., and David, J. R. (1971). *In* "Cell Interactions and
    Receptor Antibodies in Immune Response" (O. Makela, A. Cross, and T. U.
    Kosunen, eds.), p. 411. Academic Press, New York.
Rice, F. A. H., and Ciavarra, R. (1971). *Proc. Soc. Exp. Biol. Med.* **137,** 567.
Riggio, R. R., Schwartz, G. H., Bull, F. G., Stenzel, K. H., and Rubin, A. L.
    (1969). *Transplantation* **8,** 689.
Rocklin, R. E., Meyers, O. L., and David, J. R. (1970). *J. Immunol.* **104,** 95.
Rocklin, R. E., Winston, C., and David, J. R. (1971). *Clin. Res.* **19,** 730.
Roelants, G. E., and Goodman, J. W. (1969). *J. Exp. Med.* **130,** 557.
Rounds, D. E. (1970). *Cancer Res.* **30,** 2847.
Salas, J., and Green, H. (1971). *Nature New Biol.* **229,** 165.

Salvin, S. B., and Nishio, J. (1971). *In* "In Vitro Methods in Cell-Mediated Immunity" (B. R. Bloom and P. R. Glade, eds.), p. 159. Academic Press, New York.

Schachter, M. (1968). *Fed. Proc.* **27**, 49.

Schellekens, P. T. A., and Eijsvoogel, V. P. (1971). *Clin. Exp. Immunol.* **8**, 187.

Scheurlen, R. G., Pappas, A., and Ludwig, T. (1968). *Klin. Worschr.* **46**, 483.

Schmidt, M., Davis, S., and Rubin, A. D. (1972). *In* "Proceedings of the Sixth Leukocyte Culture Conference" (M. R. Schwartz, ed.), p. 297. Academic Press, New York.

Silk, M. (1967). *Cancer* **20**, 2088.

Simon, H. B., and Sheagren, J. N. (1971). *J. Exp. Med.* **133**, 1377.

Smith, R. T., Bausher, J. A. C., and Adler, W. H. (1970). *Amer. J. Pathol.* **60**, 495.

Smith, R. S., McMillan, R. L., Reid, R. T., and Craddock, C. G. (1971). *Int. Arch. All. Appl. Immunol.* **40**, 631.

Snyderman, R., Shin, H. S., and Hausman, M. H. (1971). *Proc. Soc. Exp. Biol. Med.* **138**, 387.

Sonozaki, H., and Cohen, S. (1971). *Cell. Immunol.* **2**, 341.

Spitler, L. E., and Fudenberg, H. H. (1970). *J. Immunol.* **104**, 544.

Spitler, L. E., and Lawrence, H. S. (1969). *J. Immunol.* **103**, 1072.

Stjernholm, R. L., Wheelock, E. F., and Van Den Noort, S. (1970). *J. Reticuloendothel. Soc.* **8**, 334.

Svejcar, J., Johannovsky, J., and Pekarek, J. (1971). *In* "In Vitro Methods in Cell-Mediated Immunity" (B. R. Bloom and P. R. Glade, eds.), p. 263. Academic Press, New York.

Terasaki, P. I., Mottironi, V. D., and Barnett, E. V. (1970). *New England J. Med.* **283**, 724.

Trubowitz, S., Masek, B., and DelRoasario, A. (1966). *Cancer* **19**, 2019.

Tubergen, D. G., Feldman, J. D., Pollocka, E. M., and Lerner, R. A. (1972). *J. Exp. Med.* **135**, 255.

Uhr, J. W., and Moller, G. (1968). *Advan. Immunol.* **8**, 81.

Valentine, F. T., and Lawrence, H. S. (1968). *J. Clin. Invest.* **47**, 989.

Valentine, F. T., and Lawrence, H. S. (1969). *Science* **165**, 1014.

Valentine, F. T. (1971). As summarized in "Progress in Immunology" (B. Amos, ed.), p. 1527. Academic Press, N.Y.

Viza, D. C., Degani, O., Dausset, J., and Davies, D. A. L. (1968). *Nature* (*London*) **219**, 704.

Wallis, C., Ver, B., and Melnik, J. L. (1969). *Exp. Cell Res.* **58**, 271.

Ward, P. A., Remold, H. G., and David, J. R. (1969). *Science* **163**, 1069.

Ward, P. A., Remold, H. G., and David, J. R. (1970). *Cell Immunol.* **1**, 162.

Ward, P. A., Offen, C. D., and Montgomery, J. R. (1971). *Fed. Proc.* **30**, 1721.

Whitfield, J. F., Perris, A. D., and Youdale, T. (1968). *J. Cell. Physiol.* **73**, 203.

Whittaker, M. G., Rees, K., and Clark, C. G. (1971). *Lancet* **i**, 892.

Williams, T. W., and Granger, G. A. (1969). *J. Immunol.* **103**, 1970.

Wilson, D. B., and Nowell, P. C. (1970). *J. Exp. Med.* **131**, 391.

Wilson, D. B., Silvers, W. K., and Nowell, P. C. (1967). *J. Exp. Med.* **126**, 655.

Wolstencroft, R. A. (1971). *In* "Cell Mediated Immunity. In Vitro Correlates" (J. P. Revillard, ed.), p. 130. Karger, Basel.

Wolstencroft, R. A., and Dumonde, D. C. (1970). *Immunology* **18**, 599.

Wolstencroft, R. A., Matthew, M., Oates, C., Maini, R. N., and Dumonde, D. C. (1971). *In* "The Role of Macrophage Lymphocytes in the Immunological Response" (D. C. Dumonde, ed.), p. 28. Springer-Verlag, Heidelberg.

10

POSSIBLE FEEDBACK INHIBITION OF LEUKEMIC CELL GROWTH: KINETICS OF SHAY CHLOROLEUKEMIA GROWN IN DIFFUSION CHAMBERS AND INTRAPERITONEALLY IN RODENTS*

Philip Ferris, Joseph LoBue, and Albert S. Gordon

* The authors' original research reported in this chapter was supported by research grants from the National Heart and Lung Institute (5-R01-HL03357-15) and the National Cancer Institute (1-R01-CA12815-01) of the U.S. Public Health Service.

I. INTRODUCTION

This chapter describes investigations designed to determine if there
are any host influences—particularly feedback inhibition—upon the proliferation of implanted tumors derived from the hematopoietic system.
Using the Shay chloroleukemia grown in Wistar rats, the cell cycle and
subcycle times were investigated by using the fraction (percent) labeled
mitoses autoradiographic technique (FLM) of Quastler and Sherman
(1959) and data for leukemic cells grown both within and out of diffusion chambers were compared. An attempt was also made to determine
whether any differences existed for chamber-grown tumor cells cultured
in normal as opposed to tumor-bearing rodents.

II. MATERIALS AND METHODS

A. Tumor

The chloroleukemia used was obtained from Drs. Eugene and Evelyn
Handler in 1967 from a colony originally established by Drs. B. S. Dornfest, J. LoBue, and A. S. Gordon at New York University with rats
donated by Dr. H. Shay. It has been serially transplanted by subcutaneous injection into 15-day-old rats on a 2-week schedule for over
3 years with 100% success.

B. Diffusion Chamber Cultivation

Diffusion chambers were constructed by a method similar to that
of Nettesheim et al. (1966). Two nylon reinforced Millipore filters (Millipore Filter Co., Bedford, Mass.) of 0.45 μm porosity were cemented
to a 1-holed Lucite ring and sterilized under ultraviolet light for 24–48
hours. The chloroleukemic cells were harvested as required and passed
through a sterile tissue press into a petri dish containing 5–8 ml of
media 199 (with 200 units of penicillin and streptomycin). The cells
were placed into a sterile test tube and briefly centrifuged to separate
the cellular debris and connective tissue from the free cells in the super-

natant fluid. The cells were then transferred to a sterile vial using a 10-ml syringe attached to a Swinny filter containing a stainless steel screen fitted with a 13 gauge trochar. All chambers received 0.1 ml of the cell suspension and were then sealed with MF cement. The filled chambers were floated in sterile media 199 in petri dishes until they were implanted into rats. Sterility was tested by culturing 0.05 ml of the tumor suspension in thioglycolate media for 48 hours at 37°C. Metaphane (Pittman-Moore Co., Indianapolis, Ind.) was used as the anesthetic and the chambers (usually 5 per rat) were introduced into the rat's peritoneal cavity by laparotomy. Two ml of media 199 were injected intraperitoneally after closure of the wound to reduce development of adhesions around the chamber (Johnson *et al.*, 1967). Chamber cell populations were carefully counted so that each experiment was initiated with the same number of cells. The times that the chamber cells were permitted to grow in any host were also kept constant for all experiments.

C. Cell Cycle Analysis (FLM Technique)

The fraction labeled mitosis (FLM) curve is a useful device developed by Quastler and Sherman (1959). It provides a reasonably good method of determining cell cycle parameters in ideal, homogeneous populations. The technique makes use of two "phase" markers, mitosis and S-phase (see Fig. 1). As initially tritiated-thymidine ([^{3}H]Tdr) pulse-labeled S-phase cells progressively enter and leave mitosis, a characteristic wavelike alteration in the percentage of labeled mitotic figures will be observed. Under the most favorable circumstances (when all cells uni-

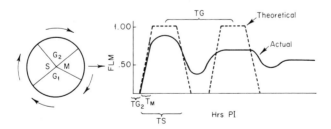

Fig. 1. Graphic illustration of the cell cycle and the FLM technique of cell cycle analysis. The cell cycle, M, mitosis (duration abbreviated, T_m) phase during which karyo- and cytokinesis occurs; G_1. (T_{G1}) the "postmitotic, presynthetic rest period"; S,S-phase (T_s) period during which DNA is replicated as a prerequisite for cell division; G_2:(T_{G2}) the "postsynthetic, premitotic rest period." PI refers to time postinjection of tritiated thymidine. (From, LoBue, 1973, with permission of W. B. Saunders Co., Philadelphia.)

formly proceed through the cell cycle) a perfectly symmetrical, regularly repeating curve will be obtained ("Theoretical" curve, Fig. 1). In practice, this does not occur because phase durations of individual cells vary and a certain amount of "diffusion" and merging of subpopulations occurs ("Actual" curve, Fig. 1). Thus the regular wave form tends to dampen as the population randomizes such that the curve is ultimately transformed into a straight line usually approximating the initial pulse labeling index. From such curves estimates of the generation time (T_g) and its subphase durations may be determined as indicated in Fig. 1.

D. Determination of Cell Cycle Times of Chloroleukemia

Diffusion chambers containing approximately 8×10^6 cells obtained from similarly aged hosts were implanted into normal 100 gm rats. On day 3 after implantation the rats were injected intraperitoneally (IP) with 0.5 μCi [³H]Tdr (specific activity 1.9 Ci/mmole per gram body weight) and sacrificed at 10 minutes, 30 minutes, 1 hour, and at 2-hour intervals thereafter up to 36 hours for estimation of the FLM. For each time interval, chambers were removed, the outsides cleaned with a moist cotton swab, and incubated in pronase for 30 minutes at room temperature to ease removal of adherent cells. The cells were removed and washed twice with sterile rat serum and then spread on clean microslides for autoradiography.

Chloroleukemic cells were also grown in rats IP without chambers. Each rat received 8×10^6 cells and the tumor was allowed to grow for 10–12 days. A pulse label of 1 μCi/gm body weight of [³H]Tdr was then injected IP. Rats were sacrificed following the previous schedule up to 36 hours. The entire tumor was removed and minced to free the cells. The mince was suspended in sterile normal rat serum and slides prepared for autoradiography.

E. Determination of Cell Cycle Time of Chloroleukemic Chamber Cells Grown in Chloroleukemic Rats

Rats were divided into two groups; half were injected IP with 10×10^6 cells and the remainder constituted the uninjected controls. Diffusion chambers containing 8×10^6 chloroleukemic cells were implanted into both groups. In those rats growing tumors IP, the tumor was allowed to progress for 6 days before chamber implantation. Once in the host, they were removed after 3 days growth. Both groups were pulse labeled with 0.5 μCi of [³H]Tdr per gram body weight and slides were made at the previously specified intervals.

F. Cell Cycle Determinations on *Ehrlich* Ascites Carcinoma (EAC) Grown in Chambers within Rats

EAC cells were grown in C_3H mice, aspirated after 7 days growth, and placed into diffusion chambers. The chloroleukemic recipient rats were permitted to grow the leukemia for 6 days before implantation of the chambers. In all rats (leukemic and normal controls), the chambers were left in their hosts for 5 days before removal for cell cycle analysis. Five $\times$ 10^6 cells were inoculated into each chamber.

G. Cell Cycles of Chloroleukemic Cells Grown in Chambers within Swiss Mice with EAC

Diffusion chambers inoculated with 7×10^6 chloroleukemic cells were placed into two groups of Swiss mice. One group comprised the untreated controls, whereas the other group was injected with 0.2 ml of a 1:10 dilution of EAC cells aspirated from a donor mouse. After 5 days growth, the chambers were removed and slides prepared as indicated above.

Labeled mitoses were always enumerated on coded slides so that the sample time interval any slide represented was not known at the time the counts were performed. In addition, all autoradiograms had to conform to the following criteria before the data were used to construct FLM curves: (1) all smears had to contain cells which were morphologically identical to those prepared from the solid tumor growing *in vivo;* (2) there had to be at least 100 mitotic figures for each time interval; (3) the background fogging had to be low enough so that cells containing 5 grains or more could be considered labeled; and (4) in those experiments in which chambers were grown in leukemic animals, all hosts had to have IP tumors for all of the sacrifice time intervals. When some of the IP tumors failed to grow, the entire experiment was excluded from the results.

III. RESULTS

A. FLM Cell Cycle Analysis

Table I summarizes the results obtained from analyses of chloroleukemia grown (1) in diffusion chambers or IP in normal rats, (2) in chambers in chloroleukemic rats, and (3) in chambers in Swiss mice with and without EAC. Also shown are the cell cycle analyses of EAC

TABLE I
SUMMARY OF GROWTH KINETIC PARAMETERS[a]

Method of tumor growth	T_{G2}			T_s		T_m (min.)	T_g (avg.)
	Min.	Avg.	Max.	Min.	Avg.		
Chloroleukemic chambers in normal rats	0.5	1.0	3.0	7.5–8.5	10.5	1.0	15.5
Chloroleukemic cells grown IP in rats	1.0	3.0	5.5	11.0	12.2	—	18.5
Chloroleukemic chambers in chloroleukemic rats	1.0	3.0	6.0	11.0–12.0	13.0	—	18.0
Chloroleukemic chambers in normal Swiss mice	0.5	1.0	3.0	7.0–8.0	9.0	1.0	16.0
Chloroleukemic chambers in EAC Swiss mice	0.5	1.0	3.0	7.0–8.0	10.0	1.0	16.0
EAC in chambers in normal rats	0.5	4.0	6.0	5.3	9.0	—	—
EAC in chambers in chloroleukemic rats	0.5	5.0	6.0	5.3	9.0	—	—

[a] Results are given in hours. Min., avg., max. refer to minimum, average, and maximal estimates, respectively.

grown in chambers in normal and chloroleukemic rats. Presentations of the actual FLM curves from which these data were obtained are given in Figs. 2, 3, and 5–9. These will now be considered individually.

B. Chloroleukemia Cells Grown in Chambers in Normal Rats

In normal rats (Fig. 2, Table I), the first labeled mitoses were observed within 30 minutes after pulse label and the data suggested a minimum duration of G_2 (T_{G_2}) of less than 0.5 hours, an average T_{G_2} of 1 hour, and a maximum T_{G_2} of approximately 3 hours. The minimum DNA-synthesis time (T_s) was approximately 7.5–8.5 hours and averaged about 10.5 hours; mitotic time (T_m) was about 1 hour and the average generation time (T_g), 15.5 hours.

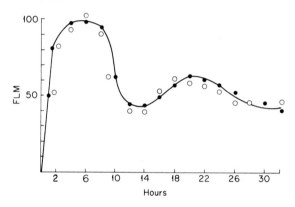

Fig. 2. FLM curve for chloroleukemia grown in diffusion chambers implanted into normal Wistar rats. For Figs. 2, 3, 5–9 each set of points represents a separate experiment and an FLM derived from at least 100 mitotic figures per point.

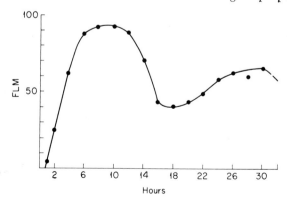

Fig. 3. FLM curve for chloroleukemia grown intraperitoneally in Wistar rats.

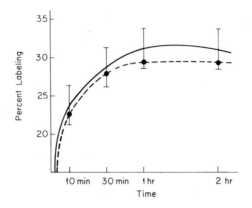

Fig. 4. Availability time for [³H]Tdr incorporation. Solid curve represents uptake by chloroleukemic cells in diffusion chambers in normal rats and dashed curve represents chloroleukemic cells growing intraperitoneally in rats. Vertical lines indicate ±1 standard error of the mean.

C. Chloroleukemic Cells Grown Intraperitoneally

When the chloroleukemic cells were grown IP, (Fig. 3, Table I), the minimum T_{G2} was about 1 hour, averaging 3 hours, with a maximum of 5.5 hours. The minimum T_s was 11–12 hours with an average of 13 hours, and the T_g averaged 18.5 hours. It is important to note that the pulse labeling index and availability time for both the chamber cells and the IP tumor were identical (Fig. 4).

D. Chloroleukemic Cells Grown in Chambers Implanted into Chloroleukemic Hosts

The FLM curves for chloroleukemic cells grown in diffusion chambers and cultured in chloroleukemic rats showed cycle and subcycle durations which were in close agreement with those observed in rats with IP tumors only (Fig. 5, Table I). The minimum T_{G2} was about 1 hour, averaging 3 hours, with a maximum of 6 hours. Minimum T_s was 11–12 hours with an average of 13 hours and an average T_g of 18 hours. These similarities are more easily visualized in Fig. 6 in which the values from Figs. 3 and 5 have been superimposed. When the data for the chamber cells in normal rats are compared by a similar graphic superimposition (Fig. 7), there is a strong indication that the cycle and subcycle durations are different. It would appear, therefore, that the growth pattern of chloroleukemia may be altered when grown in leukemic hosts. This difference seems to be related to the fact that the same kind of

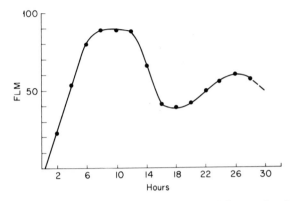

Fig. 5. FLM curve for chloroleukemic cells grown in diffusion chambers implanted in chloroleukemic Wistar rats.

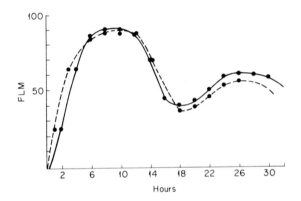

Fig. 6. FLM curves from Figs. 3 and 5 superimposed to indicate the close similarity. Solid line represents intraperitoneal tumor and dashed curve represents chamber cells grown in chloroleukemic hosts.

tumor was growing in the peritoneum, since the presence of the chloroleukemia in chambers in mouse hosts with a different IP tumor (EAC) apparently had no effect on the cycle times of the chloroleukemia (Fig. 8, Table I).

E. The Effect of Intraperitoneal Chloroleukemia on EAC Grown in Diffusion Chambers

The time parameters for the Ehrlich ascites carcinoma cells grown in both normal and leukemic rats, show almost identical FLM curves (Fig. 9). Values for T_{G2}, T_s, and T_g exhibited no significant differences and are similar in duration for values previously described by Baserga

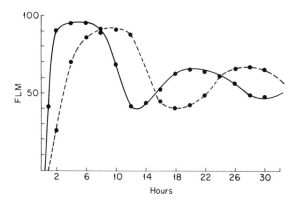

Fig. 7. FLM curves from Figs. 2 and 3 superimposed to indicate differences. Solid curve represents chloroleukemic cells grown in diffusion chambers in normal rats; dashed curve represents chloroleukemic cells grown intraperitoneally.

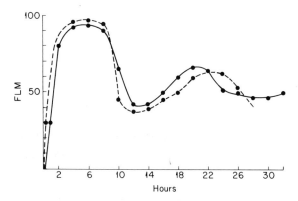

Fig. 8. FLM curves for chamber-grown chloroleukemic cells implanted in Swiss mice. Solid curve represents chamber cells in normal mice and dashed curve, in mice with Ehrlich's ascites carcinoma.

and Lisco (1963). The chambers in many leukemic hosts were surrounded either by ascitic fluid laden with chloroleukemic cells or by solid tumor. As can be seen, the average T_s was between 9 and 10 hours, and the T_g approximately 16 hours.

F. Effect of EAC on Cell Cycle Time of Chloroleukemia Grown in Diffusion Chambers in Swiss Mice

The FLM curves for chloroleukemic cells grown in diffusion chambers in normal and EAC Swiss mice, alluded to earlier, are given in Table I and also illustrated in Fig. 8. In both cases, cell cycle and subcycle parameters resemble each other remarkably well. Moreover, the values

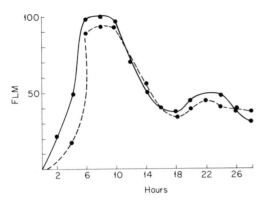

Fig. 9. FLM curves of Ehrlich's ascites cells grown in diffusion chambers implanted in Wistar rats. Solid curve represents chambers grown in normal rats and dashed curve represents chambers grown in chloroleukemic rats.

are almost identical to those seen in Fig. 2 in which the chamber cells were grown in normal rats.

IV. DISCUSSION

Comparison of the data for chloroleukemic cells grown in chambers within normal rats with those for tumors grown IP, suggest a shift to a longer T_{G2}, T_s, and T_g in the latter experiments. When chambers containing chloroleukemic cells were implanted into leukemic hosts, the FLM curve for this population conformed remarkably well to that of the solid tumor. There were several possible reasons for this: (1) the physiological state of the hosts may have been severely impaired by leukemic disease since in almost all cases, rats growing the tumor IP were in a debilitated state; (2) the chamber surface may have been obstructed by solid tumor often observed around them or the outer surface of the membranes coated by tumor cells from the ascitic fluid persistently seen surrounding them; (3) the uptake of [³H]Tdr was diminished in chambers growing in animals with solid tumors when compared to chamber cells grown in normal rats; (4) the growth of the solid tumor in some manner inhibited the growth of the chamber population, reflected by the changes in the cell cycle.

Possibilities (1) and (2) must be discarded from results obtained when EAC cells were grown in diffusion chambers in both normal and chloroleukemic rats. The EAC was selected for several reasons. This tumor had been successfully grown in diffusion chambers (Amos and

Wakefield, 1958) and several cell cycle studies have already been performed (Baserga, 1963, 1965; Baserga and Kisieleski, 1962; Kisieleski
et al., 1961). Moreover, the tumor is easily transplanted into many strains
of mice and the cells grow in an ascitic cell suspension *in vivo*, facilitating
quantitation. When EAC was grown in diffusion chambers in both normal and leukemic rats, *there was no difference in cell cycle time.* To
insure that this was not merely a characteristic of the EAC cells, and
to determine whether chloroleukemic cell growth would be affected
by the simultaneous growth of another tumor in the peritoneum, chambers with chloroleukemic cells were implanted into Swiss mice, one
group of which was normal, and the other bearing EAC. Despite
the observation that the chambers in the tumor-bearing mice were completely surrounded by ascitic fluid laden with EAC cells, both chamber
populations showed remarkable similarity in their FLM curves and the
temporal parameters were well within the range of chloroleukemic cells
grown in diffusion chambers in normal rats.

A third alternate explanation, that there may have been a difference
in the uptake of [³H]Tdr between chamber grown cells and the solid
tumor, was shown to be untenable by the data recorded in Fig. 4. It
is readily seen that there was no significant difference in the [³H]Tdr
uptake between these two tumor systems. Hence any differences in FLM
curves could not be ascribed to differences in [³H]Tdr availability.

Thus, there may be validity to the fourth suggestion, namely, *that
the presence of chloroleukemic cells can somehow retard the growth
of a chamber population of identical cells placed in the same host.*
This may have been nothing more than a manifestation of competitive
inhibition for specific metabolites. On the other hand, the occurrence
of mitotic and DNA synthetic inhibition in normal and malignant tissues
by cell feedback via humoral chalone and chalonelike substances or
specific macromolecules is well documented (Bullough, 1965, 1973;
Bullough and Lawrence, 1968; Houck *et al.*, 1971; Mohr *et al.*,
1968). In fact, Rytomaa and Kiviniemi (1968) have actually postulated
the existence of a chloroleukemic chalone which retards chloroleukemic
cell proliferative activity. Clearly the existence of such a specific humoral
feedback control to explain the data presented herein and its potential
significance for leukemic cell growth will require further investigation.

V. SUMMARY

The cell cycle of chloroleukemic cells was analyzed using the fraction
labeled mitosis technique for cells growing both in diffusion chambers

and intraperitoneally (IP) in rodents. Durations of T_{G2}, T_s and T_g were greater for chloroleukemic cells grown IP than for identical cells grown within diffusion chambers in normal hosts. Similar decreased proliferative activity was also found for chloroleukemic cells in diffusion chambers grown in leukemic rats. The data presented suggest that differences in generation times and subcycle phase durations may involve a feedback inhibition of leukemic cell growth since no such differences were found when Ehrlich ascites carcinoma (EAC) was grown in diffusion chambers in either normal or leukemic rat hosts. In addition, there were no observed differences in cycle time parameters of chloroleukemic cells grown in diffusion chambers in either normal mice or those with EAC.

REFERENCES

Amos, B. D., and Wakefield, J. D. (1958). *J. Nat. Cancer Inst.* 21, 657.
Baserga, R. (1963). *A.M.A. Arch. Pathol.* 75, 156.
Baserga, R. (1965). *Cancer Res.* 25, 581.
Baserga, R., and Kisieleski, W. E. (1962). *J. Nat. Cancer Inst.* 28, 331.
Baserga, R., and Lisco, E. (1963). *J. Nat. Cancer Inst.* 31, 1559.
Bullough, W. S. (1965). *Cancer Res.* 25, 1683.
Bullough, W. S. (1973). This monograph.
Bullough, W. S., and Lawrence, E. B. (1968). *Nature (London)* 220, 134.
Johnson, L. I., Chan, P-C., LoBue, J., and Gordon, A. S. (1967). *Exp. Cell Res.* 47, 201.
Houck, J. C., Irausquin, H., and Leikin, S. (1971). *Science* 173, 1139.
Kisieleski, W. E., Baserga, R., and Lisco, H. (1961). *Atompraxia* 7, 81.
LoBue, J. (1973). *Med. Clin. No. Amer.* 57, 265.
Mohr, V., Althoff, J., Kinzell, V., Suss, R., and Volm, M. (1968). *Nature (London)* 220, 138.
Nettesheim, P. T., Makinodan, T., and Chadwick, C. T. (1966). *Immunology* 11, 427.
Quastler, H., and Sherman, F. G. (1959). *Exp. Cell Res.* 17, 420.
Rytomaa, T., and Kiviniemi, K. (1968). *Nature (London)* 220, 136.

III

HUMORAL CONTROL OF ORGANS AND TISSUE GROWTH

11

THE NERVE GROWTH FACTOR

Ruth Hogue Angeletti, Pietro U. Angeletti, and Rita Levi-Montalcini

I. INTRODUCTION

During the more than 20 years since its discovery, the nerve growth factor (NGF) has indeed been demonstrated to be an essential protagonist in the growth and development of the sympathetic nervous system (Levi-Montalcini and Angeletti, 1968). The NGF-stimulated growth response has been amply described from both metabolic and ultrastructural viewpoints, while, more recently, attention has been focused on the relation of these biological properties to the chemical structure of the NGF molecule.

The earliest observation that implants of mouse sarcoma 180 into 3-day chick embryos became innervated with sensory nerve fibers (Bueker, 1948) led to the discovery that this and other tumors produce a proteinaceous nerve growth factor which causes marked hypertrophic responses in sympathetic and embryonic sensory neurons (Levi-Montalcini and Hamburger, 1951; Levi-Montalcini, 1952; Cohen and Levi-Montalcini, 1956). Further studies revealed that NGF activity is present in larger amounts in the venoms of all three families of poisonous snakes (Cohen, 1959). The most abundant source of NGF so far elucidated is the adult male mouse submandibular gland (Cohen, 1960). This discovery further documented the well-known sexual dimorphism of this tissue (Caramia *et al.*, 1962), for the amount of NGF found in the glands of the adult male is approximately 10-fold higher than in those of the adult female. Because of the availability of NGF from this source, most of the biological, metabolic, and chemical studies to be described below were carried out with this material.

II. STRUCTURAL PROPERTIES OF NGF

In order to facilitate studies on the mechanism of action of NGF, the characterization of the physicochemical properties was begun. The NGF first isolated from adult male mouse salivary glands by Cohen (1960) was a protein moiety with an apparent molecular weight of 44,000. This preparation contained trace amounts of several hydrolytic activities, which did not, however, appear to be responsible for the NGF response. With the advent of more sophisticated techniques of ion-exchange chromatography and gel filtration, it became possible to obtain NGF samples of sufficient purity to permit structural analysis. Two distinct approaches to the purification of NGF have been made.

It has been established that NGF activity can be isolated from salivary gland homogenates in association with a high molecular weight species having a sedimentation coefficient of 7.1 S (Varon *et al.*, 1967a, b; Smith *et al.*, 1968). Upon incubation at mildly acidic or alkaline conditions, this 140,000 molecular weight complex dissociates into three classes of smaller molecular weight subunits: α, β, and γ. The β subunit is the only one possessing inherent NGF activity. Of the other two classes of subunits, the γ is an arginine esterase, while no precise function has yet been determined for the acidic α species. Whereas the β subunit appears to be a uniform, 26,000 molecular weight dimer of identical polypeptide chains (Greene *et al.*, 1971), within the α and γ subunits there exists a marked microheterogeneity (Shooter *et al.*, 1971). The acidic α subunit is isolated as a 29,000 molecular weight moiety and appears to contain two nonidentical polypeptide chains. Slight differences in these component chains, possibly the result of carbohydrate moieties seem to be indicated as the source of the microheterogeneity. The structure of the arginine esterase, or γ subunit, is not yet defined but it, too, has an apparent molecular weight of 30,000. The three subunits, α, β, and γ, can be recombined under conditions of neutral pH to yield only 7 S protein material. The stoichiometric proportions of the three classes of subunits present has not yet been established (Shooter *et al.*, 1971).

By a modification of the procedure of Cohen (1960), Bocchini and Angeletti (1969) obtained in two steps a biologically active moiety with a sedimentation coefficient of 2.5 S and an apparent molecular weight of 30,000. Experiments by Zanini *et al.* (1968) indicated that the NGF could be separated both by gel filtration and by ion-exchange chromatography into two fractions of 28,000 and 14,000 molecular weight. Sedimentation equilibrium analyses in the presence and absence of denaturing agents demonstrated that the NGF as isolated is composed of two subunits of nearly equal molecular weight, associated by noncovalent bonds (R. H. Angeletti *et al.*, 1971). A column fingerprint analysis of the tryptic digest of [¹⁴C]S-carboxymethyl NGF unequivocally established that the native molecule is a dimer of two identical polypeptide chains, each containing three intrachain disulfide bridges. The analysis of the peptides from these experiments, plus those from five other enzymatic digests, permitted the construction of a tentative, self-consistent, covalent structure of the nerve growth factor, as shown in Fig. 1 (R. H. Angeletti and Bradshaw, 1971). The fundamental unit is composed of 118 amino acids with a resultant molecular weight of 13,259, possessing amino terminal serine and carboxy terminal arginine. The native dimer, therefore, has a molecular weight of 26,518, as compared to the molecular weight of 29,000 determined by sedimentation equilibrium measurements (R. H. Angeletti *et al.*, 1971). The amide content of the structure, 6 out of 11 aspartate and 2 out of 8 glutamate residues, is in excellent agreement with the observed isoelectric point of 9.3, determined by isoelectric focusing (Bocchini, 1970). It is of interest to note the distribution of charged residues within the polypeptide chain. The amino terminal portion of the molecule is considerably less basic than the carboxyl terminal region. Conversely, the carboxyl terminal portion of the molecule is more basic in character and, as judged by alignment of the disulfide bonds, exists in a more rigid conformation. The most distinctive feature of the molecule as shown in Fig. 1 is, in fact, the 14-residue loop that results from the formation of two of the disulfide bonds. The disulfide bridges apparently impart a particularly rigid and resistant nature to the NGF molecule, as indicated by its striking resistance to enzymatic, chemical, and heat denaturation (Zanini and Angeletti, 1971).

An unusual feature of the NGF molecule was revealed during sequenator analysis of the intact, S-carboxymethylated protein (R. H. Angeletti

Fig. 1. A schematic representation of the amino acid sequence and disulfide bond pairing of mouse submaxillary gland nerve growth factor. (From Angeletti and Bradshaw, 1971.)

et al., 1973). Samples of NGF as prepared by the Bocchini and Angeletti method (1969), appear to contain equimolar amounts of two polypeptide chains, one of which lacks the first eight amino terminal residues. The mode of generation of these chains and the physiological significance of the phenomenon is not yet understood.

The functional relationship of the NGF's purified by different methods remains to be completely elucidated. However, recent evidence suggests that the primary structure of the β-NGF subunit is very similar to that of the "2.5 S" subunit NGF (personal communication, E. M. Shooter). Preparations of β-NGF, however, appear to be uniformly of the 118 residue length, in contrast to the amino terminal heterogeneity which exists in preparations of 2.5 S NGF. This suggests that the latter may indeed result from a proteolytic and not a genetic event (Bradshaw et al., 1972). Full comparison of the two types of NGF preparations awaits the determination of the complete primary structure of β-NGF. Although exogenous 2.5 S NGF is completely effective in producing the in vivo and in vitro biological effects and, conversely, antiserum to 2.5 S NGF effectively produces immunosympathectomy, the question remains whether it is the monomer, the dimer, or the 7 S moiety which is the physiologically active unit. Although the stoichiometry of the 7 S molecule is not yet known, there is convincing evidence that it can be specifically reconstituted from the component α, β, and γ subunits even in the presence of crude submaxillary gland extract (Shooter et al., 1971). The possibility that the 7 S NGF is, in part, a biosynthetic complex and that NGF may be released from a larger precursor has been suggested by Cohen (1970), in view of the fact that NGF possesses carboxyl terminal arginine, and the associated γ subunit is an arginine esterase.

NGF has also been isolated from the venoms of several representatives of all families of snakes (Cohen, 1959; Banks et al., 1969; R. H. Angeletti, 1970). All are of molecular weight between 20,000 and 28,000. No extensive physicochemical characterization has been performed, but immunochemical analyses indicate that all venom NGF's are cross-reactive among themselves. In addition, there is a 10% cross reaction between Naja naja NGF and mouse salivary gland NGF. The elucidation of the structure of venom NGF and its comparison to mouse NGF may prove useful to understanding the structural requirements for biological activity.

III. DISTRIBUTION OF NGF

The highest concentration of NGF appears in the male mouse submaxillary gland, where it approaches 1% of the total soluble protein (Cohen, 1960). Using a double-faced radioimmunoassay technique, Hendry et al. (1972) reconfirmed this and the sexual dimorphism noted above. As seen in Table I, the amount of NGF protein found in adult male salivary glands by this method is about 3-fold higher in males than in females, and a 100-fold higher than that of prepuberal mice. Small amounts of NGF have also been detected by the bioassay method in several mouse tissues, human blood and extracts of embryonic tissues (Levi-Montalcini and Angeletti, 1961; Bueker et al., 1960).

Although indirect evidence has been presented that NGF is synthesized in the salivary glands (Levi-Montalcini and Angeletti, 1968), the

TABLE I

RADIOIMMUNOASSAY OF NGF CONTENT
OF MOUSE SALIVARY GLANDS[a]

Source of tissue	β-NGF concentration μg/gm wet weight
Prepubertal male and female mice	0.6 ± 0.02
Adult female mice	60.0 ± 3.0
Adult male mice	17.0 ± 24.0

[a] Submaxillary glands were dissected from Swiss Albino mice and homogenized in 10 volumes of veronal buffer and serial dilutions were prepared in this buffer; aliquots of the diluted homogenates were used for radioimmunoassay of β-NGF content (duplicate assay). Results are mean values ±sem for groups of four animals. (From Hendry et al., 1972.)

physiological mode of its release and interaction with sympathetic nerve cells in vivo is as yet unknown. Recent experiments in which peripheral tissues were homogenized in isotonic sucrose buffers and subjected to subcellular fractionation procedures have shown that the NGF noted previously in peripheral tissues appears to be concentrated in the microsomal fractions isolated (P. U. Angeletti and Vigneti, 1971). These results were confirmed both by bioassay and microcomplement fixation techniques. Table II shows the relatively high concentration found in the microsomal pellet from adult mouse heart, spleen and kidney. The

TABLE II

RELATIVE CONCENTRATIONS OF NGF IN THE MICROSOMAL
FRACTIONS FROM MOUSE TISSUES[a]

Tissue	Expt. no.	NGF (ng)	Microsomal fraction (ng)	NGF concentration (mg %)
Heart	1	6	2400	0.25
	2	4.5	3300	0.14
	3	7	1400	0.5
	4	5.4	3200	0.17
Spleen	1	7	3500	0.20
	2	4.5	3200	0.14
	3	5.4	1800	0.30
Kidney	1	4.5	3800	0.12
	2	5.4	2200	0.24
	3	6.4	4500	0.14

[a] (From Angeletti and Vigneti, 1971.)

possibility that NGF is associated with or acts via the adrenergic terminals is being investigated, along with the role this topical location might play in the maintenance and function of sympathetic neurons. Preliminary experiments with radioactively labeled NGF indicate that NGF not only selectively accumulates in the sympathetic ganglia (R. H. Angeletti *et al.*, 1972a), but that the optic adrenergic nerve terminals are capable of taking up and transporting NGF retrogradely to the cell body (Sjöstrand, in preparation).

IV. BIOLOGICAL PROPERTIES OF NGF

NGF evokes a growth response both *in vivo* and *in vitro* from sympathetic nerve cells throughout development and maturation, and from embryonic sensory nerve cells during a very limited period of development (Levi-Montalcini *et al.*, 1972).

The *in vitro* bioassay of NGF uses the 8-day chick embryonic sensory ganglia. After 18–24 hours of incubation at 37°C in plasma clot, hanging drop cultures, an optimal NGF concentration (ng = 1 biological unit) elicits the uniform outgrowth of a "halo" of nerve fibers as shown in Fig. 3. Lesser amounts of NGF cause fewer and longer fibers to appear, gradually diminishing to none, as is the case when NGF is not present in the medium (Fig. 2). Increased concentration of NGF in the culture medium results in a paradoxical effect, namely the decrease in length of the axons of the fibrillar halo, as shown in Figs. 4 and 5, which depict, respectively, the effects of 10 (Fig. 4) and 100 (Fig. 5) biological units of NGF. Dissociated nerve cells cultured in a minimum essential medium in absence of NGF, undergo disintegration and death within 24 hours (Fig. 6), while they survive if the NGF is added at a concentration of 1–10 biological units (Fig. 7) (Levi-Montalcini and Angeletti, 1968). Although used for the bioassay technique, it is not clearly understood why the sensory ganglia, in fact only the mediodorsal cells of the sensory ganglia, are sensitive from just the seventh to the twelfth day of embryonic life. Indeed, why they are sensitive to NGF at all, considering the otherwise apparently high specificity of NGF toward sympathetic nerve cells, is an important, unanswered question.

The response of sympathetic embryonic ganglia in culture is essentially the same as that of the sensory ganglia (Partlow and Larrabee, 1971). *In vitro*, however, they are responsive during all developmental stages. *In vivo*, the NGF effect can be clearly demonstrated by the 3- to 4-fold enlargement of sympathetic ganglia after injection of adult mice with

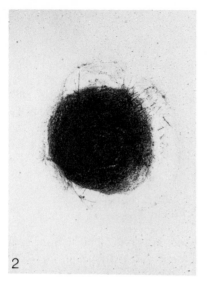
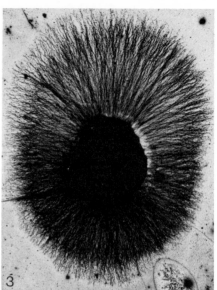

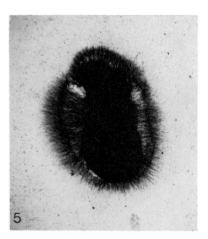

Fig. 2. Whole mount of control 8-day sensory ganglion incubated 24 hours in a control semisolid medium. (From Levi-Montalcini and Angeletti, 1968.)

Fig. 3. Whole mount of 8-day sensory ganglion incubated 24 hours in semisolid medium with one biological unit of NGF. (From Levi-Montalcini, 1964.)

Figs. 4 and 5. Whole mount of 8-day sensory ganglia incubated for 24 hours in semisolid medium in presence of 10 (Fig. 4) and 100 (Fig. 5) biological units of NGF. Note progressive decrease in length and increase in thickness of fibrillar halo.

NGF (10 μg/gm body weight per day), compared to the 10- to 12-fold enlargement of ganglia seen in newborn animals (Levi-Montalcini and Booker, 1960a, b) (Figs. 8–10). It has been shown that the latter is due to an increase both in size and number of sympathetic cells, whereas the former is the result of only cell size increase, the adult cells no longer being subject to mitosis. In addition, a study of the peripheral distribution of sympathetic nerve fibers showed a marked hyperneurotization of the viscera and increased supply of sympathetic nerve fibers around hair bulbs and in the external tunica of blood vessels in the NGF-treated animals as compared to controls (Levi-Montalcini and Booker, 1960a, b). Furthermore, Olson (1967) demonstrated that there is an increased density of the adrenergic fibrillar network in the iris, submaxillary and parotid glands, and in the intramural ganglionic plexuses of the intestinal tract.

Although earlier studies indicated that there was no effect of NGF on higher brain centers when injected peripherally (Levi-Montalcini and Angeletti, 1966), recent studies by Björklund and Stenevi (1972) reveal that NGF injected in the caudal hypothalamus intraventricularly will augment the formation and growth of adrenergic fibers of previously transected axons. It thus appears confirmed that NGF is not just a component in the growth control of the peripheral nervous system, but is most likely an important determinant in the growth and development of parts of the central nervous system as well.

V. IMMUNOSYMPATHECTOMY

Immunosympathectomy, the selective destruction of sympathetic nerve cells by antiserum to NGF, emphasizes the central role played by NGF in the life of these cells (Cohen, 1960; Levi-Montalcini and Angeletti, 1961). Whereas in adult animals the sympathetic ganglia are only reduced in size, in newborn mice, 90–95% of all sympathetic cells are destroyed. Although the precise mechanism by which the antiserum functions is not known, the cytotoxic effects have been described in detail. Within 2 hours of anti-NGF treatment, the superior cervical ganglia of newborn mice already show marked alterations in the fine structure of the nucleoli, the chromatin begins to clump, and the ribosomes appear to become more disorganized (Levi-Montalcini et al., 1969). Within 12 hours, the nucleus shows further atrophy and the cytoplasm begins to show signs of necrosis, particularly with respect to ribosomal structure. Twenty-four hours after injection, the nuclear and cytoplasmic components become mixed. Within 1 week, the ganglia are physically destroyed

238   R. H. Angeletti, P. U. Angeletti, and Levi-Montalcini

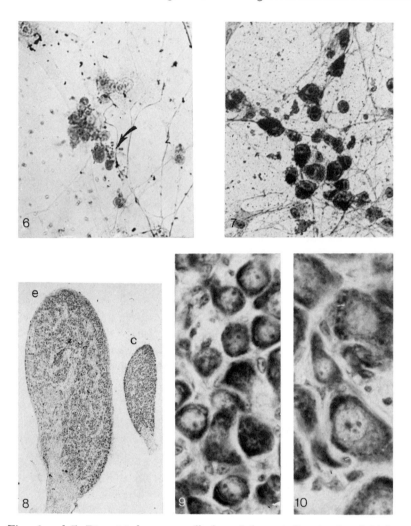

Figs. 6 and 7. Dissociated sensory cells from 8-day ganglion incubated 24 hours in control liquid medium (Fig. 6), and in presence of NGF at a concentration of 0.1 μg/ml (Fig. 7). Arrow in Fig. 6 points to necrotic neurons. (Fig. 6 from Levi-Montalcini, 1964; Fig. 7 from Levi-Montalcini, 1966.)

Fig. 8. Frontal section of experimental (e) and control (c) superior cervical ganglia of 9-day baby mouse. The experimental mouse received daily injections of the NGF from the first to the ninth day. (From Levi-Montalcini, 1966.)

Figs. 9 and 10. Comparison of size of sympathetic neurons in control (Fig. 9) and treated (Fig. 10) 19-day old mice. (From Levi-Montalcini and Cohen, 1960.)

and eliminated, probably through the action of macrophages. In adult mice, there are subtle signs of ultrastructural degeneration in the sympathetic cells within 24 hours, accompanied by impairment of catecholamine storage and uptake. However, these effects in adults appear to be reversible within a few months after cessation of treatment P. U. Angeletti *et al.*, 1971).

VI. METABOLIC EFFECTS OF NGF

The first experiments attempting to elucidate the mechanism of action of NGF were aimed toward the detection of a specific metabolic pathway stimulated by its action. However, only a general overall increase in the cells' secondary metabolic processes has been demonstrated and no isolated event has been singled out as the primary point of intervention by NGF in the nerve cell's growth processes.

Both embryonic chick sensory ganglia and chick sympathetic ganglia have been used to study the metabolic effects of NGF (Cohen, 1959; P. U. Angeletti *et al.*, 1964, 1965; Toschi *et al.*, 1965; Partlow and Larrabee, 1971). Energy is required for the outgrowth of nerve fibers (Cohen, 1959). Glucose utilization in sensory and sympathetic ganglia appears to be stimulated, but primarily through the direct oxidative pathway (P. U. Angeletti *et al.*, 1967; Partlow and Larrabee, 1971). The incorporation of [^{14}C]acetate, primarily into triglycerides, is also markedly increased within 3–4 hours (P. U. Angeletti *et al.*, 1964; Liuzzi and Foppen, 1968). No *de novo* DNA synthesis is detectable, whereas RNA synthesis is markedly enhanced after 2–4 hours. New mRNA synthesis appears to be required for the full expression of the NGF response (halo outgrowth), as shown by the effect of actinomycin D on the *in vitro* cultures. However, some fiber formation still results even in the presence of this inhibitor, indicating that this is not the primary response to NGF (Larrabee, 1972; Levi-Montalcini and Angeletti, 1971). There is a 2-fold increase of polysome formation in NGF-treated ganglia after 3 hours, but there are only slightly higher ratios of polysomes:monomers and no difference in the proportions of free and membrane-bound ribosomes over controls (Amaldi, 1971). Morphologically this is seen to be manifested by markedly increased numbers of polysomal clusters in electron microscope studies of NGF-treated ganglia (Crain *et al.*, 1964; Levi-Montalcini *et al.*, 1968).

As might be expected from the increase in polysomes, protein synthesis appears to be required for any response to NGF at all. There is a net increase in the incorporation of ^{14}C-labeled amino acids into protein after NGF treatment for 6–8 hours, with a concomitant rise in the total protein content of the ganglia. Both amino acid incorporation and nerve fiber outgrowth are eliminated by the use of the specific protein synthesis inhibitors, puromycin, and cycloheximide. Fractionation of labeled proteins extracted from ganglia incubated in the presence or absence of NGF indicates that there is a preferential increase of incorporation into acidic proteins (Gandini-Attardi et al., 1967). It is of interest to note that there is an increased number of microtubules, composed of acidic polypeptide units, present in the cytoplasm (Levi-Montalcini et al., 1968). It has recently been established by Hier et al. (1972) that this augmentation of neurotubular elements is preceded by de novo synthesis of new microtubular polypeptide subunits within 3–6 hours of incubation with NGF (Table III). This result is of particular significance in view of the report by Roisen et al. (1972) that cyclic AMP in high concentrations stimulates neurite outgrowth from sensory ganglia, similarly to NGF. Hier and his collaborators clearly demonstrated that cAMP stimulation is not followed by an increase in microtubular subunits in the cytoplasm, thus indicating that cAMP and NGF stimulated fiber outgrowth probably occur by different mechanisms. That neurofilament production is essential to the process of fiber growth was emphasized by the blockage of the NGF effect using inhibitors of neurotubule subunit

TABLE III

EFFECT OF NGF ON NEUROTUBULE
PROTEIN CONTENT OF GANGLIA
TREATED WITH VINCRISTINE[a]

Preparation	Colchicine-binding activity per ganglion (cpm)
NGF	2200; 2400
NGF + vincristine	1900; 2700
Vincristine	1000; 1500
Control	1050; 1750

[a] Ganglia were treated either with NGF (3 units/ml), vincristine (0.1 μM), or NGF plus vincristine. Control ganglia were untreated. After 18 hours at 37°, ganglia were assayed for colchicine-binding activity. Data are presented for two separate experiments. (From Hier et al., 1972.)

polymerization, colchicine, and vinblastine (Levi-Montalcini and Angeletti, 1970).

The entire cellular synthetic apparatus thus seems to be turned on, as manifested ultrastructurally by the increased number of nucleoli, the rise in free and membrane-bound ribosomes and the massive amounts of neurotubules (Figs. 11 and 12), yet none of these appears to be an early primary event. No clear answer as to whether these processes cause or result from nerve fiber growth stimulation has yet been obtained.

The stimulatory effect of NGF on sympathetic nerve cells has also been demonstrated for the pathway which distinguishes their functional activity, i.e., the synthesis of adrenergic neurotransmitter. The total content of norepinephrine in sympathetic ganglia from normal and NGF-treated mice was markedly increased and, in addition, the ratio of norepinephrine to total protein was also significantly higher (Crain and Wiegand, 1961). Studies of the enzymes in the biosynthetic pathway of norepinephrine have been carried out by Thoenen *et al.* (1971). Superior cervical ganglia from 7-day old rats treated since birth were analyzed for several enzymatic activities (Table IV). Tyrosine hydroxylase activity was 15- to 20-fold greater in NGF-treated ganglia, with a concomitant 3- to 4-fold apparent increase in specific activity. While the K_m of this rate-limiting enzyme remained constant, the V_{max} was markedly increased. Dopamine β-hydroxylase, a membrane-bound enzyme, also appears to be selectively induced about 10-fold. On the contrary, the authors could account for the slight increase noted for dopa decarboxylase and monoamine oxidase by a relative increase in the ratio of neuronal versus satellite cells. No direct effect of NGF on these enzymes could be demonstrated *in vitro,* nor was evidence found for control by NGF of activators or inhibitors of the above enzymes.

Further experimentation by Thoenen *et al.* (1972) revealed that, in addition to the increase of noradrenergic synthesizing enzymes, there is a corresponding increase in choline acetyltransferase activity of the superior cervical ganglion of the rat. The authors showed that this enzyme was indeed localized in the cholinergic sympathetic nerve terminals innervating the superior cervical ganglion. It was noted, however, that cholinergic terminals innervating other cholinergic neurons or other tissues, such as in the heart, do not show a similar increase in choline acetyltransferase. The NGF, therefore, does not appear to have a direct action on the cholinergic cells or on this enzyme, but achieves this effect transsynaptically through the adrenergic-cholinergic cell interaction.

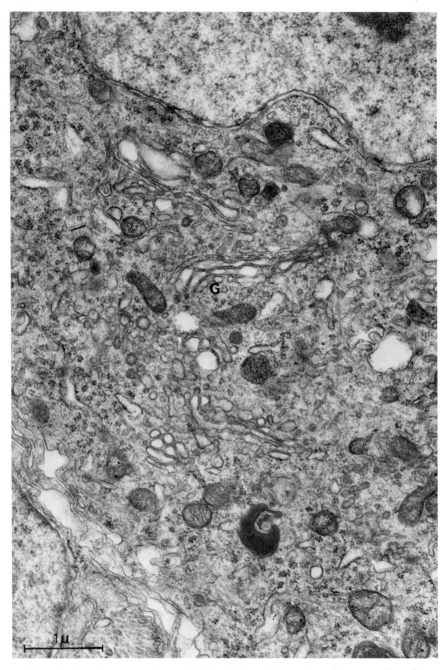

Fig. 11. Dorsal root ganglion of 8-day chick embryo cultured in control medium for 4 hours. The Golgi membranes (G) are prominent.

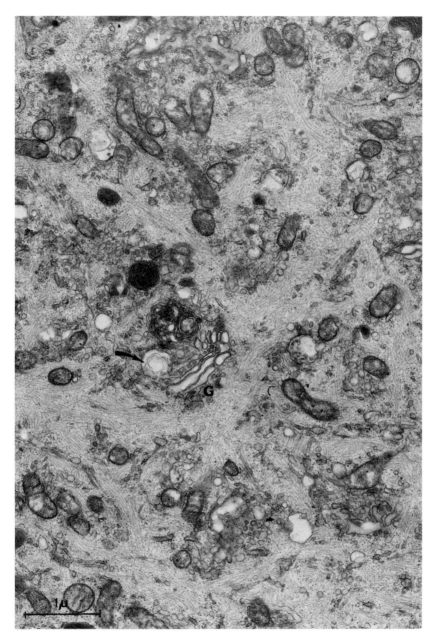

Fig. 12. Ganglion of 8-day chick embryo cultured for 12 hours in presence of NGF. Neurotubules and neurofilaments fill practically all the cytoplasmic area. The Golgi apparatus (G) is prominent with numerous cytoplasmic vesicles and vacuoles. Arrow points to a coated vesicle. (From Levi-Montalcini et al., 1968.)

TABLE IV

Effect of NGF on Specific and Total Activities of Enzymes in Superior Cervical Ganglia of Newborn Rats

Enzyme[a]	Controls		NGF-treated	
	Specific	Total	Specific	Total
Tyrosine hydroxylase	1.9 ± 0.06	0.10 ± 0.004	8.4 ± 0.9	1.9 ± 0.02
Dopamine β-hydroxylase	11.8 ± 0.1	1.1 ± 0.04	42.0 ± 2.6	13.0 ± 0.6
Dopa decarboxylase	0.32 ± 0.03	0.02 ± 0.001	0.51 ± 0.03	0.13 ± 0.07
Monoamine oxidase	160 ± 8	23.5 ± 2.2	189 ± 12	57.8 ± 8.5

[a] Activities (mean ± SE, n = 6–8) are amounts of product formed per hour per milligram protein (specific) and amounts of product formed per hour per pair of ganglia (total); tyrosine hydroxylase, nmoles of dopa; dopamine β-hydroxylase, pmoles of octopamine; dopa decarboxylase, μmoles of dopamine; monoamine oxidase, nmoles of indoleacetic acid. (From Thoenen et al., 1971.)

VII. CONCLUDING REMARKS

Recent advances in the analysis of the structure of NGF and in the more detailed investigation of its metabolic effects on the target nerve cells have not solved the question of the fundamental mechanism of action of NGF, but have produced stimulating insights into means of approaching this problem. Of particular interest is the observation that the covalent structure of NGF displays a striking homology to the primary and secondary structure of proinsulin (Frazier *et al.*, 1972). The possibility that NGF first interacts with the sympathetic nerve cell through a membrane-bound receptor as has been demonstrated for insulin by Cuatrecasas (1969, 1972), is currently being investigated in several laboratories. If this is indeed the case, then studies on the molecular interaction of NGF and receptor and identification of the subsequent cellular events should provide new answers and propose new problems in the spatial and temporal control of cellular differentiation.

ACKNOWLEDGMENTS

This work has been supported in part by grants from the National Institutes of Health, USPHS (NS-03777) and from the National Science Foundation (GB-16330X).

REFERENCES

Amaldi, P. (1971). *J. Neurochem.* 18, 827.
Angeletti, P. U., and Vigneti, E. (1971). *Brain Res.* 33, 601.
Angeletti, P. U., Liuzzi, A., Levi-Montalcini, R., and Gandini-Attardi, D. (1964). *Biochim. Biophys. Acta* 90, 445.
Angeletti, P. U., Liuzzi, A., and Levi-Montalcini, R. (1965). *Biochim. Biophys. Acta* 84, 778.
Angeletti, P. U., Calissano, P., Chen, J. S., and Levi-Montalcini, R. (1967). *Biochim. Biophys. Acta* 147, 180.
Angeletti, P. U., Levi-Montalcini, R., and Caramia, F. (1971). *J. Ultrastruct. Res.* 36, 24.
Angeletti, R. H. (1970). *Proc. Nat. Acad. Sci. U.S.* 65, 668.
Angeletti, R. H., and Bradshaw, R. A. (1971). *Proc. Nat. Acad. Sci. U.S.* 68, 2417.
Angeletti, R. H., Bradshaw, R. A., and Wade, R. G. (1971). *Biochemistry* 10, 463.
Angeletti, R. H., Angeletti, P. U., and Levi-Montalcini, R. (1972a). *Brain Res.* 46, 421.

Angeletti, R. H., Hermodson, M., and Bradshaw, R. A. (1973). *Biochemistry* **12**, 100–115.
Banks, B. E. C., Banthorpe, D. U., Berry, A. R., Davies, H. S., Doonan, S., Lamont, D. M., and Vernon, C. A. (1969). *Biochem. J.* **108**, 157.
Björklund, A., and Stenevi, U. (1972). *Science* **175**, 1251.
Bocchini, V. (1970). *Eur. J. Biochem.* **15**, 127.
Bocchini, V., and Angeletti, P. U. (1969). *Proc. Nat. Acad. Sci. U.S.* **64**, 787.
Bradshaw, R. A., Frazier, W. A., and Angeletti, R. H. (1972). *In* "The Chemistry and Biology of Peptides" (J. Meienhofer, ed.), Ann Arbor Science Publ., Ann Arbor pp. 423–439.
Bueker, E. D. (1948). *Anat. Rec.* **102**, 369.
Bueker, E. D., Schenkein, I., and Bane, J. L. (1960). *Cancer Res.* **20**, 1220.
Caramia, F., Angeletti, P. U., and Levi-Montalcini, R. (1962). *Endocrinology* **70**, 915.
Cohen, S. (1959). *J. Biol. Chem.* **234**, 1129.
Cohen, S. (1960). *Proc. Nat. Acad. Sci. U.S.* **46**, 302.
Cohen, S. (1970). *Proc. Nat. Acad. Sci. U.S.* **67**, 164.
Cohen, S., and Levi-Montalcini, R. (1956). *Proc. Nat. Acad. Sci. U.S.* **42**, 571.
Crain, S. M., and Wiegand, R. G. (1961). *Proc. Soc. Exp. Biol. Med.* **107**, 571.
Crain, S. M., Benitez, H., and Vatter, A. E. (1964). *Ann. N.Y. Acad. Sci.* **118**, 206.
Cuatrecasas, P. (1969). *Proc. Nat. Acad. Sci. U.S.* **63**, 450.
Cuatrecasas, P. (1972). *Proc. Nat. Acad. Sci. U.S.* **69**, 318.
Frazier, W. A., Angeletti, R. H., and Bradshaw, R. A. (1972). *Science* **176**, 482.
Gandini-Attardi, D., Calissano, P., and Angeletti, P. U. (1967). *Brain Res.* **6**, 367.
Greene, L. A., Varon, S., Piltch, A., and Shooter, E. M. (1971). *Neurobiology* **1**, 1.
Hendry, I., Addison, G. M., and Iversen, L. L. (1972). *In* "Nerve Growth Factor and its Antiserum" (E. Zaimis, ed.), pp. 262–270. Athlone Press, London.
Hier, D. B., Arnason, B. G. W., and Young, M. (1972). *Proc. Nat. Acad. Sci. U.S.* **69**, 2268.
Larrabee, M. G. (1972). *In* "Nerve Growth Factor and its Antiserum" (E. Zaimis, ed.), pp. 71–88. Athlone Press, London.
Levi-Montalcini, R. (1952). *Ann. N.Y. Acad. Sci.* **55**, 330.
Levi-Montalcini, R. (1964). *Science* **143**, 105.
Levi-Montalcini, R. (1966). *Harvey Lect.*, Ser. **60**, 217.
Levi-Montalcini, R., and Angeletti, P. U. (1961). *In* "Regional Neurochemistry" (S. S. Kety and J. Elkes, eds.), pp. 362–376. Pergamon, Oxford.
Levi-Montalcini, R., and Angeletti, P. U. (1966). *Pharmacol. Rev.* **18**, 69.
Levi-Montalcini, R., and Angeletti, P. U. (1968). *In* "Growth of the Nervous System" (G. E. W. Wolstenholme and M. O'Connor, eds.), pp. 126–147. Churchill, London.
Levi-Montalcini, R., and Angeletti, P. U. (1970). *Proc. Nat. Acad. Sci. U.S.* **67**, 7A.
Levi-Montalcini, R., and Angeletti, P. U. (1971). *In* "Hormones in Development" (M. Hamburgh and E. J. W. Barrington, eds.), pp. 719–730. Appleton, New York.
Levi-Montalcini, R., and Booker, B. (1960a). *Proc. Nat. Acad. Sci. U.S.* **46**, 373.
Levi-Montalcini, R., and Booker, B. (1960b). *Proc. Nat. Acad. Sci. U.S.* **46**, 384.
Levi-Montalcini, R., and Cohen, S. (1960). *Ann. N.Y. Acad. Sci.* **85**, 324.

Levi-Montalcini, R., and Hamburger, V. (1951). *J. Exp. Zool.* 116, 321.
Levi-Montalcini, R., Caramia, F., Luse, S. A., and Angeletti, P. U. (1968). *Brain Res.* 8, 347.
Levi-Montalcini, R., Caramia, F., and Angeletti, P. U. (1969). *Brain Res.* 12, 54.
Levi-Montalcini, R., Angeletti, R. H., and Angeletti, P. U. (1972). In "Structure and Function of Nervous Tissue" (G. H. Bourne, ed.), Vol. 5, pp. 1–38. Academic Press, New York.
Liuzzi, A., and Foppen, F. H. (1968). *Biochem. J.* 107, 191.
Olson, L. (1967). *Z. Zellforsch. Mikrosk. Anat.* 81, 155.
Partlow, L., and Larrabee, M. G. (1971). *J. Neurochem.* 18, 2101.
Roisen, F. J., Murphy, R. A., Pichichero, M. E., and Braden, W. G. (1972). *Science* 175, 73.
Shooter, E. M., Bamburg, J., Perez-Polo, J. R. Piltch, A., and Strauss, D. (1971). *Abstr. Int. Meeting Int. Soc. Neurochem., 3rd* p. 428.
Sjöstrand, J. (in preparation).
Smith, A. P., Varon, S., and Shooter, E. M. (1968). *Biochemistry* 7, 3259.
Thoenen, H., Angeletti, P. U. and Levi-Montalcini, R. (1971). *Proc. Nat. Acad. Sci. U.S.* 68, 1598.
Thoenen, H., Saner, A., Angeletti, P. U., and Levi-Montalcini, R. (1972). *Nature* (*London*) 236, 26.
Toschi, G., Dore, E., Angeletti, P. U., Levi-Montalcini, R., and de Haen, C. H. (1965). *J. Neurochem.* 13, 539.
Varon, S., Nomura, J., and Shooter, E. M. (1967a). *Biochemistry* 6, 2202.
Varon, S., Nomura, J., and Shooter, E. M. (1967b). *Proc. Nat. Acad. Sci. U.S.* 57, 1782.
Zanini, A., and Angeletti, P. U. (1971). *Biochim. Biophys. Acta* 229, 724.
Zanini, A., Angeletti, P., and Levi-Montalcini, R. (1968). *Proc. Nat. Acad. Sci. U.S.* 61, 835.

12

HUMORAL ASPECTS OF LIVER REGENERATION

Frederick F. Becker

I. INTRODUCTION

The adult mammalian hepatocyte is structurally and functionally complex. It is mitotically quiescent, yet it retains the capacity to divide in response to numerous stimuli. The alteration which most frequently provokes hepatocyte division is a loss of other hepatocytes, whether such loss is focal or widespread. It has been suggested that humoral mechanisms control or participate in the panhepatic mitotic activity which follows amputation, toxic or infectious "removal" of a threshold number of hepatocytes. Following such cell loss the residual hepatocytes enter the mitotic cycle and remain mitotically active until the normal hepatocyte number is approximated, after which they return to a non-mitotic state. This complex phenomenon, which has been termed liver regeneration, has been reviewed (Bucher, 1967, 1971; Becker, 1970).

The stimulus most frequently used in the experimental analysis of liver regeneration is that of surgical amputation of liver parenchyma (hereafter referred to as partial hepatectomy). The results of this proce-

250   *Frederick F. Becker*

dure appear to be similar in mouse and man. The pivotal aspects of
the response to partial hepatectomy can be summarized as follows:

a. The alterations which take place following partial hepatectomy
do not represent wound-healing in the usual sense. They occur equally
throughout all of the residual parenchyma and require that a threshold
cell number be removed before they take place.

b. Although a period of morphological simplification of the residual
hepatocytes occurs soon after the operation, no major functional deficit
has been reported. Indeed, these cells may perform normal functions
at a heightened level during their preparation to divide (Teebor *et
al.*, 1967; Asofsky and Becker, 1972).

c. The metabolic alteration which best characterizes the hepatocytes' preparations for division is the appearance of ribonucleotide re-
ductase. This enzyme, absent from the nondividing cell, signals the appearance of the metabolic apparatus responsible for DNA synthesis.

d. Concurrent with the impulse to divide, to undergo hyperplastic
increase, the hepatocytes also enlarge and undergo hypertrophy alteration. It is the hyperplastic thrust of the residual hepatocytes which de-
fines the regenerative activity of the liver. It is imperative, therefore,
that this response alone be quantitated in examining mechanisms which
may influence liver regeneration, since it has been clearly demonstrated
that hypertrophy results from a separate set of stimuli (Becker, 1963;
Sigel *et al.*, 1967). Indeed, cell division can occur in the presence
of a rapidly decreasing liver weight. The hyperplastic response appears
to be related to interaction of blood-components with hepatocytes, while
hypertrophy or atrophy appear to result from blood flow:hepatocyte
relationships.

e. Following several mitotic waves, which result from one to two
hepatocyte divisions, mitotic quiescence ensues.

Over the years, many stimuli of liver regeneration have been suggested
and these can be grouped roughly as follows: injury, blood flow, humoral
factors and/or summated physiological alteration. There is now no evidence to support either injury or blood flow as obligatory participants
in this phenomenon. The last suggestion, summated alteration which
will be discussed later, has scant supporting evidence.

There is a considerable body of evidence in the literature that humoral
mechanisms participate in liver regeneration after partial hepatectomy.
Stimulation of "liver-derived cells" in culture by post-hepatectomy serum
(Wrba *et al.*, 1962; Grisham *et al.*, 1967; Hays *et al.*, 1969), stimulation
of grafted liver tissue (Sigel *et al.*, 1963; Virolainen, 1964; Leong *et
al.*, 1964), and of parabiotic normal livers (Moolten and Bucher, 1967;
Sakai, 1970), have suggested participation of blood-borne factors.

Three types of humoral controls have been suggested.

A. *Appearance of a mitotic stimulant in response to partial hepatectomy.* This proposal suggests that the liver, or another organ, responds to partial hepatectomy by the secretion of a mitotic stimulant which affects the hepatocytes. This model corresponds roughly to that of erythropoietin in which the stimulant has an origin in one organ and its target in another; or could be the result of a liver produced stimulant acting upon liver.

B. *Stimulation resulting from imbalance between normal blood components and hepatocyte number.* The operative feature of this proposal is the interaction of a normal component of blood with hepatocytes despite the absence of any alteration in its (absolute) level after partial hepatectomy. In this instance the severe reduction in hepatocyte number would create a relative increase in the level of the stimulant per hepatocyte exposed; this imbalance between it and its target cell would result in cell division.

C. *Diminution of circulating inhibitor.* This hypothesis is comparable to that of epidermal chalone. It is based on the proposal that one or more of the immumerable liver export substances, such as plasma proteins, act(s) as a mitotic inhibitor. Following partial hepatectomy the plasma levels of this substance would decline (especially if its turnover was rapid) and the hepatocytes would be "released" to divide.

In the majority of experiments reported it has been impossible to distinguish between proposals A and C. Until a specific humoral factor is isolated, this problem will persist.

II. EXPERIMENTAL RESULTS

In the complex area of humoral control of liver regeneration few experimental results have withstood the test of time, and occasionally the time involved has been embarrassingly short. A limited number of results remain generally accepted, such as the finding that the removal of many of the major endocrine glands does not significantly alter the regenerative response (Becker, 1970). The normal liver of parabiotic rats has been consistently stimulated by partial hepatectomy of its partner, and its mitotic response is roughly proportion to the extent of amputation (Moolten and Bucher, 1967; Fisher *et al.*, 1971a). In our own hands, removal of the spleen, thymus, or kidney had no effect on regeneration nor did exteriorization of hepatic lymph (Becker, 1970), while germfree rats responded in a normal fashion (Asofsky and Becker, 1972).

It is with a resounding thud that any concept of general acceptance of experimental findings must end after this brief list. Let us try then to group the various conflicting experimental results according to the problems which they attempted to solve.

Proposal A. A circulating stimulant might arise in the residual hepatocytes after partial hepatectomy. This possibility seemed totally refuted
by the work of Fisher *et al.* (1971a) which achieved maximal mitotic
stimulation of the normal liver after 100% hepatectomy of a parabiotic
partner. Clearly, the "residual hepatocytes" did not exist and could not
secrete anything. Yet, Levi and Zeppa (1971) have been able to stimulate
DNA synthesis in isolated, normal livers by cross-perfusion with the
residual tissue of a partially hepatectomized liver. Further, they could
evoke a similar response in the normal liver exposed only to perfusate
which had been exposed to the residual hepatocytes. Their conclusion,
that the liver *does* secrete a mitotic stimulant in response to partial
hepatectomy. In their analysis of Fisher's 100% hepatectomy-parabiotic
stimulation, they concluded that the normal liver itself, hooked into
two rats, was the "residual" liver; the equivalent of a 50% hepatectomy.

The work of Levi and Zeppa (1971), however, is also vulnerable
to questioning. After an extremely brief exposure to the "prepared" perfusate or to cross-perfusion with partially hepatectomized liver, DNA
synthesis is detectable in the hepatocytes of the normal liver. Since
DNA synthesis requires the presence of deoxyribonucleotides as well
as DNA polymerase this response would involve their synthesis and
transport to the nucleus at an incredible rate. This dichotomy will require further clarification.

Proposal B. Imbalance of components producing stimulation. As a
result of other experiments, Fisher *et al.* (197b) have concluded that
the stimulus producing the regenerative response of liver is a factor
in normal portal blood. The diversion of portal blood from a partially
hepatectomized, portacaval shunted rat to a parabiotic partner caused
maximal mitotic response in the normal liver. These experiments further
suggested that hepatocytes inactivate this factor so that passage of portal
blood through any liver tissue decreases its stimulatory activity propor-
tionate to the number of hepatocytes traversed. This is certainly in keeping with the finding of Sigel *et al.* (1968) that the stimulus to mitotic
activity of residual hepatocytes resided in portal blood; and that stimulation diminished progressively as the blood traversed the liver lobule.
This was true whether the ingress of portal blood was via central or
portal vein. As a result of their own studies, Fisher's group concluded
that stimulation of liver regeneration results from an imbalance between
a factor found in normal portal blood and the number of residual
hepatocytes.

Other investigators, however, have less evidence for the inactivation
by liver of this portal blood factor or of any stimulatory factor. The
work of Moolten and Bucher (1967) and the impressive interaortic ex-

change experiments of Sakai (1970) suggest very significant stimulation of the normal parabiotic liver despite prior passage of portal blood through the operated liver. In response to this possibility, Fisher *et al.* (1971a) suggest an additive effect of portal factor which eludes liver inactivation in the partially hepatectomized liver plus that of the portal blood of the normal liver.

However, the failure of Moolten and Bucher (1967) to detect a greater than normal mitotic response in parabiotic rats, where both were partially hepatectomized (indeed, it was 50% less that of a single operated rat) lessens the probability that parabiotic liver was responding to the combined level of portal blood factors. In all parabiotic studies, the total mitotic response of the livers is invariably less than that which would be expected by an additive effect of portal blood factors.

More difinitively damaging to the concept of stimulation by a portal blood factory is the demonstration by Price *et al.* (1971) that a normal regenerative response took place in the absence of all portal organs. This study apparently refutes any possible intervention by a portal factor.

Proposal C. A circulating inhibitor. The work of Fisher *et al.* (1971b) did strongly decrease the likelihood that the regenerative response is the result of a diminished level of circulating inhibitor. A 30% liver transplant in the presence of a normal host liver remained capable of a normal regenerative response.

Until this time the work of Grisham *et al.* (1966) had suggested the presence of a mitotic inhibitor in normal rat blood. The exchange-transfusion of normal blood into partially hepatectomized rats delayed the mitotic response of the latter. More importantly, when the transfused blood was obtained from partially hepatectomized rats inhibition did not occur. This conflict of results cannot be explained.

Levi and Zeppa (1971) concluded that the liver-secreted stimulatory factor was the result of the prolonged, heightened metabolic load upon the liver during the first 12 postoperative hours. This suggestion is similar to that previously offered: that the initiating factor in liver regeneration was a summated physiological challenge (Becker, 1970). However, both aspects of this concept are incompatible with the work of Fisher *et al.* (1971b) in which the 30% transplant regenerated in the presence of a normal liver (a total presence of 130% of functional liver).

III. CONCLUSION

This then is the confusing and frustrating picture of the humoral control of liver regeneration. The tremendous complexity of most of the experimental models evokes our admiration for the skills of the investigators but inhibits attempts of other groups to reproduce the studies. Almost all of these recently reported experiments suggest that humoral substances control hepatocyte division. There has been, however, no progress in identifying a stimulant or an inhibitor involved in this process.

Several obstacles loom which may long prevent the isolation and identification of the "control" factor(s). It appears likely that such factors exist at low concentration, that they are labile and rapidly catabolized, and that for these, or other reasons, the hepatocytes must be exposed to them for prolonged periods.

It is impossible, at this time, to reconcile much of the experimental data from different laboratories and to propose a unified theory for the humoral control of liver regeneration. I cannot accept the concept that the conflicting results are the result of species or sex differences. Whenever a broad principal of biological control has been revealed it has proven to be common to all species, i.e., erythropoietin, chalone, trophic hormones, etc. I would suggest certain conceptual proposals that might enable us to explore these differences more rationally.

Schema for Humoral Control of Liver Regeneration

There is a strong possibility that the regenerative response is multiphasic in relation to humoral control as it is biochemically and cytologically. Thus, an initiation phase would take place during the first 12 hours after partial hepatectomy; the response phase in the next 24 hours, during which DNA synthesis commences and then a control phase, that period in which mitosis ceases.

1. INITIATION PHASE

Initiation might result from absolute or relative alterations in substances already present in the blood. As has been suggested, the decrease in level of a rapidly catabolized, liver-synthesized plasma protein might result in release of the hepatocytes from mitotic inhibition. Similarly, increased exposure of hepatocytes to a normal blood component, due to the reduction in their number, rather than the substance's absolute increase, could achieve the same effect.

We suggest that a combination of these events would also result in hepatocyte stimulation. Thus, the diminution of a plasma protein which normally inhibits the stimulatory action of the normal blood factor would result in hepatocyte mitosis. The reduction of a plasma protein which normally acted as a binding transport vehicle for the latter would allow it to circulate in an *elevated, unbound* and possibly stimulatory form.

2. RESPONSE PHASE

There is every possibility that this phase results from a secretory stimulant arising from the initiated liver. This material may well result in DNA synthesis, and it remains possible that it is required for mitotic division to ensue (Levi and Zeppa, 1971). The interaction of this substance with other(s) of normal blood is possible.

3. CONTROL PHASE

During this phase of liver regeneration DNA synthesis and mitotic activity cease. It is most likely that cessation results from reversal of one or more of the control interactions implicated above. Or, in view of the reported refractory period during which recently mitotic hepatocytes cannot be further stimulated, it is possible that the stimulatory mechanisms are no longer effective. One further hypothesis, without experimental foundation, is the possibility that the liver now secretes yet another protein which results in mitotic inhibition. The heightened synthesis of at least one interesting plasma protein has been reported by Asofsky and Becker (1972) during this period.

In view of the enormously complex physiological alterations induced by partial hepatectomy and our knowledge that many different challenges may instigate hepatocyte division, we must remain exceedingly cautious in accepting the results of highly complicated biological experiments as "revealing" the single factor involved. In view of the conflicting results of experiments in this field it remains possible that the residual hepatocytes "sense" the loss of others in a manner akin to the intercellular communication of Loewenstein and Penn (1967) or by means as yet hidden. Perhaps the liver "knows."

REFERENCES

Asofsky, R., and Becker, F. F. (1972). *Cancer Res.* **32**, 914–920, 1972.
Becker, F. F. (1963). *Amer. J. Pathol.* **43**, 497.
Becker, F. F. (1970). *In* "Progress in Liver Disease" (H. Popper and F. Schaffner, eds.), Vol. III, pp. 60–76. Grune and Stratton, New York.
Bucher, N. L. R. (1967). *New England J. Med.* **227**, 686, 738.
Bucher, N. L. R. (1971). *In* "Regeneration of Liver and Kidney" (N. L. R. Bucher, and R. A. Malt, eds.). Little, Brown, Boston, Massachusetts.
Fisher, B. Szuch, P., Levine, M., and Fisher, E. R. (1971a). *Science* **171**, 575.
Fisher, B., Szuch, P., and Fisher, E. R. (1971b). *Cancer Res.* **31**, 322.

Grisham, J. W., Leong, G. F., Albright, M. L., and Emerson, J. D. (1966). *Cancer Res.* **26**, 1476.

Grisham, J. W., Kaufman, D. G., and Alexander, R. W. (1967). *Fed. Proc.* **26**, 624.

Hays, D. M., Tedo, I., and Matsushima, Y. (1969). *J. Surg. Res.* **9**, 133.

Leong, G. F., Grisham, J. W., Hole, B. V., and Albright, M. L. (1964). *Cancer Res.* **24**, 1496.

Levi, J. U., and Zeppa, R. (1971). *Ann. Surg.* **174**, 364.

Loewenstein, W. R., and Penn, R. D. (1967). *J. Cell Biol.* **33**, 235.

Moolten, F. L., and Bucher, N. L. R. (1967). *Science* **158**, 272.

Price, Jr., J. B., Takeshige, K., Max, M. H., and Voorhees, A. B., Jr. (1971). *Gastro* **60**, 749.

Sakai, A. (1970). *Nature (London)* **228**, 1186.

Sigel, B., Dunn, M. R., and Butterfield, J. (1963). *Surg. Forum* **14**, 72.

Sigel, B., Baldia, L. B., Menduke, H., and Feigl P. (1967). *Surg. Gynecol. Obstet.* **125**, 95.

Sigel, B., Baldia, L. B., Brightman, S. A., Dunn, M. R., and Price, R. I. M. (1968). *J. Clin. Invest.* **47**, 1231.

Teebor, G. W., Becker, F. F., and Seidman, I. (1967). *Nature (London)* **216**, 396.

Virolainen, M. (1964). *Exp. Cell Res.* **33**, 588.

Wrba, H., Rabes, H., Ripoll-Gomez, M., and Rany, H. (1962). *Exp. Cell Res.* **26**, 70.

13

RENAL GROWTH FACTOR

Ronald A. Malt

I. INTRODUCTION

Neural and hemodynamic stimuli may modulate compensatory growth of the mammalian kidney, but the initial and major regulators appear to be carried in the blood (Bucher and Malt, 1971). Whether the mediating factor is a stimulant (or stimulants) added to the circulation or an inhibitor (or inhibitors) removed as a result of loss of one kidney is not known. Whatever its cause, compensatory renal growth probably does not arise simply from one kidney doing the work formerly done by two.

II. DENERVATED, TRANSPLANTED KIDNEY

If, at first approximation, compensatory growth is considered only an increase in the volume of the solitary kidney, good evidence for a humoral mediator is the enlargement of a kidney transplanted to an alien site in a compatible recipient. During its removal from the donor this kidney is removed from central neurogenic stimuli as a consequence of division of the external nerve fibers and the nerves that run within the renal artery and vein, although intrinsic parasympathetic nerves may survive (Norvell et al., 1970). Adrenergic connections across the new vascular unions are not identifiable until a month after transplantation in man (Gazdar and Dammin, 1970) and until about 3 months in dogs (Weitsen and Norvell, 1969; Norvell, 1970). In the meantime, not only is the transplanted kidney virtually without extrinsic innervation, but in its new site in flank or pelvis it is divorced from changes in blood flow, blood pressure, and pulse contour present at the normal renal pedicle.

Despite its isolation from normal connections, the transplanted kidney grows in man and in the rat (Starzl, 1964; Hume, 1968; Fletcher et al., 1969; Klein and Gittes, 1972). In the dog, a kidney from a small mongrel transplanted to an immunosuppressed large dog seems to grow proportionately more than a kidney from a large dog transplanted to a small one (Cohn et al., 1967).

Functional as well as morphological compensation occurs in the denervated kidney. Although the metabolic upsets following transplantation preclude certainty about how soon functional compensation takes place, clearly within 6–7 weeks the transplanted kidney can maintain sodium balance and regulate renin secretion in response to varying loads of sodium ion equally with the solitary, normally innervated kidney in the donor (Blaufox et al., 1969). Over the longer term, most renal functions are parallel in recipient and donor, especially if the pair are genetically similar (Bricker et al., 1956; Krohn et al., 1966; Donadio et al., 1967; Ogden et al., 1967; Flanigan et al., 1968).

III. KIDNEYS IN VASCULAR PARABIOSIS

A. Background

Despite the ways in which the transplanted kidney mimics many responses of the kidney remaining in the donor, behavior of the transplanted organ cannot definitively settle the nature of the stimulus to compensatory growth because of the variables produced by surgery, by the immune response, and by the state of preservation. Adaptation of the vascular parabiosis devised by Moolten and Bucher (1967) for studying humoral factors in hepatic regeneration serves this purpose better. Cross-circulation of blood between rats shows that loss of 50% of renal mass evokes in the remaining kidneys almost the same compensatory increase in renal weight and in the quantity of RNA compared with the quantity of DNA (RNA/DNA) as loss of 50% of renal mass in a single animal (see Section III,B).

Since there is substantial agreement among laboratories about the kinetics of renal weight and RNA/DNA during compensatory hypertrophy, these parameters appear to be the surest indicators of renal hypertrophy in response to contralateral nephrectomy. Normally, unilateral nephrectomy in the adult rat causes a 3–11% rise in the mass or protein content of the other kidney within the first postoperative day and an approximately equal response the next day (Halliburton and Thomson, 1965; Johnson and Vera Roman, 1966; Coe and Korty, 1967; Threlfall *et al.*, 1967; Kurnick and Lindsay, 1968a; Tomashefsky and Tannenbaum, 1970; Janicki and Lingis, 1970; Katz, 1970; Dicker and Shirley, 1971). The mouse has only about a 5% increase in mass after 1 day and a 10% increase at 2 days (Malt and Lemaitre, 1968).

In both the rat and the mouse, RNA/DNA rises 20–40% after 2 days and remains at that plateau for the next month (Halliburton and Thomson, 1965; Threlfall *et al.*, 1967; Malt and Lemaitre, 1968; Kurnick and Lindsay, 1968a; Dicker and Shirley, 1971) (Fig. 1). The increased RNA/DNA results from an accretion of RNA in the proximal tubules without an appreciable increase in the content of DNA (Vančura *et al.*, 1970).

B. Experiments

The similarity of the increase in renal mass and RNA/DNA in cross-circulated rats with half their renal complement compared with the

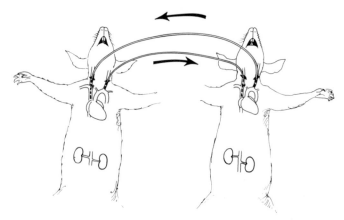

Fig. 1. Cross-circulation between the carotid arteries and jugular veins of rats, after Moolten and Bucher (1967). (From van Vroonhoven *et al.*, 1972, by permission.)

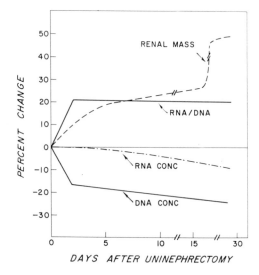

Fig. 2. Idealized changes in RNA/DNA with respect to renal mass during compensatory hypertrophy. (From Malt, 1969, by permission.)

increase in RNA/DNA and mass in unilaterally nephrectomized single rats strongly suggests that the stimulus to early compensatory hypertrophy is carried in the blood. In vascular parabiosis (Fig. 2) with rapid mixing of blood between a rat with both kidneys removed and a rat with both kidneys intact, compared with a sham-nephrectomized preparation, the kidneys of the intact partner grew 14.5% in weight in 1 day and 16% ($p < 0.001$) after 2 days; RNA/DNA increased 7.5% after 1 day and 20% ($p < 0.025$) after 2 days (van Vroonhoven *et al.*, 1972; Soler-Montesinos *et al.*, unpublished). Values in the solitary kidneys of unilaterally nephrectomized rats maintained under similar conditions were practically the same as these.

When cross-circulation was stopped after 2 days, hypertrophic kidneys returned almost to normal in another 2 days, showing that the early changes of compensatory hypertrophy were not permanent, but that compensation could operate in either direction to achieve balance. Further evidence for compensation in response to magnitude of loss was the 50% increase in renal mass and the 60% increase in RNA/DNA when three kidneys were removed and only a single kidney served the cross-circulated pair for 2 days. In every instance, the protein content remained about 17–18% of the wet weight, just as in normal kidneys or as in the solitary kidney undergoing compensatory hypertrophy.

Kurnick and Lindsay (1968b) had earlier demonstrated that after isologous mice made parabiotic through the peritoneal cavities for 20 days were deprived of a portion of the renal substance for a further 10 days, the weight of remaining kidneys increased. Removal of the left kidney from one mouse produced a 36% increment in weight of the right kidney in the same animal compared with the usual 10% preponderance of the right; observations on sham-operated animals were not reported. In the parabiotic preparation the effect of this left nephrectomy was to increase in the intact partner the mass of the left kidney by 11% compared with normal ($p = 0.05$) and the mass of the right kidney 21% compared with the normal 10% increment over the left kidney ($p = 0.01$). Trinephrectomy produced a 55% increase in residual renal mass. Considered per unit time, the limited augmentation of renal mass may have been a consequence of nutritional disturbances during the lengthy experiments, conducted in mice that tended to be small for age at the onset. The 30% mortality following binephrectomy and the larger compensatory response in the same animal than in the parabiont suggests either incomplete mixing of the two circulations or an unstable humoral mediator.

The experiments of Kurnick and Lindsay and of van Vroonhoven *et al.*, extend both Braun-Menéndez's observation (1958) of renal enlargement in rats after cutaneous parabiosis with anephric partners and Steuart's note (1958) stating that removal of four kidneys among a parabiotic triplet elevated mitotic activity in the two remaining kidneys. Failure to detect humoral transmission in other parabiotic preparations has probably been the result either of inadequate mixing or of using labeling of DNA as a parameter (see Section IV,B) (Lytton *et al.*, 1969; Johnson and Vera Roman, 1968).

IV. PROBLEMS IN ASSAYS

A. Renal Weight

The more that assays of compensatory renal hypertrophy deviate from the relative simplicity of renal weight or of protein content toward the complexities of biochemistry, the more subject to error they become. And even weight and protein content are not immune from extraneous influences. When body weight and renal weight change but slightly in comparisons among normal animals, sham-nephrectomized animals, and nephrectomized animals during experiments covering a few days, the results can validly be normalized in terms of milligrams of renal weight per gram of body weight, provided that the body weights measured do not vary widely. Perhaps a 5% acute variation could be allowed before a linear relation of body weight:renal weight is not accepted. But for long-term experiments and certainly for those with variations of more than 5% in body weight, pair-feeding programs are essential. Poor diet and prolonged recovery from operation depress the hypertrophic response (Malt and Lemaitre, 1968; Halliburton, 1969; Goldman, 1972), and a high-protein diet enhances it (MacKay *et al.*, 1938; Halliburton, 1969; Dicker and Shirley, 1971).

Indirectly, long-continued malnutrition may depress pituitary function, thereby inhibiting augmentation of renal mass—not by suppressing compensatory hypertrophy itself, but by causing a diminution in the size of many organs, including kidney, and thereby counterbalancing the compensatory growth (Bates *et al.*, 1964; Ross and Goldman, 1970). In the mouse the maintenance of comparable levels of sex hormones among experimental animals is essential, since estrogen sharply diminishes the weight of kidneys in male mice (Shimkin *et al.*, 1963), and androgens strongly promote both renal mass and protein and nucleic acid synthesis (Kochakian *et al.*, 1972; Dofuku *et al.*, 1971). Transfer of either male or female hormones into organ cultures may thus also affect *in vitro* assays adventitially.

B. DNA Synthesis and Mitosis

Still more qualifications are introduced into assays using DNA synthesis and mitosis as the indices of renal growth, whether *in vivo* or *in vitro*. DNA synthesis and mitosis are unnecessary for renal compensation, they are dissociable from it, and they are readily influenced by extraneous circumstances. The early stages of compensatory growth are apparently normal even when DNA synthesis or mitosis is suppressed with hydroxyurea or azathioprine (Janicki and Lingis, 1970; Cobbe *et al.*, 1970). Mitoses are easily inhibited by abnormal health or diet (Williams, 1962; Konishi, 1962; Reiter, 1965; Wachtel and Cole, 1965; Sharipov, 1967; Connolly *et al.*, 1969) and are promoted by high-protein diets (Halliburton, 1969). Minor degrees of contralateral or ipsilateral renal trauma, which may be related even to the route of injection of a radioactive precursor, can elevate the rates of DNA synthesis and mitosis in the absence of compensatory hypertrophy (Argyris and Trimble, 1964; Malamud *et al.*, 1972). Intense DNA synthesis may be provoked without a response in renal weight or in RNA synthesis, as by a single injection of isoproterenol (Malamud and Malt, 1971; Burns *et al.*, 1972). Finally, the appearance of a peak in renal DNA synthesis and mitosis in the whole animal 40–48 hours after contralateral nephrectomy (reviewed by Bucher and Malt, 1971) is no assurance that the times of the expected maxima will be the same *in vitro*. To the contrary, with the overall acceleration of renal cell division in culture (Lieberman *et al.*, 1963; Lee *et al.*, 1970), a further enhancement in proliferation from addition of substances to be tested for renotropic activity should be sought much earlier; yet 40–48 hours has been the point chosen in most such experiments. The influence of thymidine triphosphate pools and thymidine degrading enzymes on radioactive labeling is not known, although there is evidence that thymidine kinase activity is increased about 2.5-fold (Mayfield, *et al.*, 1967).

Using the rate of DNA synthesis in liver as an internal control to identify factors that stimulate mitosis, but are not specific for kidney, may be open to question because of the possibility that nephrectomy inhibits DNA synthesis in liver (Arasimowicz, 1967).

C. RNA and Protein Synthesis

Even though renal RNA and protein synthesis are better parameters of compensatory growth than DNA synthesis is, they have not been so widely used. The list of problems associated with the assay of renal RNA and protein synthesis is therefore short, not because more do not exist, but because they have not knowingly been encountered or studied.

As with DNA synthesis and mitosis, a deficient diet will inhibit RNA and protein synthesis (Bucher and Malt, 1971). In addition, studies of the kinetics of RNA and protein synthesis employing, of necessity, precursors that are less specific than a short exposure to radioactive thymidine is for DNA, introduce not only problems in compensating for specific activities of the final precursor pools (Bucher and Swaffield, 1966; Bucher and Malt, 1971), but in ascertaining that the radio-active product isolated is specifically the one that is sought, not a product of some other metabolic pathway (Baserga and Malamud, 1969).

With respect to RNA synthesis, data seem to exist only for the sizes of the ATP, ADP, and UTP pools—all of which remain constant during the first week of compensatory hypertrophy—and for the specific activity of the UTP pool with respect to labeling of RNA following injection of (^{3}H-5)uridine (Hill et al., unpublished). The availability of these data suggests that more valid information about RNA synthesis in kidney could be derived using radioactive uridine as a precursor than using other radiochemicals. Even under those circumstances there is no way to assess compartmentation of the nucleotide precursor pools, a problem that probably exists in liver (Bucher and Swaffield, 1965). With respect to protein synthesis, even though our unpublished experiments detected no changes in the free amino acid pools of kidney and showed no change in the kinetics of labeling of the leucine pool during compensatory hyper-trophy, they, too, could not evaluate compartmentation of amino acids (Rosenberg et al., 1963). The increased concentration of α-aminobutyric acid (a model short-chain polar amino acid) in mouse kidney during compensatory hypertrophy emphasizes the need for considering pre-cursor pools (Ross et al., 1973).

Assays of growth factors relying upon the processing of rapidly labeled species of RNA (Willems et al., 1969), and on labeling of proteins (Coe and Korty, 1967; Tomashefsky and Tannenbaum, 1970; Malt and Baptiste, unpublished), especially, will all be plagued by the problem of specific activities and compartmentation in precursor pools. Assays that depend upon enhanced labeling of ribosomes with [^{3}H]orotic acid for short periods (Halliburton, 1969)—short in comparison with the hour that it takes for finished ribosomes first to appear in the cyto-plasm—might be reevaluated in light of our inability to confirm this phenomenon with [^{3}H]uridine as the precursor (Hill, et al., unpublished).

V. TRANSFER OF SERUM

Because of these problems, most attempts to identify serum factors presumably responsible for compensatory hypertrophy by their effect on renal cell division and RNA and protein synthesis are open to challenge. The study with the largest number of variables controlled *in vitro* showed a 29% increase in the specific activity of DNA labeled with [³H]thymidine (p < 0.05) and a 30% increase in specific activity of RNA labeled with [¹⁴C]uridine (p < 0.05) when slices from male rats were incubated in the presence of serum taken from male rats 24 hours after unilateral nephrectomy compared with serum from sham-nephrectomized rats (Preuss *et al.*, 1970). Serum from bilaterally nephrectomized rats produced no change, and DNA and RNA labeling in liver slices was depressed or unchanged, respectively, under these conditions. Investigation of the role of enzymes for thymidine degradation in these results was not a part of the protocol.

Earlier experiments supporting the presence of serum stimulators in compensatory growth had used either mitotic index of the thymidine-labeling as the parameter. In the first such experiment, serum from rats two days after unilateral nephrectomy doubled the mitotic activity of cultured specimens of rat outer medulla (Ogawa and Nowinski, 1958). The stimulating activity of the serum was stable after being heated at 56°C for 30 minutes and was not affected by dialysis; it was absent from serum drawn 15 days after unilateral nephrectomy. Preliminary corroboration was later reported (Lowenstein and Lozner, 1966). Although it was organ-specific for kidney, the factor described by Ogawa and Nowinski was not species specific, for serum from renoprival puppies reproduced the phenomenon. Likewise, the increased labeling index of the kidneys of rats given intraperitoneal injections of serum from human beings soon after unilateral nephrectomy argues against species specificity of the factor that stimulates cell proliferation (Schiff and Lytton, unpublished).

Confirmation of stimulated cell proliferation *in vivo* has not been general. On the one side, intraperitoneal injections of serum from rats 48 hours after unilateral nephrectomy into the other rats twice daily for 4 days raised the labeling index of kidneys in the recipient animals from 12 ± 0.9 (S.E.)/1000 cortical cells in sham-injected recipients to 20 ± 2/1000 cells (p < 0.02) in those injected with postnephrectomy serum (Lowenstein and Stern, 1963). Renal weights were not reported. The report of Vichi and Earle (1970) was similar. To the contrary, Kurnick and Lindsay (1967) attributed such findings to differential weight gains among the groups of animals and felt that the stimulation was nonspecific. Responses of female mice given serum from male donors (Silk *et al.*, 1967) are difficult to interpret because of the renotropic effect of androgens.

VI. TISSUE EXTRACTS

The concept of self-regulation of organ size by feedback from its own products has guided many of the efforts that have shown an inhibition of renal cell proliferation by parenterally administered extracts of kidney. With some reservation (Roels, 1969), all of these experiments may have their effects on mammalian kidney explained by resultant disturbances in nutrition and health (Williams, 1962; Goss, 1963a; Royce, 1963; Speilhoff, 1971). Evidence for intrinsic tissue regulators may be stronger in the mesonephric chick kidney (Andres, 1955) and in the pronephric kidney of Xenopus (Chopra and Simnett, 1971).

VII. GROWTH REGULATORS VERSUS WORK HYPERTROPHY

For several decades contention has centered about whether organ-specific stimulatory (or inhibitory) substances modulate compensatory renal hypertrophy, or whether compensation is a response to the need for one kidney to do the regulatory work in excretion formerly done by two (work hypertrophy). At present, the existence of specific modulators for renal mass seems likely. But if advocates of specific regulators had to defend their cause, their strongest argument might be to retreat to the axiom of physics that, unless certain events are specifically forbidden, they must exist.

"Work hypertrophy," though not explicitly forbidden, is tacitly discouraged. Its original foundation—that added work is done by a single kidney in excreting urea and other products—is not valid. Tubular reabsorption of sodium ions from the glomerular filtrate is the process actually responsible for 99% of renal work (Johnson, 1969) and studies in rats (Katz and Epstein, 1967; Katz, 1970; Weinman et al., 1973) and in dogs (Kiil and Bugge-Asperheim, 1968) showed that increases in renal mass after contralateral nephrectomy preceded measurable changes in the reabsorption of sodium ion. Therefore, by these criteria compensatory hypertrophy precedes an increase in renal work. The converse situation, diversion of one canine ureter into the peritoneal cavity to produce major increases in sodium reabsorption of the intact kidney, did not increase mitosis in the way that unilateral nephrectomy did (Bugge-Asperheim and Kiil, 1968). Because of the reservations about mitotic index as an assay of growth and because of the difficulty in controlling for an inhibition of mitosis produced by the operation, however, this preparation cannot be considered strong refutation of the work-hypertrophy theory.

The findings of Katz and Epstein and of Weinman *et al.*, were confirmed by *in vitro* measurements of renal work in terms of oxygen consumption, which is stoichiometrically related to reabsorption of sodium ion, and in terms of utilization of the major energy substrates of the cortex and medulla. No increase in the observed capacity for aerobic or anaerobic metabolism was detectable until the interval between the first and second days after unilateral nephrectomy, perceptibly later than the first accretion of RNA and protein (Vančura and Malt, 1973).

Perhaps the best argument against work hypertrophy is that it is illogical when considered in biochemical terms. Since no biosynthesis can take place without biochemical work being done, all growth in every system must be work hypertrophy, and there is really nothing to argue about. As it leads to nonproductive thought, the concept of work hypertrophy should be abandoned.

VIII. SEARCH FOR THE FACTOR

A. Vasoactive Substances

Because speculation becomes simpler and analyses more reasonable if a single substance is held responsible for the control of renal growth from start to finish, the earliest sign of compensatory hypertrophy may be a key to identification of the controlling mechanism. Hyperemia of the renoprival kidney is the earliest documented event, although there is no evidence that it is an obligatory antecedent to compensatory growth. In the mouse the solitary kidney becomes plethoric within a minute (personal observation), and in the dog, renal artery blood flow rises 16–50% within the first few minutes (Payer and Siman, 1967; Krohn *et al.*, 1970; Roding *et al.*, 1971), with larger increments thereafter.

Since glomerular filtration rate does not appear to change at these early intervals in the rat (Peters, 1963; Katz, 1970), dog (Kiil and Bugge-Asperheim, 1968), and rabbit (Fajers, 1957), either the intraglomerular filtration pressure is diminishing or the larger amount of blood is being distributed among glomeruli with a smaller filtering surface (Barger and Herd, 1971). This hypothetical change in filtration area may even occur in a larger number of glomeruli than normal, since 1 day after unilateral nephrectomy in the rabbit, the number of perfused glomeruli increases from 44–78% of the total to 91–99% of the total (Moore,

1929), and in the rat 4–7 days after nephrectomy the rise is 39% in young animals and 17% in older ones (Hartman and Bonifilio, 1959).

Although rapid vascular responses such as these could implicate kinins, prostaglandins, or angiotensins, the stable intrarenal levels of cyclic adenosine monophosphate (Robison and Regen, unpublished; Achar and Nowinski, 1972), through which some of the actions of these vasoactive substances are mediated, suggests that other influences may be at work.

B. Endocrine Glands

Many experiments have shown what the stimulus is not. It is not a primary pituitary hormone or an endocrine secretion regulated by the pituitary, since compensatory renal hypertrophy occurs in hypophysectomized rats (Ross and Goldman, 1970). It is not aldosterone, since there is no evidence that aldosterone secretion in physiological doses reproduces the proper events (discussed by Vančura et al., 1971). It is not testosterone or its metabolites since mice lacking the nuclear receptor for testosterone undergo normal compensatory hypertrophy (unpublished). It probably is not insulin, since the ability of rats with stereptozotocin-induced diabetes to compensate is normal (Ross and Goldman, 1971), but reservations must be held in view of the potentiality of compensatory growth having been produced in response to a nephrotoxicity induced by streptozotocin.

C. Other Humoral Substances

It does not seem to be a substance secreted into the renal vein and ultimately inactivated by the liver, for renal mass and RNA synthesis are not promoted by unilateral diversion of the renal-vein effluent into the portal circulation (Bump and Malt, 1970), and it may not be related to immune competence (Fox and Wahman, 1968) because compensatory hypertrophy in mice is unchanged by neonatal thymectomy (Bump and Malt, 1969). Compensatory growth of normal magnitude proceeds for at least two days in rats that are abdominally eviscerated and maintained on intravenous alimentation (Ross et al., unpublished).

Failure of maternal nephrectomy in the pregnant rat to affect weight, mitotic activity, and RNA/DNA in fetal kidneys (Goss, 1963b; Malt and Lemaitre, 1969) is more likely to be a consequence of inability of fetal kidneys to respond than to failure of the maternal stimulus to cross the placenta; to the contrary, maternal nephrectomy at the start of the last trimester of pregnancy may inhibit the development of fetal rat kidneys (Škreb et al., 1971). Despite the limited growth of fetal rat kidneys after unilateral fetal nephrectomy (Goss and Walker, 1971), compensatory hypertrophy in the human kidney does not begin until after birth. The one normal-size kidney of human fetuses with congenital unilateral renal dysplasia begins compensatory growth only after birth (Laufer and Griscom, 1971). What is not known is whether the solitary kidney fails to respond *in utero* because it cannot or because the maternal kidneys are adequate for homeostasis.

D. Chemicals

If their physiological validity could be confirmed, several substances would be attractive candidates as humoral stimulators. Excess ribonuclease in the serum after unilateral or bilateral nephrectomy could easily be imagined to release controls on macromolecular synthesis (Rabinovitch, 1959; Royce, 1967; Rosso *et al.*, 1973), but its presence after unilateral nephrectomy has not been widely substantiated (see discussion in Nowinski and Goss, 1969). The circulating α-globulin produced by the male rat and present after bilateral nephrectomy (Royce, 1968) has not yet been demonstrated after unilateral nephrectomy, and the preliminary report of increased levels of an inhibitor to serum complement in the dog after unilateral nephrectomy (Delgado and Nathan, 1971) has to be balanced by the 4-fold increase in complement also reported (Babaeva and Sokolova, 1970). Stimulatory effects of folate and vitamin B_{12} on renal growth are more likely results of tubular blockage than of a direct stimulatory effect (Taylor *et al.*, 1968; Baserga *et al.*, 1968).

E. Speculation

I like to believe that the regulator of renal compensatory hypertrophy is a substance of extrarenal origin, perhaps a prohormone, that is inactivated by renal tissue (Braun-Menéndez, 1958). The direct effect within the kidney of this substance could be mediated by a substance as simple as potassium ion, which radically influences the proliferation and differentiation of kidney in culture (Orr *et al.*, 1972; Crocker and Vernier, 1970).

IX. SUMMARY

The primary regulator of compensatory renal hypertrophy is carried by the blood and can operate independent of neural and hemodynamic influences. Strong evidence for this view derives from the hypertrophic responses of the kidney transplanted to a heterotopic site in an anephric recipient and of kidneys in vascular parabiosis with an anephric animal. Assays of presumptive serum regulators that depend for their validity upon techniques of labeling with radioactive precursors are difficult to interpret because of variables difficult to control. Compensatory hypertrophy does not occur as a consequence of the solitary kidney doing more regulatory work. Rather, the regulator (or regulators) of renal compensation are likely to be organ specific. The identity of the renal growth factor is entirely unknown.

ACKNOWLEDGMENT

This work was supported by the National Institutes of Health (AM-12769) and by the Shriners Burns Institute.

I thank Drs. Daniel Malamud, Nancy L. R. Bucher, and Jeffrey S. Ross for helpful criticism.

REFERENCES

Achar, S. B., and Nowinski, W. W. (1972). *Fed. Proc.* 31, 824Abs.
Andres, G. (1955). *J. Exp. Zool.* 130, 221.
Arasimowicz, C. (1967). *Nature (London)* 215, 756.
Argyris, T. S., and Trimble, M. E. (1964). *Anat. Rec.* 150, 1.
Babaeva, A. G., and Sokolova, E. V. (1970). *Byull. Eksperim. Biol. Med.* 70, 91.
Barger, A. C., and Herd, J. A. (1971). *New England J. Med.* 284, 482.
Baserga, R., and Malamud, D. (1969). "Autoradiography: Techniques and Application." Harper, New York.
Baserga, R., Thatcher, D., and Marzi, D. (1968). *Lab Invest.* 19, 92.
Bates, R. W., Milkovic, S., and Garrison, M. M. (1964). *Endocrinology* 74, 714.
Blaufox, M. D., Lewis, E. J., Jagger, P., Lauler, D., Hickler, R., and Merrill, J. P. (1969). *New England J. Med.* 280, 62.
Braun-Menéndez, E. (1958). *Circulation* 17, 696.
Bricker, N. S., Guild, W. R., Reardon, J. B., and Merrill, J. P. (1956). *J. Clin. Invest.* 35, 1364.
Bucher, N. L. R., and Malt, R. A. (1971). "Regeneration of Liver and Kidney." Little, Brown, Boston, Massachusetts.
Bucher, N. L. R., and Swaffield, M. N. (1965). *Biochim. Biophys. Acta* 108, 551.
Bucher, N. L. R., and Swaffield, M. N. (1966). *Exp. Mol. Pathol.* 5, 443.
Bugge-Asperheim, B., and Kiil, F. (1968). *Scand. J. Clin. Lab. Invest.* 22, 255.
Bump, S., and Malt, R. A. (1969). *Transplantation* 8, 750.
Bump, S., and Malt, R. A. (1970). *J. Appl. Physiol.* 28, 682.
Burns, R. E., Scheving, L. E., and Tsai, T-H. (1972). *Science* 175, 71.
Chopra, D. P., and Simnett, J. D. (1971). *Exp. Cell Res.* 64, 396.

Cobbe, S. M., Herbertson, B. M., and Houghton, J. B. (1970). *Transplantation* **10**, 443.

Coe, F. L., and Korty, P. R. (1967). *Amer. J. Physiol.* **213**, 1585.

Cohn, R., Eckert, D., and Knudsen, D. F. (1967). *Amer. Soc. Nephrol. Abstr.* **1**, 13.

Connolly, J. G., Demelker, J., and Promislow, C. (1969). *Can. J. Surg.* **12**, 236.

Crocker, J. F. S., and Vernier, R. L. (1970). *Science* **169**, 485.

Delgado, O. M., and Nathan, P. (1971). *Fed. Proc.* **30**, 602Abs.

Dicker, S. E., and Shirley, D. G. (1971). *J. Physiol.* **219**, 507.

Dofuku, R., Tettenborn, U., and Ohno, S. (1971). *Nature New Biol.* **232**, 5.

Donadio, J. V., Farmer, C. D., Hunt, J. C., Tauxe, W. N., Hallenbeck, G. A., and Shorter, R. G. (1967). *Ann. Int. Med.* **66**, 105.

Fajers, C.-M. (1957). *Acta Pathol. Microbiol. Scand.* **41**, 34.

Flanigan, W. J., Burns, R. O., Takacs, F. J., and Merrill, J. P. (1968). *Amer. J. Surg.* **116**, 788.

Fletcher, E. W. L., Chir, B., and Lecky, J. W. (1969). *Brit. J. Radiol.* **42**, 892.

Fox, M., and Wahman, G. E. (1968). *Invest. Urol.* **5**, 521.

Gazdar, A. F., and Dammin, G. J. (1970). *New England J. Med.* **283**, 222.

Goldman, J. K. (1972). *Proc. Soc. Exp. Biol. Med.* **138**, 589.

Goss, R. J. (1963a). *Cancer Res.* **23**, 1031.

Goss, R. J. (1963b). *Nature (London)* **198**, 1108.

Goss, R. J., and Walker, M. J. (1971). *J. Urol.* **106**, 360.

Halliburton, I. W. (1969). *In* "Compensatory Renal Hypertrophy" (W. W. Nowinski and R. J. Goss, eds.), pp. 101–130. Academic Press, New York.

Halliburton, I. W., and Thomson, R. Y. (1965). *Cancer Res.* **25**, 1882.

Hartman, M. E., and Bonfilio, A. C. (1959). *Amer. J. Physiol.* **196**, 1119.

Hume, D. M. (1968). *In* "Human Transplantation" (F. T. Rapaport and J. Dausset, eds.), pp. 110–150. Grune and Stratton, New York.

Janicki, R., and Lingis, J. (1970). *Amer. J. Physiol.* **219**, 1188.

Johnson, H. A. (1969). *In* "Compensatory Renal Hypertrophy" (W. W. Nowinski and R. J. Goss, eds.), pp. 9–27. Academic Press, New York.

Johnson, H. A., and Vera Roman, J. M. (1966). *Amer. J. Pathol.* **49**, 1.

Johnson, H. A., and Vera Roman, J. M. (1968). *Cell Tissue Kinet.* **1**, 35.

Katz, A. I. (1970). *Yale J. Biol. Med.* **43**, 164.

Katz, A. I., and Epstein, F. H. (1967). *Yale J. Biol. Med.* **40**, 222.

Kiil, F., and Bugge-Asperheim, B. (1968). *Scand. J. Clin. Lab. Invest.* **22**, 266.

Klein, T. W., and Gittes, R. F. (1972). *J. Urol.*, **109**, 19.

Kochakian, C. D., Dubovsky, J., and Broulik, P. (1972). *Endocrinology* **90**, 531.

Konishi, F. (1962). *J. Gerontol.* **17**, 151.

Krohn, A. G., Ogden, D. A., and Holmes, J. H. (1966). *J. Amer. Med. Ass.* **196**, 110.

Krohn, A. G., Peng, B. B. K., Antell, H. I., Stein, S., and Waterhouse, K. (1970). *J. Urol.* **103**, 564.

Kurnick, N. B., and Lindsay, P. A. (1967). *Lab. Invest.* **17**, 211.

Kurnick, N. B., and Lindsay, P. A. (1968a). *Lab. Invest.* **18**, 700.

Kurnick, N. B., and Lindsay, P. A. (1968b). *Lab. Invest.* **19**, 45.

Laufer, I., and Griscom, N. T. (1971). *Amer. J. Roentgenol.* **113**, 464.

Lee, M. J., Vaughan, M. H., and Abrams, R. (1970). *J. Biol. Chem.* **245**, 4525.

Lieberman, I., Abrams, R., and Ove, P. (1963). *J. Biol. Chem.* **238**, 2141.

Lowenstein, L. M., and Stern, A. (1963). *Science* **142**, 1479.

Lowenstein, L. M., and Lozner, E. C. (1966). *Clin. Res.* 14, 383.

Lytton, B., Schiff, M., and Bloom, N. (1969). *J. Urol.* 101, 648.

MacKay, L. L., Addis, T., and MacKay, E. M. (1938). *J. Exp. Med.* 67, 515.

Malamud, D., and Malt, R. A. (1971). *Lab. Invest.* 24, 140.

Malamud, D., Paddock, J., and Malt, R. A. (1972). *Proc. Soc. Exp. Biol. Med.* 139, 28.

Malt, R. A. (1969). *New England J. Med.* 280, 1446.

Malt, R. A., and Lemaitre, D. A. (1968). *Amer. J. Physiol.* 214, 1041.

Malt, R. A., and Lemaitre, D. A. (1969). *Proc. Soc. Exp. Biol. Med.* 130, 539.

Mayfield, E. D., Liebelt, R. A., and Bresnick, E. (1967). *Cancer Res.* 27, 1652.

Moolten, F. L., and Bucher, N. L. R. (1967). *Science* 158, 272.

Moore, R. A. (1929). *J. Exp. Med.* 50, 709.

Norvell, J. E. (1970). *New England J. Med.* 283, 261.

Norvell, J. E., Weitsen, H. A., and Sheppek, C. G. (1970). *Transplantation* 9, 169.

Nowinski, W. W., and Goss, R. J. (eds.) (1969). "Compensatory Renal Hypertrophy." Academic Press, New York.

Ogawa, K., and Nowinski, W. W. (1958). *Proc. Soc. Exp. Biol. Med.* 99, 350.

Ogden, D. A., Porter, K. A., Terasaki, P. I., Marchioro, T. L., Holmes, J. H., and Starzl, T. E. (1967). *Amer. J. Med.* 43, 837.

Orr, C. W., Yoshikawa-Fukada, M., and Ebert, J. D. (1972). *Proc. Nat. Acad. Sci.* 69, 243.

Payer, J., and Siman, J. (1967). *Urol. Int.* 22, 250.

Peters, G. (1963). *Amer. J. Physiol.* 205, 1042.

Preuss, H. G., Terryi, E. F., and Keller, A. I. (1970). *Nephron* 7, 459.

Rabinovitch, M. (1959). *Proc. Soc. Exp. Biol. Med.* 100, 865.

Reiter, R. J. (1965). *Lab. Invest.* 14, 1636.

Roding, B., Williams, B. T., and Schenk, W. G. (1971). *J. Trauma* 11, 263.

Roels, F. (1969). In "Compensatory Renal Hypertrophy" (W. W. Nowinski and R. J. Goss, eds.), pp. 69–85. Academic Press, New York.

Rosenberg, L. E., Berman, M., and Segal, S. (1963). *Biochim. Biophys. Acta* 71, 664.

Ross, J., and Goldman, J. K. (1970). *Endocrinology* 87, 620.

Ross, J., and Goldman, J. K. (1971). *Endocrinology* 88, 1079.

Ross, J. S., Vančura, P., and Malt, R. A. (1973). *Proc. Soc. Exp. Biol. Med.* 142, 632.

Rosso, P., Diggs, J., and Winick, M. (1973). *Proc. Nat. Acad. Sci. U.S.A.* 70, 169.

Royce, P. C. (1963). *Proc. Soc. Exp. Biol. Med.* 113, 1046.

Royce, P. C. (1967). *Amer. J. Physiol.* 212, 924.

Royce, P. C. (1968). *Amer. J. Physiol.* 215, 1429.

Sharipov, F. K. (1967). *Byull. Eksper. Biol. Med.* 64, 86.

Shimkin, M. B., Shimkin, P. M., and Andervont, H. B. (1963). *J. Nat. Cancer. Inst.* 30, 135.

Silk, M. R., Homsy, G. E., and Merz, T. (1967). *J. Urol.* 98, 36.

Škreb, N., Domazet, Z., Luković, G., and Hofman, L. (1971). *Experientia* 27, 76.

Speilhoff, R. (1971). *Proc. Soc. Exp. Biol. Med.* 138, 43.

Starzl, T. E. (1964). "Experience in Renal Transplantation." Saunders, Philadelphia, Pennsylvania.

Steuart, C. D. (1958). *Carnegie Inst. Wash. Year Book* 57, 347.

Taylor, D. M., Threlfall, G., and Buck, A. T. (1968). *Biochem. Pharmacol.* 17, 1567.

Threlfall, G., Taylor, D. M., and Buck, A. T. (1967). *Amer. J. Path.* **50**, 1.
Tomashefsky, P., and Tannenbaum, M. (1970). *Lab. Invest.* **23**, 190.
Van Vroonhoven, T. J., Soler-Montesinos, L., and Malt, R. A. (1972). *Surgery* **72**, 300.
Vančura, P., Miller, W. L., Little, J. W., and Malt, R. A. (1970). *Amer. J. Physiol.* **219**, 78.
Vančura, P., and Malt, R. A. (1973). *Amer. J. Physiol.* In press.
Vančura, P., Sharp, G. W. G., and Malt, R. A. (1971). *J. Clin. Invest.* **50**, 5431.
Vichi, F. L., and Earle, D. P. (1970). *Proc. Soc. Exp. Biol. Med.* **135**, 38.
Wachtel, L. W., and Cole, L. J. (1965). *Radiat. Res.* **25**, 78.
Weinman, E. J., Renquist, K., Stroup, R., Kashgarian, M., and Hayslett, J. P. (1973). *Amer. J. Physiol.* **224**, 565.
Weitsen, H. A., and Norvell, J. E. (1969). *Circ. Res.* **25**, 535.
Willems, M., Musilova, H. A., and Malt, R. A. (1969). *Proc. Nat. Acad. Sci.* **62**, 1189.
Williams, G. E. G. (1962). *Nature (London)* **196**, 1221.

14

HORMONAL INFLUENCE ON SKELETAL GROWTH AND REGENERATION

Edgar A. Tonna*

I. INTRODUCTION

The skeletal system is a member of the more extensive connective tissue system which includes tendons, fasciae, ligaments, joint capsules,

* Supported by N.I.H. grants HD-03677 and DE-03014.

dermis, septae, capsules and interstitial tissue of essentially all organs, adipose and mucous tissue, parts of the cardiovascular tissue, dental pulp, dentin, and cementum. All are derivatives of the mesenchyme. It is comprised of a variety of cell compartments which have been determined under normal circumstances to function in the capacity of producing bone and cartilage matrices. Although matrix production is a characteristic feature of connective tissues in general, it is outstanding with regard to the skeleton whose matrices are mineralizable. This peculiar feature accounts for the obvious function of the skeleton, namely, structural support of the organism, the site for muscle attachments, in addition to serving as the vestibule for bone marrow and red blood cell production. By virtue of the property of mineralization, the skeleton becomes the store house for the essential minerals, calcium, phosphorus, sodium, and magnesium, and important organic substances such as citrate. The constant need of these substances for the very viability and functional integrity of all the other cell systems which make up the organism calls for a dynamic biological container system capable of both the rapid storage and release of the needed mineral substances. In addition, a sensitive regulatory mechanism is essential which can respond effectively to the subtle mineral requirements and overload by regulating the precise quantitative flow of minerals in or out of the skeleton. This is achieved by the existence of a feedback mechanism which is at the heart of the homeostatic regulation of these substances. Like any other tissue system the skeleton contains cell subcompartments whose purpose is to supply and replenish the required number of functional cell members essential to its general viability through growth, development, and repair of the system. In addition, the skeleton possesses a unique osteochondroclastic cell subcompartment capable of bone and cartilage resorption, which is called upon throughout the life of the organism to participate in the remodeling of the skeletal structures.

When taken individually, all of the vital functions to which the skeletal system is committed appear to operate independently. The truth of the matter is, however, that the vital functions are intimately interrelated and their individual expressions form part of a concerted activity necessary to the total function and utility of the system. Within set limits, its potential quantitative expressions are subject to functional demands. The adaptability of the system, in which its cellular and matrical compartments operate in concerted effort to meet with the demands of the organism, is under hormonal influence and regulation. Numerous hormones are known to exert their influence on skeletal growth and differentiation; these include growth hormone, androgens, estrogens, adrenocortical hormones, thyroid hormone, parathormone, thyrocalcitonin, insulin, and possibly parotin. Of these hormones, only growth hormone of the anterior lobe of the pituitary is claimed to exert a specific effect upon bone growth which continues until cessation of skeletal growth occurs. The significance of endocrine factors to skeletal growth, development, repair, maintenance, and aging is well recognized through clinical and experimental experiences, and a plethora of documentation exists. Much is known about the chemical nature of hormones and their varied amino acids, peptides, proteins, and steroids. However, little is known of the transmission of the effect of the circulating hormones to their peripheral targets. The nature of the stimulation or inhibition is entirely unknown. Above all, the action is mainly directed to cells, but extracellular substances are also effected (Asboe-Hansen, 1966). Hormones influence enzyme systems, the synthesis and activity of enzymes, coenzymes, activators, and/or inhibitors (Miner and Henegan, 1951). In reaching their target site, numerous vascular and cellular membranes must be traversed. Their influence on the subcellular structures and cellular organelles which possess the very enzyme systems on which hormone action depends is largely unknown, except for the prestigious work on articular cartilage by the Silberbergs' and their colleagues at Washington University School of Medicine. It is known, however, that the endocrine secretion of one gland affects the activity of others, just as the effect on one enzyme system influences other enzyme systems. Hormonal influence may be exerted (1) by means of changes in tissue enzyme concentration; (2) by the hormone functioning as a component of an enzyme system, or (3) by direct or indirect affect on accelerators and/or inhibitors of enzyme systems. Such action or actions assuredly influence cell membranes and membrane systems. With the progressive increase in knowledge about hormones one factor appears to stand out, i.e., that the classical textbook description of hormone action is indeed simplistic. The portrayed sequence involving origin, secretion, pathway, and specific target relationships may be demonstrated *in vitro;* however, the *in vivo* sequences and interrelationships are far more involved and complex than hitherto realized. The hormonal influence on skeletal growth, development, maintenance, repair, and aging is certainly no less complex or less significant than occurs elsewhere.

An in-depth treatise of the ramifications and interrelationships of the influence of numerous hormones on skeletal growth and regeneration are beyond the scope of the present chapter and the knowledge at hand. Despite this limitation, however, an attempt will be made to outline the influence of individual endocrine secretions, especially at the cellular level where this is feasible in the light of more recent information. Although the following brief discussions will focus on the normal influence on skeletal structure, this does not imply that this influence is either the main or the exclusive effect of the hormone, e.g., somatotropic hormone is usually thought of in terms of its capacity to stimulate skeletal growth; however, it has many other functions, including effects on metabolism of proteins, fats and carbohydrates, on the immune response and also serves as a synergist in enhancing the effects of other hormones (Li, 1969b; Wilhelmi and Mills, 1969; Turner and Bagnara, 1971).

II. THE HIERARCHY OF SKELETAL CELL COMPARTMENTS

In view of the varying terminology which has been used in the literature for the hierarchy of cells which comprise the skeletal system, it was deemed pragmatic to define the specific cell types and their interrelationships in the hope of establishing a lucid comprehension of cell nomenclature used in this chapter and better insight into the cellular aspects of hormonal influence.

As can be seen in Fig. 1, three main cell compartments exist, i.e., fibrous, bone, and cartilage cell compartments. Specialized cells in each

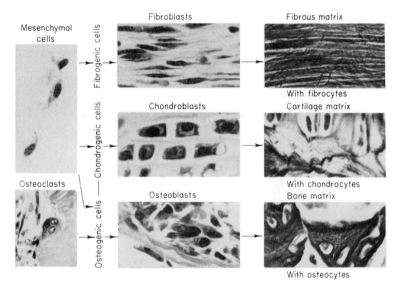

Fig. 1. The hierarchy of skeletal cell compartments is illustrated with their respective matrices at the light microscopic level. Note that all cell compartments are originally derived from the mesenchymal cell pool via precursor cells which give rise to the individual cell compartments. Osteogenic cells, however, can give rise to both the osteoblastic and chondrogenic cell lines, as well as, to osteoclasts following osteogenic cell fusion. Mouse tissues; hematoxylin stained.

compartment are capable of matrix production. The matrices consist largely of the fibrous protein, collagen, embedded in a ground substance containing mucoproteins. The hexosamine polysaccharides called muco-polysaccharides (glycosaminoglycans) are a characteristic component of the ground substance. These include hyaluronic acid, chondroitin sulfates A, B, and C, keratosulfate, and other less known sulfated fractions (Meyer, 1956; Meyer *et al.*, 1958). The matrices differ in their collagen content, types of mucopolysaccharides, and in their concentrations which are subject to age variations (Stidworthy *et al.*, 1958, Sinex, 1968). These differences may well account, in part, for the normal mineralizability of bone and cartilage and not fibrous tissue. Unfortunately, the biologic role of the different mucopolysaccharides remains unsettled and little understood.

Fibroblasts, osteoblasts, and chondroblasts are involved with matrix production by cellular synthesis of protein and mucopolysaccharide matrical precursors. Each cell type is derived via functional differentiation from the respective progenitive elements, the fibrogenic, osteogenic, and chondrogenic cells. It has been demonstrated autoradiographically using [³H]thymidine as a tracer, that osteogenic cells are not fully determined to produce bone cells exclusively, since cartilage cells are also produced (Tonna, 1961; Tonna and Pentel, 1972). Consequently, these cells are in reality osteochondrogenic cells. Whether these cell precursors give rise to differentiated bone or cartilage cells appears to depend largely upon the vascular integrity of the surrounding matrix (Ham, 1930). There exists no experimental proof that chondrogenic cells can or do give rise to osteogenic cells. It may well be that a chondrogenic cell is more differentiated than an osteogenic cell; on the other hand, both cell types appear to be able to be derived from a similar cell precursor which retains more of the pluripotential properties of mesenchymal cells from which the various skeletal cell compartments are derived. Figures 2–8 illustrate the ultrastructural characteristics of the different skeletal cell compartments and their respective matrices. Since this chapter emphasizes, where possible, the effects of hormones on skeletal cell ultrastructure, these figures will serve as general controls so that normal features may be compared with alterations induced by hormonal administration or withdrawal.

After a widely varied period of active matrix precursor synthesis (Tonna, 1965, 1971; Tonna and Pavelec, 1971) by the functional members of each major cell compartment, the cells in response to diminished physiological demands for growth and aging reveal a concomitant reduction in ultrastructural components essential to synthesis. In the case of bone, functional osteoblasts in growing animals become osteocytes

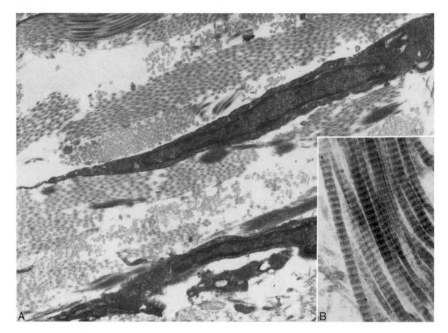

Fig. 2. The electron micrograph shows typical mouse fibroblasts embedded in a matrix of laminated collagen. Collagen lamina are oriented in various directions. The general ultrastructural detail of collagen is shown in B. A, ×13,000; B, ×46,150.

as they are surrounded by matrix which subsequently mineralizes. These cells, as well as those which remain at the surface, also go through a series of diminishing ultrastructural changes in response to aging. Chondroblasts also become chondrocytes; however, they are usually destroyed during growth at both the articular and epiphyseal plate regions. Those which remain in older animals are seen to go through similar age changes as do fibroblasts and fibrocytes. Consequently, bone cells may respond with different intensity to hormones at varying ages.

Osteoclasts constitute another cell type which is essential to bone growth, development, and repair. The cells are found at the perichondrial zone of the periosteum and at metaphyseal and epiphyseal trabecular surfaces. Here, osteoclasts are responsible for bone remodeling through osteoclasis. Osteoclasts at the epiphyseal plate and at the base of the articular cartilage are called chondroclasts because they exhibit chondroclasis. However, the name seems inappropriate since the origin and morphology of these cells is similar to osteoclasts, i.e., resulting from the fusion of osteogenic and osteoblastic cells (Tonna, 1960). Only the material substrate of their activity differs, i.e., cartilage instead of bone.

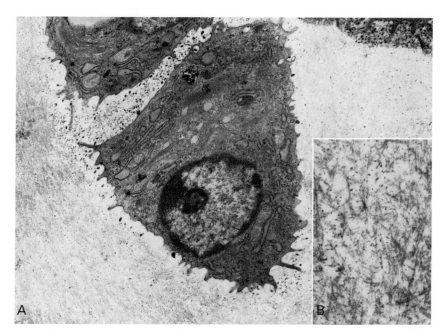

Fig. 3. A typical cartilage cell of the proliferative zone of mouse femoral epiphyseal plate is shown in the electron micrograph (A) embedded in its matrix. An enlargement of the matrix is illustrated in (B). The cells reveal a well-developed, round nucleus and Golgi complex, few mitochondria, some rough endoplasmic reticulum and a few footlets. A, ×13,000; B, ×32,500.

In addition to the skeletal cell compartments per se, i.e., fibrogenic, chondrogenic and osteogenic, the skeletal system consists of other matrices, nervous, vascular elements, peripheral blood, bone marrow cells, and associated muscle. It must be borne in mind that the skeletal response to hormones involves orders ranging from the cell ultrastructure to that of the system, so that the individual cell reaction is but a part of the complex interrelated response to hormones, even if it be the initial responsive tissue unit.

III. SOMATOTROPIC HORMONE (GROWTH HORMONE, STH)

At least nine hormones have been obtained from the hypophysis, namely: somatotropin (STH, or growth hormone), corticotropin (ACTH), thyrotropin (TSH), prolactin (LTH, lentotropin or lactogenic hormone), follicle-stimulating hormone (FSH), luteotropin hormone (LH, or interstitial cell-stimulating hormone ICSH), and melanophore-

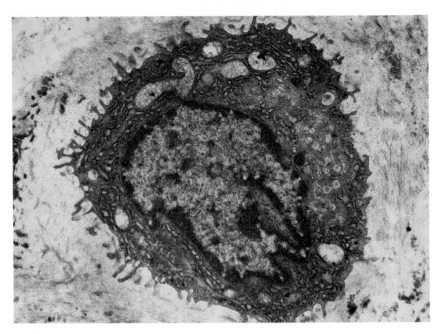

Fig. 4. A typical chondrocyte at the upper portion of the hypertrophic zone of mouse femoral epiphyseal plate is shown in the electron micrograph. The nucleus is folded, while the cytoplasm reveals numerous mitochondria, well-developed Golgi apparatus, and an abundance of rough endoplasmic reticulum. Cytoplasmic granules and footlets are more numerous. Compare with Fig. 3. ×14,000.

stimulating hormone (MSH, or intermedin) of the adenohypophysis, and in addition, the various oxytocins and vasopressins of the neuro-hypophysis. Recent evidence suggests that β-lipotropin (β-LPH), is yet another hormone belonging to this list (Li, 1969a). All of the hormones are proteins or polypeptides and several (TSH, FSH, and LH) also contain carbohydrates (Turner and Bagnara, 1971).

Of the many hormones which are known to influence skeletal growth including androgens, estrogens, thyroxine, parathormone, thyrocalcitonin, insulin, and adrenocortical hormones, only somatotropic hormone of the anterior pituitary gland exhibits a specific effect on bone growth. STH is a protein hormone; the average molecular weight of somatotropins from all species studied is approximately 22,000. Considerable variations, however, exist in the isoelectric points, ranging from pH 4.9 for human to pH 6.8 for sheep and ox. Immunochemical differences are also highly variable, despite the similarity in amino acid composition of various somatotropins. The amino acid sequence of human STH is known, ex-

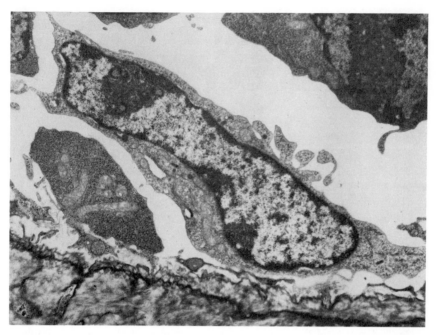

Fig. 5. A 1-year-old mouse osteoblast is shown in the electron micrograph adjacent to decalcified bone matrix of a femoral metaphyseal trabeculum. Although the cell reveals a well developed nucleus, few mitochondria are seen. The Golgi complex is small while little endoplasmic reticulum is present; ribosomes are free and scattered. The cell shows little evidence of involvement in matrix production. Compare with Fig. 6. Decalcified ×9750.

hibiting at least two disulfide bridges (Li, 1969b). Cleavage of these bonds apparently does not affect the growth-promoting activity of the hormones.

The specific skeletal effects of STH are exerted on cells of the epiphyseal plate. The plate constitutes the growth apparatus which is responsible for longitudinal bone growth. The hormone influence continues until cessation of longitudinal bone growth occurs, resulting from closure of the epiphyseal plate. Plate closure constitutes the deposition of bone on its superior and inferior aspects terminating cartilage cell proliferation and the invasion by capillaries and osteogenic cells essential to diaphyseal growth. STH effect is not, however, exclusively limited to the growth plate cartilage, but involves articular cartilage as well (Silberberg and Silberberg, 1957). Any interference with the growth apparatus by either physical or chemical means during the growth phase, in time reveals itself upon the size of the organism.

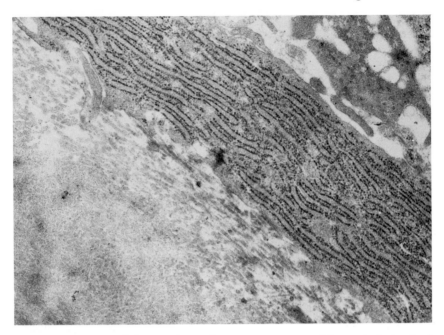

Fig. 6. A portion of the cytoplasm of a young mouse osteoblast actively involved in matrix production is shown in the electron micrograph. Note the abundance of rough endoplasmic reticulum lined by well formed ribosomes. Young, recently formed, loose collagen, constitutes the matrix adjacent to the cell. Compare with Fig 5. Decalcified. ×29,250.

Hypophysectomy of the young rat results in a rapid cessation of growth in length (Fig. 9) and additional gain in protein weight (Walker *et al.*, 1952; Asling and Evans, 1956). This results primarily from the loss of STH which has a stimulatory effect on the formation of cartilage and bone. The effect is mainly on the germinal and proliferative zones of epiphyseal and articular cartilage, wherein, inhibition of mitosis and cell size reduction are seen. Consequently, the cell columns become significantly shorter than normal in growing animals (Becks *et al.*, 1949), taking on the appearance of old rats. Concomitantly, vascular progression on the growth side of the cartilage is arrested. The osteoblasts associated with the spongiosa (trabecular bone) become spindle-shaped, resembling inactive cells and the spongiosa itself starts to disappear. Eventually the epiphyseal plate becomes "sandwiched" by bone lamina marking the cessation of growth.

Administration of growth hormone to a hypophysectomized animal during the state of growth plate quiescence, promptly reactivates skeletal growth and increases body weight. Replacement therapy is effective

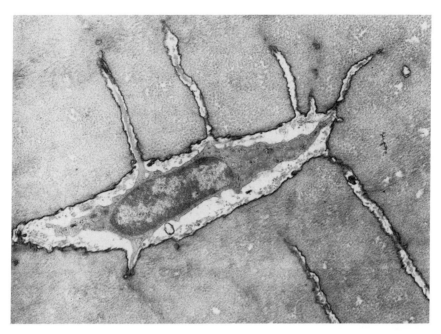

Fig. 7. The electron micrograph shows a mouse femoral cortical bone osteocyte residing within a lacuna lined by a dark osmiophilic lamina. Cytoplasmic projections make contact with other osteocytes via numerous canaliculi. Decalcified. ×9000.

in repairing the growth defects in hypophysectomized rats (Asling and Evans, 1956). The animals are responsive even if hypophysectomy was performed over a year earlier (Becks, *et al.*, 1946). It is most interesting to point out that the age of the animal at the time of hypophysectomy is not a factor, since when animals 6 days (Walker *et al.*, 1952) to 6–7 months of age were used (Moon *et al.*, 1951), response to STH administration was positive. At the articular cartilage, not only restorative growth was observed, but also increased incidence of degenerative joint disease normally encountered in older animals (Silberberg and Silberberg, 1957). Coincidental with this point, it is significant to note that collagen aging occurs, but is reduced in hypophysectomized rats (Verzár and Spichtin, 1966). STH apparently influences the aging rates of certain tissue compartments. The epiphyseal plate and adjacent spongiosa reveal resumed activity as the bony seals are resorbed, allowing for the reestablishment of the vascular invasion of the epiphyseal plate essential to growth. The growth plate widens; however, the width attained is proportional to the hormone dosage used (Becks *et al.*, 1941). Widening of the plate results from reactivated cell proliferation and

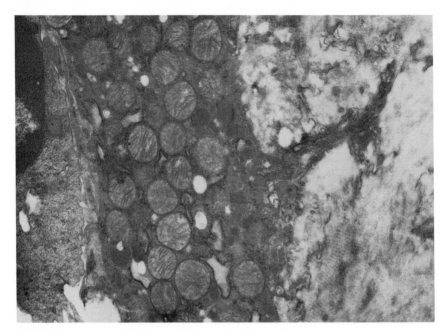

Fig. 8. A cytoplasmic portion typical of an osteoclast is shown in the electron micrograph. The cytoplasm is characterized by large numbers of mitochondria and scattered lysosomal bodies. Cytoplasmic footlets (brush border) are also present juxtaposition to the bone surface. Mouse femoral metaphyseal trabeculum. Decalcified. ×19,500.

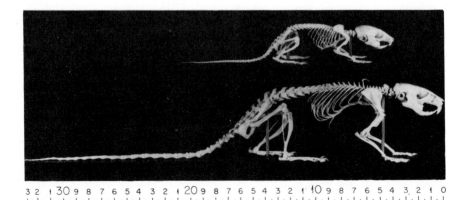

Fig. 9. A skeletal comparison is shown at 60 days between a rat hypophysectomized at 6 days of age (above) and a normal rat (below). (Reproduced with the courtesy of the authors and publisher from Walker *et al.*, 1952.)

the production of numerous chondrocytes. Associated osteoblasts undergo morphological changes resulting in the resumption of their functional activity.

Rigal (1964) reported interesting results which are significant to our awareness of the complexity of the relationship of hormones of different species to cells and to their internal milieu. When bovine growth hormone was administered to rabbits *in vivo*, a marked increase in cartilage cell proliferation at the epiphyses was observed in the germinal and proliferative zones of the epiphyseal growth cartilage and the corresponding cells of the articular cartilage and cartilage anlage. On the other hand, when the hormone was administered to normal tissue explants *in vitro*, stimulation of cell proliferation did not occur. The reason for the difference in response is unknown, however, a number of suggestions were postulated, namely, (1) growth hormone may require modification within the animal prior to mimicking the effect of endogenous rabbit growth hormone, and (2) perhaps growth hormone does not exert its effect directly upon cartilage cells, but acts via an intermediate enzymatic, hormonal mechanism, or enhancement of nutritional or vascular supply of the region. Earlier studies by Daughaday and Mariz (1962a) showed similar results using [^{14}C]proline as a hydroxyproline tracer. STH addition *in vitro* was without effect, while hydroxyproline production increased *in vivo* in hypophysectomized rats. However, addition of normal rat serum to the incubating medium resulted in increased hydroxyproline formation.

Similar histological and cytological changes can be elicited in nonhypophysectomized animals which have normally reached the advanced stage of diminished growth or growth cessation by STH therapy. Growth hormone administration rapidly restores the histological appearance of epiphyseal cartilage plates to that characterizing actively growing young bones. It is noteworthy that the effect of STH on skeletal growth can be augmented markedly through the synergistic action of thyrotropic hormone (Marx *et al.*, 1942). In fact, it proved impossible to produce skeletal gigantism in thyroid deficient rats by continued administration of STH; however, in intact or hypophysectomized rats this was achieved (Asling *et al.*, 1965). Synergistic activity was also revealed when both hormones were administered to rats simultaneously and individually during fracture repair (Udupa and Gupta, 1965).

The ultrastructure of the articular cartilage of hypopituitary dwarf mice (dwdw) and its deviation from the normal architecture of cells was reported by R. Silberberg *et al.* (1966a). In other electron microscopic investigations, the degree to which hypophyseal imbalance contributed to the subcellular picture was studied following STH administration (M. Silberberg *et al.*, 1965, 1966a). In general, STH treatment resulted in advanced organellar development, and the effect was manifest as early as 2 hours after its administration. Although STH counteracted many of the cellular defects present in dwarf mice, complete ultrastructural restoration to the normal nondwarf chondrocyte structure was not attained due to the ultrastructural defects induced by other hormonal imbalances in dwarf mice (Figs. 10–12). STH treatment did result in increasing the granular endoplasmic reticulum essential to protein synthesis, improved Golgi complex development essential to numerous cellular secretions including mucopolysaccharides, and increased the size and number of mitochondria essential to cellular respiration and energy supply. Ultrastructural results essentially similar to those of M. Silberberg *et al.* (1966a) were reported for rat epiphyseal cartilage by Fahmy (1968). The ultrastructural changes unquestionably point.to a significant conversion and improvement in the cell's functional capacity required for cartilage and bone matrix production. Support for this statement can be derived from the earlier work of Daughaday and Marix (1962b) who reported increased hydroxyproline formation by rat cartilage in response to growth hormone. Hydroxyproline is specific to collagen structure via the conversion of proline. Conversely, hypophysectomy reduced the rate of collagen synthesis. While there is insufficient evidence to establish the definitive mechanism of STH action, one important aspect of its function is to promote the transfer of extracellular amino acids across cell membranes (Knobil, 1961). In an investigation of the subcellular localization of ^{3}H-acetyl-labeled human growth hormone in the liver of hypophysectomized rats, it was noted that the initial binding sites of the hormone appear to be the microsomal and mitochondrial fractions (Maddaiah *et al.*, 1970). A remarkable increase in histochemically demonstrable mucopolysaccharides of rat cartilaginous tissues was reported by Tinacci *et al.* (1962) following STH administration. Increase in alkaline phosphatase and small increases in lipids and lipase were also observed. In this regard, somatotropin encourages the movement of unesterified fatty acids from fat reserves (Swislocki and Szego, 1965).

In man afflicted with hypopituitarism or in primordial dwarfism of genetic origin, mature skeletal proportions are observed, generally commensurate with the age of the individual. A marked retardation in ossification time of the epiphyses is seen in hypopituitary dwarfs, while primordial dwarfs usually exhibit an essentially normal bone age. In hypopituitary dwarfs epiphyseal fusions do not occur or appear very late in life. Normal fusion, however, occurs in primordial dwarfs. The bones remain small and delicate in pituitary dwarfs, probably due to relatively poor muscle development. Some individuals are slow in all

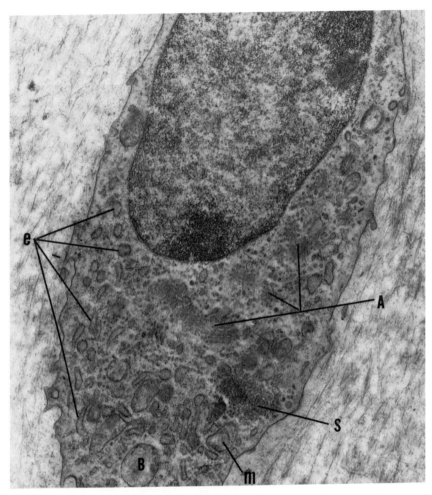

Fig. 10. An electron micrograph is shown of a deep midzonal chondrocyte of the articular cartilage from the femoral head of an untreated dwarf mouse. Note the few small footlets, widespread Golgi system (A), small deposits of glycogen (S), interrupted endoplasmic reticulum (e), few mitochondria (m) and multivesicular body (B). ×20,000. (Reproduced with the courtesy of the authors and publisher from Silberberg *et al.*, 1966a.)

their growth and development throughout childhood exhibiting a late onset of puberty. Eventually, they mature into entirely normal, although, sometimes short adults, and are considered constitutional instances of delayed growth and development (Wilkins, 1955). Since the three types of individuals present no characteristic differences in their habitus, it

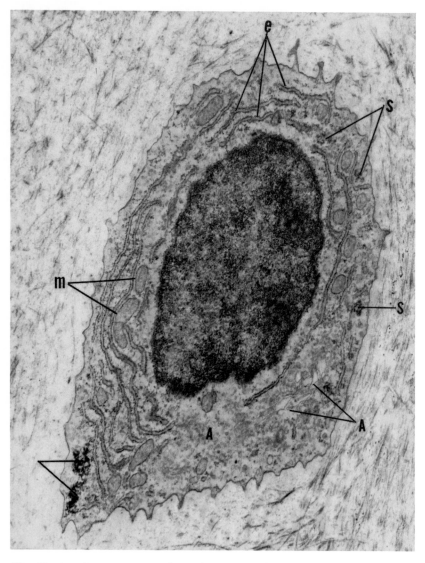

Fig. 11. An electron micrograph is shown of a deep midzonal chondrocyte of the articular cartilage from the femoral head of a dwarf mouse treated with STH for three days. Note the well-developed granular endoplasmic reticulum (e), Golgi system (A), more numerous mitochondria (m), and scattered small deposits of glycogen (S). Compare with Fig. 10. ×20,000. (Reproduced with the courtesy of the authors and publisher from Silberberg et al., 1966a.)

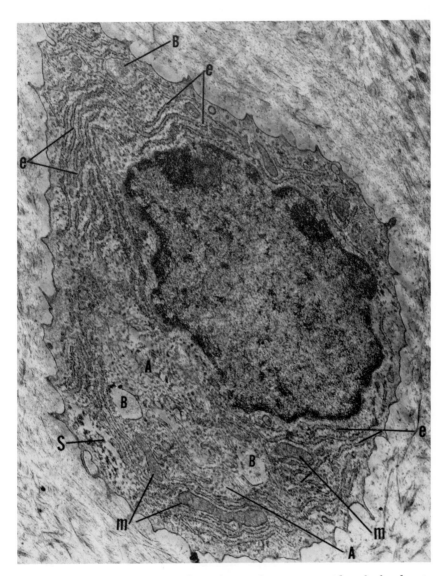

Fig. 12. An electron micrograph is shown of an upper midzonal chondrocyte of the articular cartilage from the femoral head of an untreated nondwarf mouse. Note the abundance of cytoplasmic footlets, extensive granular endoplasmic reticlum (e), numerous multivesicular bodies (B), scattered mitochondria (m) and glycogen (S). Compare with Figs. 10 and 11. ×20,000. (Reproduced with the courtesy of the authors and publisher from Silberberg *et al.*, 1966a.)

is difficult in childhood to distinguish between hypopituitary dwarfism, primordial dwarfism (although such individuals often exhibit associated congenital anamolies), and constitutionally delayed growth and development. In the assessment of the effects of hormones on skeletal growth and maturation a number of indices are used, namely: (1) the ratio of upper to lower skeletal segments, (2) development of the naso-orbital configuration, and (3) ossification time of the carpal and tarsal cartilages and the epiphyseal centers of long bones. In addition, a variety of biochemical criteria, e.g., urinary nitrogen output and retention, calcium absorption, urinary sodium and potassium output are also used.

To date, human administration of growth hormone derived from bovine and porcine sources has failed to fulfill its expectations based on animal experimentation. Hormone preparations derived from human and primate pituitary glands have demonstrated the importance of species specificity, since human beings and monkeys both respond to primate growth hormones, but not those from any other vertebrate (Turner and Bagnara, 1971). Purified human growth hormone administered intramuscularly in minute amounts resulted in stimulated linear growth, the appearance of open, active epiphyses, increased calcium resorption, and decreased urinary nitrogen, sodium, and potassium (Henneman et al., 1960). It is of interest to note that for some yet unknown reason, the guinea pig is unresponsive to any known somatotropin, even its own (Turner and Bagnara, 1971). Collipp et al. (1966) reported that injected labeled human growth hormone appeared in comparable amounts in the tissues of rats and guinea pigs and concluded that the data suggest that the unresponsiveness of guinea pigs does not result from abnormal tissue distribution, or from excessive loss in urine or bile.

Basically, the effects of STH on growth and development of the skeletal system appear to be complex involving numerous parameters. These result from the stimulation of germinal cell proliferation, the promotion of amino acid transfer across cell membranes and the cellular maintenance of the ultrastructural functional status, allowing for normal matrix production in response to growth, development, and reparative needs of the organism in terms of both bone and cartilage. The latter involves, in addition, SO_4 uptake and the sulfation factor.

IV. THYROID HORMONE

The most important protein present in the colloid of the thyroid follicle is thyroglobulin, an iodized glycoprotein with a molecular weight of approximately 680,000. This is considered to be the storage form of the thyroid hormone; consequently, thyroglobulin does not appear in circulation. Normally, this glycoprotein is enzymatically hydrolyzed yielding a number of iodinated amino acids. Of the iodothyronines that are secreted into the circulation, only 3,5,3'-triiodothyronine and thyroxine (3,5,3',5'-tetraiodothyronine) are known to exhibit biological activity. The former substance is known to be over 7 times more active than thyroxine, and to produce its effects more quickly. Although triiodothyronine and thyroxine are natural products, thyroxine may not represent the active form of the hormone. It may require conversion to triiodothyronine or some other form before affecting peripheral tissues. The kidney may play a role in this event since kidney slices can deiodinate thyroxine into triiodothyronine. The deaminated analogs, tetraiodothyroacetic acid and triiodothyroacetic acid, have been demonstrated only in peripheral tissues. Each have qualitatively different biological effects than thyroxine and triiodothyronine. It is not known what form of the thyroid hormone acts on peripheral tissues, and it may well be that a complex of thyroid compounds operate at the tissue level (Lardy et al., 1957; Larson and Albright, 1958). It is believed that as the hormones circulate through the tissues, they are released from protein carriers e.g., albumin and α-globulin, and pass through capillary walls affecting tissue cells (Tata, 1958; Hamolsky et al., 1961).

Attainment of normal adult form and dimensions in the absence of thyroid secretions cannot be achieved by most vertebrates. Thyroidectomy in young rats leads to a significant reduction in growth. The small skeleton of such animals grossly resembles those of hypophysectomized rats, with the exception that the epiphyseal cartilages remain open. Thyroid hormone administration allows for the resumption of normal growth.

Strong experimental evidence exists to show that thyroid hormone and somatotropin act synergistically in promoting normal skeletal growth (Marx et al., 1942; Geschwind and Li, 1955; Udupa and Gupta, 1965; Riekstniece and Asling, 1966). A dose of thyroxine so minute as to elicit no detectable response alone, is able to increase the sensitivity of the "tibial line" assay for growth hormone (Geschwind and Li, 1955). Thyroxine injected into normal rats does not cause an increase in body length. However, Riekstniece and Asling (1966) reported that thyroxine induces a slight but discernible unsustained elongation of the skeleton in hypophysectomized rats—a transient effect which is independent of the action of STH. Administration of thyroid hormone prevents the arrest of endochondral ossification in hypophysectomized growing rats. In these animals (Fig. 13A, B) chondrogenesis is not maintained; thyroxine permits continued epiphyseal plate cartilage resorption and bony replace-

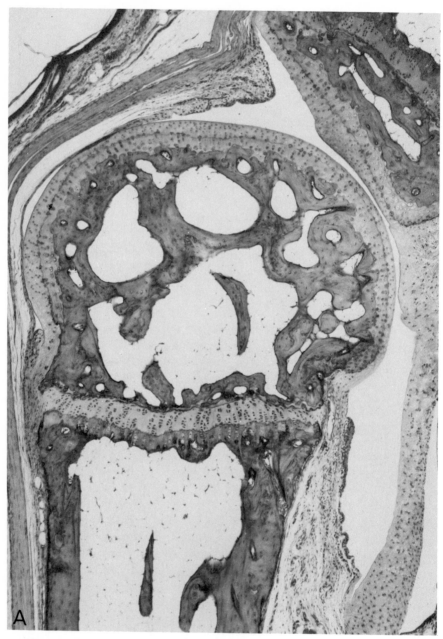

A

Fig. 13. The photomicrographs show the distal end of the third metacarpal of hypophysectomized female rats. A, Taken from an untreated rat one year post surgery. Note the retarded maturation of bone by the persistence of atrophic epiphyseal plates which normally close at about 100 days of age. B, Taken from

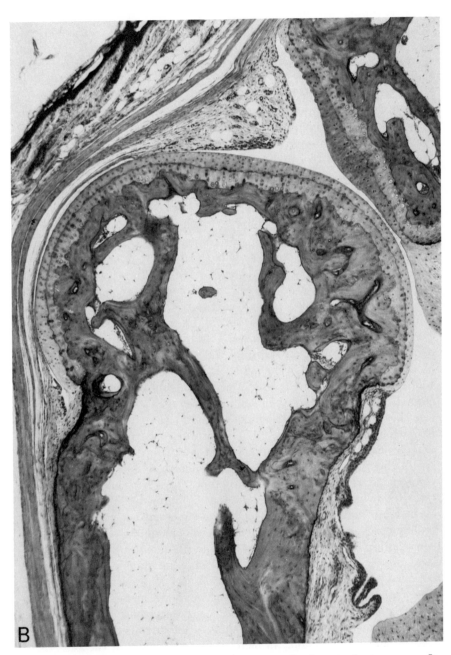

a similar animal, but which received treatment with thyroxin for one year. In this animal, the epiphyseal plate was resorbed. ×39. (Reproduced with the courtesy of the authors and publisher from Asling *et al.*, 1954.)

ment, curtailing skeletal linear growth (Asling et al., 1954). Therefore, thyroxine can induce the appearance of new epiphyseal ossification centers and their subsequent fusion with the diaphysis through epiphyseal plate resorption. Apparently, these events can occur in the absence of the pituitary (Asling et al., 1954), the thyroid (Scow et al., 1949; Ray et al., 1950b), or both (Ray et al., 1954). The phenomenon is affected through continued erosion enhanced by the appearance of chondroclasts during thyroxine administration (Ray et al., 1950a). In the absence of the thyroid, the integrity of articular cartilage of mice is lost and it may, therefore, be concluded that the thyroid hormone contributes in some way to the maintenance of articular cartilage (Silberberg and Hasler, 1968). It is also of interest to point out that in a histological study by Silberberg and Silberberg (1954a) in which ^{131}I was used to induce thyroid deficiency in mice, the skeletal response to this deficiency decreased with advancing age (Figs. 14A–D). Thyroid deficiency in mice revealed a number of cytological changes at the epiphyseal plate. Retardation of columnar chondrocytes into hypertrophic cells was evident, resulting in transitory crowding of cells. Subsequently, proliferation and vascular erosion of plate cartilage were decreased. Ossification of cartilage was delayed and incomplete. Formation of the primary spongiosa was retarded and decreased, but the bone which was present persisted for an abnormal length of time due to inhibition of resorptive processes. This resulted in the appearance of an interlacing bony network at the metaphysis. Vascularization of the metaphysis and shaft was abnormal. Periosteal ossification and lacunar resorption of cortical bone were also diminished. Although the above description results from thyroid deficiency induced by ^{131}I administration, it is similar to changes observed following thyroidectomy (Silberberg and Silberberg, 1954a).

Silberberg and Hasler (1968) in an electron microscopic study of young mice, reported rapid ultrastructural changes in articular chondrocytes following a single administration of thyroxine. Within 1–8 hours, an increase in hypertrophy of cells was observed accompanied by cell proliferation and marked organellar development, especially the rough endoplasmic reticulum and increased free ribosomes indicative of stimulated protein synthesis (Fig. 15). The Golgi apparatus became conspicuous and mitochondria more numerous, some showing enlargement. The association of the Golgi with carbohydrate metabolism and mitochondria with cellular respiration and energy mechanisms, taken together with stimulated protein synthesis, point to the augmentation of cartilage cells in matrix production. An increase was also noted in vacuoles, dense bodies, and lysosomes. The intensification of the hormone effect on the cell ultrastructure proceeded to 8 hours, after which time no further

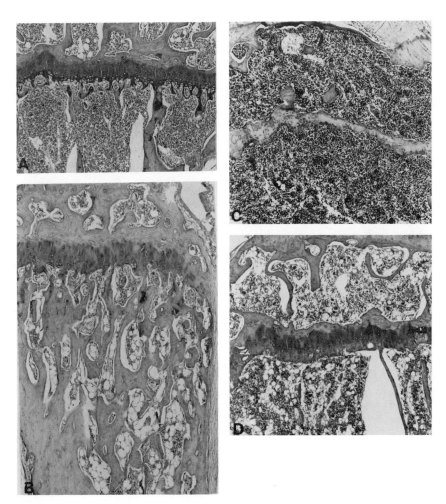

Fig. 14. The photomicrographs represent sections of the epiphyseal plates taken from the proximal ends of male mouse tibias. A, Untreated 4-month-old control mouse. Regular cartilage columns are seen separated by hyaline matrix. Short delicate trabeculae are noted at the metaphysis. B, Four-month-old animal which was treated with ^{131}I at 1 month of age. Note the wider epiphyseal growth zone and increased matrix leading to an extensive metaphyseal-trabecular framework. The bone marrow consists of numerous fat cells. C, Untreated 18-month-old control mouse. The epiphyseal plate is narrow consisting essentially of osseous material. D, Eighteen-month-old animal which was treated with ^{131}I at 1 month of age. Note the persistence of cartilaginous substance sandwiched between bony plates. ×55. (Reproduced with the courtesy of the authors and publisher from Silberberg and Silberberg, 1954a.)

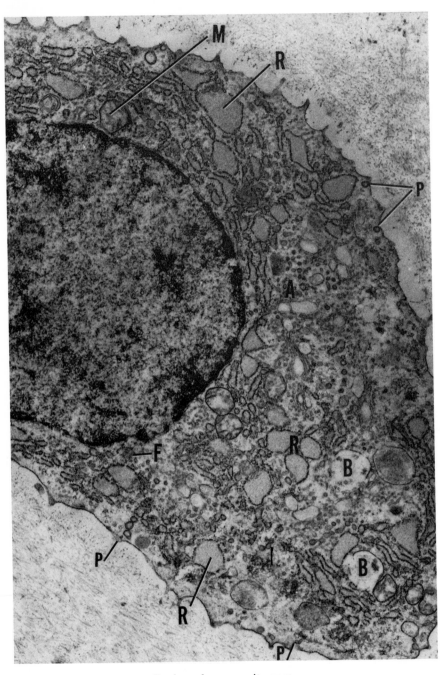

See legend on opposite page.

augmentation was seen. If thyroxine was again given at 24 hour intervals, the hormonal action was continued, manifesting itself in additional acceleration of cellular development and degeneration and in advancing mineralization of the matrix. Ultrastructural regressive changes were seen concomitant with progressive subceullular alterations. The electron microscopic study revealed hormonal stimulation of both cell growth and functional capacity followed by premature exhaustion.

It is significant to note as previously stated in the STH discussion, that it proved impossible to stimulate skeletal growth to the point of gigantism in thyroidectomized rats by prolonged administration of high doses of STH, and that this effect was, however, achieved in intact or hypophysectomized rats (Asling *et al.*, 1965). Growth hormone stimulated osteogenesis, revealing widening and thickening of bones, but not elongation which depends on chondrogenesis at the epiphyseal plates. When traces of thyroxine were supplemented full endochondral osteogenesis resumed. It would appear, therefore, that growth hormone alone supports osteogenesis, but that thyroxine is necessary for growth hormone to stimulate sustained chondrogenesis. A dose of 0.25–0.5 μg/day of l-thyroxine restored vigorous endochondral osteogenesis in thyroidectomized rats, increased their pituitary STH to normal and in hypophysectomized-thyroidectomized rats augmented the effect of growth hormone dose equal to that given alone to hypophysectomized rats (Riekstniece and Asling, 1966).

In fracture repair studies by Udupa and Gupta (1965), it was shown that supplementation of growth hormone with thyroxine augmented the formation of the cartilaginous callus, followed by a more rapid bony transformation and remodeling. The process of ossification was undisturbed and the bony tissue formed was completely normal. Similar findings were reported by Tarsoly *et al.* (1965), where the difference between controls and experimentals was the increased rate of chondrogenesis and earlier growth of trabeculae. In hypothyroid animals the rate of callus formation was reduced. The inhibition of callus formation was

---

Fig. 15. An electron micrograph is shown of a deep midzonal chondrocyte of the articular cartilage from the femoral head of a 3-week-old mouse, 8 hours after a single injection of 20 μg of thyroxine was administered. Regressive changes were promoted simultaneously with progressive changes, as can be noted in the extensive granular endoplasmic reticulum (R) showing both dilated and degenerated cisternae resulting from over stimulation. Numerous pinocytotic vesicles are encountered (P), few regular and enlarged mitochondria (M), several microvesicular bodies (B), free ribosomes (F) and Golgi system (A). Compare with Fig. 12. ×14,500. (Reproduced with the courtesy of authors and publisher from Silberberg and Hasler, 1968.)

made manifest principally through delayed chondrogenesis and prolonged ossification.

In man, the skeletal effects of hypothyroidism have long been recognized. When hypothyroidism occurs during childhood, it is always characterized by retardation of growth and all the processes of development.
There occurs not only a marked decrease in the rate of skeletal growth,
but also changes in body proportion associated with skeletal maturation.
The hypothyroid dwarf in contrast with the hypopituitary dwarf, retains
infantile skeletal proportions. Accordingly, an untreated congenital cretin
shows marked infantile characteristics which are not shown by an individual who becomes hypothyroid later in life after a more advanced
level of development (Wilkins, 1955). In stunted growth, a marked
delay in ossification time of the epiphyseal centers is always observed.
The delay is not specific to hypothyroidism for it may be observed
in cases of severe malnutrition, recently associated with hypopituitarism
and secondary hypothyroidism. When growth retardation and delayed
epiphyseal ossification is due to hypothyroidism, thyroid hormone administration induces a rapid and spectacular improvement.

Epiphyseal dysgenesis is the most specific skeletal abnormality in hypothyroidism (Wilkins, 1955). Normally, ossification is initiated in a
single center. In dysgenesis, ossification begins in several centers within
the cartilage. These grow and coalesce producing a stippled appearance.
Further calcification and bone formation is impeded unless thyroid hormone treatment is initiated.

It would appear from a plethora of experimental evidence that thyroid
hormone, like pituitary growth hormone, stimulates cell proliferation,
growth, and cell functional capacity in protein and possibly carbohydrate
synthesis, inducing general skeletal growth. The effects are mediated
via response of the epiphyseal plate, as well as articular cartilage and
bone cell compartments. Where growth hormone is essential to stimulate
osteogenesis, thyroxine is necessary for growth hormone to stimulate
chondrogenesis. Supplementation of growth hormone with small amounts
of thyroxine are essential to normal skeletal growth and development.
On this basis, thyroid hormone appears to be synergistic with growth
hormone.

V. ANDROGENS AND GONADOTROPINS

A. Androgens

Androgens are masculinizing compounds which are produced mainly
by the interstitial tissue and to a degree by seminiferous tubules of

the testes. A number of organs including the adrenal cortex, ovary, and placenta, however, possess the enzymatic system necessary to produce androgenic steroids. Thus, the endocrine difference between these organs is quantitative rather than qualitative. Testosterone and androstenedione are the main circulating androgens of testicular origin. Recent experimental evidence indicates that dihydrotestosterone (5α-androstan-17β-ol-3-one), a metabolite of testosterone, probably serves as the biologically active form of the hormone, acting on target cells. It is possible, however, that the manner in which testosterone acts at the cellular level may not be the same in all tissues.

In the growing animal, skeletal development is retarded following castration. The long bones of rats orchiectomized at 12 weeks of age are significantly shorter than controls when sacrificed at 1 year of age (McLean and Urist, 1961). Skeletal retardation is more pronounced the earlier in life the orchiectomy is performed (Silberberg et al., 1958b). In humans, the epiphyseal plates which normally close prior to puberty are least effected or noneffected. Plates such as femoral, tibial, radial, and ulnar which unite during puberty reveal retardation of epiphyseal closure, while those of the iliac crest which close postpuberty, may never unite. Delay in epiphyseal closure is followed by extension of the growth period. Growth, however, is not accelerated, but proceeds at a reduced rate (Silberberg and Silberberg, 1956). Despite the resulting body size, a disproportionate increase is observed in the length of bones whose epiphyses were open at the time of orchiectomy (Horstmann, 1949). These eunuchs exhibit long arms, hands, legs, feet, and elongated skeletal facial features resembling acromegalic individuals. The skeleton is, in general, delicate and sometimes osteoporotic (Ravault et al., 1950). The findings in a wide variety of animals are quite similar to those seen in eunuchs.

Histologically, the epiphyseal cartilage of the long bones of castrated mice reveals an increased number of columnar cells, with a delay in hypertrophy (Fig. 16A, B). Regressive modifications and provisional cartilage replacement by bone are also retarded (Silberberg and Silberberg, 1946). Interestingly enough, the frequency of degenerative joint disease and aging is also described in articular cartilage (Silberberg and Silberberg, 1954b; Silberberg et al., 1958b). In response to gonadectomy, rat cartilage and trabecular bone exhibit intense PAS (Figs. 16A, B and 17A, B) and alcian blue-PAS staining and metachromasia following toluidine blue staining (Bernick, 1970). It was concluded that the results implied that the ground substance contains carbohydrate–protein complexes in a state of lesser degree of polymerization or a relative increase in neutral hexosamine-containing polysaccharides with a de-

302   *Edgar A. Tonna*

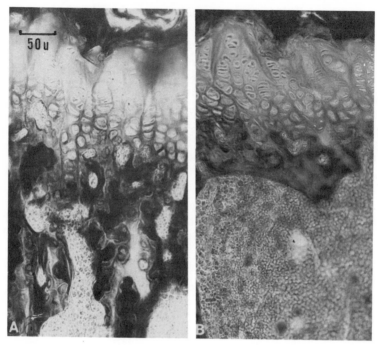

Fig. 16. The photomicrographs show sections of the epiphyseal plates taken from the proximal head of the tibias of an (A) intact rat and (B) orchiectomized rat sacrificed 8-months postsurgery. A comparison shows closure of the epiphyseal plate by bone. ×125. (Reproduced with the courtesy of the author and publisher from Bernick, 1970.)

crease in acid mucopolysaccharides. Normally, in intact animals, the matrices are initially PAS positive and metachromatic. Coincidental with mineralization, the matrix is altered losing its metachromasia. With further maturation, the formed trabeculae lose their PAS and metachromatic intensity. It may be assumed, therefore, that following gonadectomy there occurs a change in the state of aggregation of polymerization of the protein-carbohydrate complexes of the ground substance of both the epiphyseal cartilage and bone (Fig. 18A, B), and that protein and acid mucopolysaccharide alterations impede the mineralization of bone leading to premature epiphyseal plate closure (Bernick, 1970).

Interestingly enough, daily administration of testosterone to young rats causes shortening of bodies, tails, femurs, and tibias. The epiphyseal plates are thin and atrophic (Trueta, 1968). While small doses of the hormone stimulate skeletal growth (Rubenstein and Solomon, 1941b) as observed during the prepubertal growth spurt occurring in normal

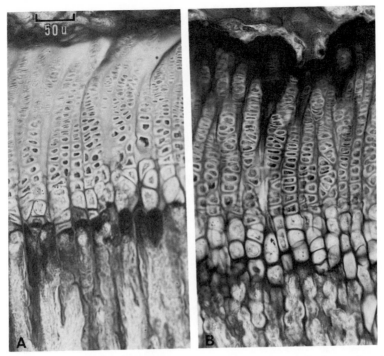

Fig. 17. The photomicrographs show sections of the epiphyseal plates taken from the proximal head of the tibias of an (A) intact rat and (B) orchiectomized rat sacrificed 1 month postsurgery. A comparison of the sections stained with the PAS method reveals the intensification of the reaction in the orchiectomized animal. ×125. (Reproduced with the courtesy of the author and publisher from Bernick, 1970.)

boys and the temporary acceleration of growth noted in sexually precocious children, large doses depress weight gain and delay linear growth (Rubenstein and Solomon, 1941a). The spurt in linear growth caused by androgens in the intact organism may be attributed to a stimulation of the growth of the epiphyseal plate, rather than to the general anabolic effect of these hormones (Silberberg and Silberberg, 1956). More recently, it was reported that testosterone increases growth hormone secretion (Illig and Prader, 1970). Whether administered in small or large doses, testosterone always hastens skeletal development, causing the premature appearance of ossification centers and epiphyseal closure. Premature epiphyseal closure accounts for the ultimate deficiency in body size of animals receiving large doses of hormone and of children with precocious sexual maturity. On the other hand, Arm-

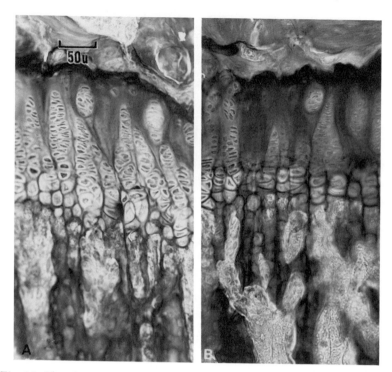

Fig. 18. The photomicrographs show sections of the epiphyseal plate taken from the proximal head of the tibias of an (A) intact rat and (B) orchiectomized rat sacrificed 5 months postsurgery. A comparison of the sections stained with the PAS method reveals at this time intensification of the reaction throughout the plate matrix and into the cartilaginous cores of the metaphyseal trabeculae. Compare with Fig. 17A, B. ×125. (Reproduced with the courtesy of the author and publisher from Bernick, 1970.)

strong (1942) reported that under strictly controlled conditions the hormone apparently had no effect on bone repair in the dog. A number of investigators, however, have reported that androgen administration enhances skeletal aging (Silberberg and Silberberg, 1946; Howard, 1962, 1963; Puche and Romano, 1968, 1971).

Testicular 17-ketosteroids have been shown to affect the levels of calcium and phosphorus in serum and bones of normal individuals (Kennedy et al., 1953). In aged humans with osteoporosis, Reifenstein and Albright (1947) reported the retention of nitrogen, calcium, and phosphorus following testosterone administration. The effect of testosterone on calcification is, however, transitory (Harris and Heaney, 1969; Puche and Romano, 1971). Testosterone was also shown to produce a shift in the bone carbohydrate metabolic pathway towards the pentose shunt; cartilage cell hyperplasia and calcification of the epiphyseal plate accompanied this shift (Puche and Romano, 1970).

Microscopic studies show that testosterone administered in large doses suppresses proliferation of cartilage, but intensifies hypertrophy, hyalinization, calcification, and ossification. Resorption of provisional cartilage and metaphyseal trabeculae is slightly inhibited (Silberberg and Silberberg, 1956). In an interesting study by Puche and Romano (1971) in which actinomycin D and testosterone were administered simultaneously, the stimulatory effect of the hormone was cancelled. Actinomycin D inhibits DNA-dependent RNA synthesis, resulting in a depression of the total hydroxyproline content of bone which is in turn indicative of reduced collagen protein synthesis. The authors believe that administration of the antibiotic for 16 days did not suppress the information already transcribed into the cell cytoplasm, but effectively cancelled new information for matrical synthesis which might be induced by androgens. Beyond 31 days, testosterone appeared to increase the rate of epiphyseal plate aging.

In a tissue culture study of chick embryo frontal bones at 12 days of development treated with testosterone and dehydroepiandrosterone sulfate, a naturally occurring steroid of adrenal origin, similar histological results were obtained (Puche and Romano, 1968). Periosteal hyperplasia and increased osteoid synthesis was observed and accompanied by enhanced alkaline phosphatase activity. Increase in alkaline phosphatase was previously reported as a response to androgen administration (Bellieni *et al.*, 1956; Rodin and Kowaleski, 1963).

At the electron microscope level, Fahmy *et al.* (1971a) reported that at the epiphyseal plate of rats, administration of supraphysiological doses of testosterone induced cell division at the proliferative zone. At the zone of maturation, cells accumulated larger amounts of secretory materials, as well as, lipid and glycogen at earlier stages of their life cycle, advancing the stage of hypertrophy. At 6 hours postinjection, an increase in nucleolar density was noted. By 24 hours, cells in the matrix secreting zone showed an increase in secretory activity and enhancement of the Golgi apparatus which was seen filled presumably with protein-polysaccharide material. The rough endoplasmic reticulum was prominent and dilated with protein precursors. Thicker and longer collagen fibers were observed in the interlacunar matrix, in addition to premature calcification.

Rubenstein *et al.* (1939) asserted that in the intact animal, large doses of testosterone may exert their effects through suppression of the hypophyseal growth hormone. Conversely, small doses stimulate skeletal growth, presumably by hypophyseal stimulation (Simpson *et al.*, 1950). In children, large doses reduce ultimate height by producing early epiphyseal closure, and small doses increase ultimate height. Growth promotion of testosterone was observed in hypophysectomized animals (Simpson *et al.*, 1944), therefore, its action is likely to be independent of the hypophysis. In hypophysectomized rats, large doses significantly depress the epiphyseal growth produced by growth hormone, suggesting also a peripheral growth inhibitory action (Reiss *et al.*, 1946). When the androgenic effects produced in tissue culture (Puche and Romano, 1968) are added to this picture, it would seem that androgens act directly on skeletal structures. The mechanism of action may well involve the current hypothesis, advanced by Hechter and Halkerston (1965), that these compounds produce their effects by regulating the genetic programming of cells in responsive tissues.

In an attempt to resolve the apparent discrepancy between the effects obtained with large versus small doses of hormone, it has been postulated (Joss *et al.*, 1963) that growth stimulation at lower doses reflects predominantly an anabolic effect in the absence of stimulation of skeletal maturation, while growth inhibition by larger doses reflects an androgenic response with marked acceleration of skeletal maturation. Furthermore, Sobel *et al.* (1956) stated that in the child, as well as, in the rat, stimulation of skeletal maturation may be proportional to the androgenic dose, while stimulation of growth rate may be relatively independent of hormone dosage.

B. Gonadotropins

The placenta, and in certain species the endometrium, produce gonadotropins which are gonad-stimulating hormones and are similar in some respects to those produced by the adenohypophysis. They differ from pituitary gonad-stimulating hormones physiologically and chemically. Chorionic gonadotropin is a glycoprotein and in primates is characteristic of pregnancy. It is secreted by the chorionic villi of the placenta and is found in the blood and in large quantities in the urine during early pregnancy. When these hormones are administered to male animals and man, they elicit skeletal growth responses resembling those of androgens. The skeletal effects of gonadotropins, by and large, result from increased production of androgenic hormones. Since the action of these hormones on skeletal tissues has been discussed in some detail in this section, the subject will not be pursued any further.

VI. ESTROGENS AND PROGESTOGENS

A. Estrogens

The ovarian structure elaborates estrogens, progestogens, androgens, and the nonsteroid hormone, relaxin. Both the mature ovarian follicle and corpus luteum produce estrogenic hormone while the latter produces, in addition, progestational hormones. The cellular source of ovarian androgens and relaxin is presently unknown. Estrogens stimulate growth and differentiation of the female reproductive tract and associated structures and also exert numerous systemic effects including those affecting the skeletal system.

Estradiol-17β, estrone, and estriol represent the predominant natural estrogens in man. It appears that the human ovary elaborates only estradiol-17β and estrone, whereas, estriol is believed to represent a degradation product. Estrogens are present in a variety of animal tissues including ovaries, testes, adrenals, and placentas.

The significant effects of ovarian deficiency have been recognized in animals and man for a number of decades. Studies have been conducted in cases of primary ovarian deficiency as occurs in the freemartin and in humans with ovarian hypoplasia or agenesis subsequent to therapeutic or surgical ovariectomy. In the growing animal, ovarian deficiency affects both endochondral and intramembranous ossification resulting in thin, delicate bones with decreased breaking strength. Retardation of epiphyseal plate and cranial suture closure occurs, while the appearance of ossification centers is delayed. Initially, linear body growth is increased due to increased tubular bone length. The tails of animals may also be increased in length. This stimulatory effect is, however, transitory and is subsequently followed by a slowdown in the growth rate (Silberg and Silberg 1956). These skeletal changes often lead to disproportionate extremities, whereas, the overall skeletal size may not exceed its normal size and may even be somewhat stunted. In the adult organism, ovarian deficiency does not exert noticeable skeletal effects. The closed epiphyses and articular cartilage react less readily. In rats and mice, osteoporosis is often observed following ovariectomy; however, it has not been determined that in women, ovarian deficiency is the cause of postmenopausal osteoporosis. Callus formation subsequent to fracture is also retarded. In young rats, Bernick (1970) reported that 1 month postovariectomy the tibial epiphyseal plates were slightly wider than those of nonoperated controls. Beyond 2 months postovariectomy, the same plates were narrowed and prematurely closed. Osteoporosis was also observed as early as 5 months following surgery. These results do not appear to differ from those of orchiectomized rats except for the fact that osteoporotic bone in male rats is not observed until 8 months following surgery. Spontaneous degenerative joint disease in the form of osteoarthritis is common in various strains of mice of both sexes. Ovariectomy performed on mice 1, 6, and 12 months of age results in delayed onset, decreased incidence and an attenuated course of osteoarthritis. The results, perhaps, may be explained on the basis of the retardation of age changes in articular cartilage following spaying (Silberberg et al., 1958a). With increasing age, the effects are less remarkable. On the other hand, intermittent administration of α-estradiol benzoate also decreases the incidence of osteoarthritis. This effect is more marked when the treatment is initiated in progressively older animals (Silberberg and Silberberg, 1970).

Histologically, during the early phase of hormone deficiency, increased numbers of cartilage cells at the epiphyseal growth zones and increased hypertrophy of cells in the cartilage columns are observed. A retardation of trabecular formation in the spongiosa is also encountered. The initial hyperplasia and hypertrophy of cartilage cells may well be due partly to growth stimulation and partly from a transitory retardation of normal degenerative changes (Silberberg and Silberberg, 1956). Cartilage and bone matrices remain strongly positive for PAS and alcian blue-PAS staining and exhibit intense metachromasia (Bernick, 1970). These reactions diminish in normal animals during maturation and mineralization of the matrices (Siffert, 1951; Van den Hooff, 1964).

Administration of estrogens results in three major skeletal changes: linear growth inhibition seen in both long and flat bones, acceleration of skeletal development, as indicated by premature epiphyseal closure and the earlier appearance of ossification centers, augmented aging and the condensation of bone (Silberberg and Silberberg, 1956; Tapp, 1966).

Microscopic studies reveal inhibition of cartilage growth and intensified hyalinization and calcification of ground substance. The erosion of provisional cartilage by invading capillaries is retarded (Silberberg and Silberberg, 1956). Chondroclasts and osteoclasts decrease in number, and vascular channels in cortical bone exhibit reduced diameters (Whitson et al., 1971). Estrogen administration exerts a profound effect on the endosteum, increasing the number of osteoblastic cells. In an [³H]thymidine autoradiographic study in which 1-month-old mice were treated with estrogen, Simmons (1963) reported that within 6–12 hours, a stimulation of osteoblast formation was observed. This stimulation persisted during the 2-week study period. In mice, trabecular bone of the metaphysis is stimulated to grow by direct elongation into the medullary canal. With proper dose and period of treatment, trabecular bone may fill the entire canal. Although no response to treatment is observed initially between 10 days of age and senility, large amounts of endosteal bone can be produced (Urist *et al.*, 1950). These events require the production of large quantities of collagen. Estrogen treatment is reported to stimulate collagen metabolism in rats (Kao *et al.*, 1967). Budy (1960) using [¹⁴C]estrone reported the selective deposition of the steroid in the endosteum, accounting for the specific effects of estrogen and/or its metabolites observed on endosteal bone.

It must be pointed out that in mice ($C_{57}Bl$) treated with estrogen (Fig. 19A–D), the cortical atrophy which is normally encountered after 17 months of age is slightly diminished if treatment is initiated from 1 month of age, and enhanced when treatment is initiated beyond 6 months of age (Silberberg and Silberberg, 1970). It would appear that this process results in extensive trabecular formation from cortical bone, in addition to the effect seen on metaphyseal bone itself. Despite the extensive new bone formation which obliterates hemopoietic elements, the normal vascular pattern of the epiphysis, metaphysis, and diaphysis persists (Brookes and Lloyd, 1961).

In the young rat (Fig. 20A, B), no new bone formation is observed; however, normal metaphyseal bone resorption is inhibited, resulting in extending the spongiosa into the medullary canal for some distance (Urist *et al.*, 1948; Budy *et al.*, 1952). The effects of estrogen are, therefore, limited in the rat to the endochondral growth apparatus; consequently, no demonstrable effects are observed in adult rats. In intramembranous bone, no effect is observed in rats at any age. Specific skeletal effects have failed to be induced in guinea pigs, hamster, rabbits, cats, and dogs (McLean and Urist, 1961). In man, the effects of estrogen administration particularly in osteoporosis has led to unequivocal results Anderson, 1950).

The effects of estrogen on avian bone is remarkable and cannot be excluded from any discussion pertaining to skeletal effects of the hormone. In birds, the increased medullary deposition of bone originates in the diaphysis rather than in the metaphysis, as seen in the long bones of mammals. In female birds, ossification of the medullary canal occurs normally prior to egg laying by the production of a secondary system of spongy bone. Bony spicules are laid down along the cortical endosteum, subsequently obliterating the marrow cavity. Once the egg is laid the excess bone is resorbed (Keys and Potter, 1934; Riddle *et al.*, 1944, 1945). Similar changes can be induced by the administration of estrogen to male birds. The overmineralization of avian long bones occurring under the influence of estrogen is related to the simultaneous appearance of hypercalcemia. Although mineralization of new bone for-

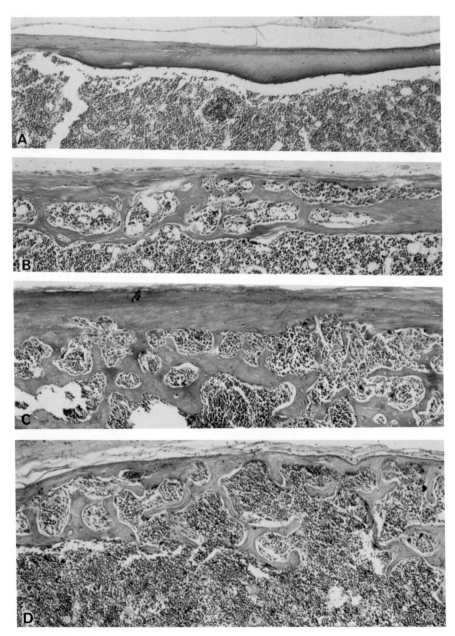

Fig. 19. The photomicrographs show regions of the femoral cortex of female mice. A, From a 2-year-old untreated mouse. B, From a 2-year-old untreated mouse exhibiting advanced osteoporosis. C, From a 16-month-old mouse treated with estrogen from the age of 1 month and sacrificed 2 months postinjection. Note excessive

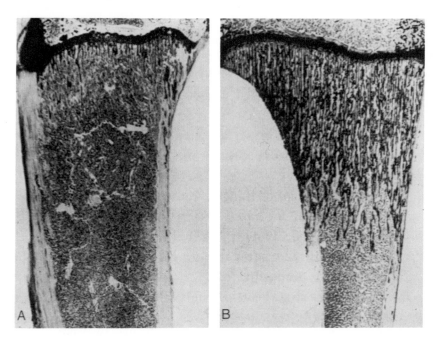

Fig. 20. The photomicrographs A and B illustrate inhibition of tibial metaphyseal-trabecular resorption following the treatment of immature rats with 8.0 mg of 17β-estradiol benzoate for a period of 5 weeks. A, control; B, treated. ×10. (Reproduced with the courtesy of the authors and publisher from McLean and Urist, 1961.)

mation is usually considered a process preparatory to egg shell formation, at the same time it may also serve as a protective mechanism against the injurious effects of hypercalcemia (Silberberg and Silberberg, 1956). It is significant to note that the fibrils of the bone matrix induced by estrogen in pigeons and ducks when observed under the electron microscope were found to lack the orderly arrangement usual to normal bone and were associated with a varying amount of cementing substance (Ascenzi et al., 1963). These findings support the view that the bone resulting from estrogen stimulation lacks mechanical function, but responds to the necessity of accumulating rapid quantities of calcium salts. It also accounts for the rapid resorption of the avian secondary spongiosa

subcortical trabecular bone formation. D, From a 19-month-old mouse treated with estrogen from the age of 1 month and sacrificed 5 months postinjection. Excess trabecular bone has been remodeled and appears to be delicate. Note the significant cortical thinning. ×90. (Reproduced with the courtesy of the authors and publisher from Silberberg and Silberberg, 1970.)

following egg laying. It is presumed that the trabeculae formed in mammals in response to estrogen are similarly devoid of any mechanical function. In addition, ^{45}Ca studies show a 50% reduction in the rate of resorption at the upper ends of rat tibias following estrogen treatment. Inhibition accounts for an effect on the secondary spongiosa while the resorption incident to remodeling is not appreciably affected by estrogen treatment (Lindquist et al., 1960).

By and large, the effects of estrogen vary depending on the time of administration, dose and duration of treatment. In addition, the osteogenic effects of estrogen vary from one part of the skeleton to another (Gardner, 1943). To a degree, estrogen also exhibits species-specificity (Urist et al., 1948, 1950; Schiff, 1966).

Investigation into the effects of estrogen on cellular biosynthesis, transport, pinocytosis, diffusion, etc., suggests that the hormone or its metabolites initially stimulate the manufacture of nucleoprotein; subsequently, cytoplasmic synthesis of protein ensues (Nichols, 1966). Hamilton (1968) postulated that accelerated nuclear RNA synthesis followed by ribosomal formation is one of several primary mechanisms of estrogen, i.e., control of genetic transcription and translation.

Electron microscopic studies reveal that in articular cartilage of mice 16 hours after estrogen treatment, chondrocytes exhibit premature development of the Golgi complex, formation of multivesicular bodies, increased glycogen deposition, in addition to increased intra- and extracellular fibrillarity. Prolonged estrogen treatment results in collagen accretion showing thicker and more densely packed fibrils (Silberberg and Silberberg, 1965; Silberberg et al., 1965). Ultrastructural observations of the rat tibial epiphyseal plate (Fahmy et al., 1971b) are reported to be similar to those observed at the articular cartilage. In both, estrogen treatment results in increased free ribosomal content, accelerated development of the Golgi complex and increased intravacuolar material. Increased fibrillogenesis is also observed in the epiphyseal cartilage of adult rats (Fahmy et al., 1969). These modifications are believed to be specific to estrogen and have not been encountered in response to testosterone (Fahmy et al., 1971b) or somatotropin (Silberberg et al., 1965; Fahmy, 1968). While cellular maturation is accelerated following administration of estrogen, testosterone, and growth hormone, distinctly different mechanisms of hormone action on chondrocytes are observed. Following estrogen treatment, the extracellular transport of protein and polysaccharide is retarded leading to hypertrophy and premature chondrocyte degeneration. Large accumulations of glycogen are also noted after testosterone and growth hormone treatment, in addition to an inordinate amount of lipid after testosterone; however, the hormones also appear to promote extracellular protein and polysaccharide transport through increased footlet formation.

The electron microscopic results of estrogen treatment, therefore, reveal an initial stimulation of RNA and DNA synthesis and subsequent protein synthesis followed by retardation of its transport as noted by the premature deposition of glycogen and by fibrillogenesis. The latter is ascribed to the depression of mucopolysaccharide production following estrogen treatment (Priest *et al.*, 1960; Minot and Hillman, 1967). Reduced collagen precursor transport in the presence of reduced polysaccharide concentration is believed to promote intracellular polymerization of collagen contributing to degeneration of chondrocytes. It must be pointed out, however, that the effect of estrogen on endosteal osteoblasts is expected to differ, since a remarkable amount of matrical production occurs in order to develop the extensive secondary spongiosa in birds and the trabecular-metaphyseal extension in some mammals.

Testosterone exerts a significant antagonistic and inhibitory effect on estrogen-induced bone formation. This has not been discussed since it is beyond the scope and purpose of this chapter; however, the reader is referred to the excellent experimental study of 167 mice by Suzuki (1958).

B. Progestogens

These steroids exhibit varied actions upon the female reproductive organs and often act synergistically with estrogens, under physiological conditions. The progestogens and estrogens are also capable of acting antagonistically, thus capable of inhibiting the actions of each other. Progesterone, 20α-hydroxypregn-4-en-3-one and 20β-hydroxypregn-4-en-3-one are known to occur naturally. These substances are present in the ovarian follicles, corpora lutea, placenta, and blood. The principal metabolites are pregnanediol and pregnanetriol which are excreted in the urine chiefly as glucuronides.

The effects of progesterone on the skeletal system, on body size, and on calcium metabolism are not remarkable. Consequently, little experimental work has been reported in this area by comparison with androgens and estrogens. Silberberg and Silberberg (1941) reported that in young guinea pigs, epiphyseal plate growth is stimulated only slightly, temporarily repressing degenerative changes and cartilage ossification. The effects resemble the initial changes observed following ovariectomy. It was concluded that the limited skeletal effects may have resulted from inhibition of other endogenous steroids which induce skeletal development, rather than through any specific action of progesterone on the skeletal system. Gardner (1936, 1940) showed that in mice progesterone administration is unable to prevent the skeletal changes induced by estrogen. When pregnant rats are treated with progesterone, however, the fetuses grow larger (Saunders and Elton, 1959) and newborn female mice exhibited masculinization and advanced skeletal maturation (Breibart *et al.*, 1963). Silberberg and Silberberg (1965a) reported that progesterone exerts a protein-catabolic effect resulting in retardation of skeletal development in growing, noncastrated mice. On the other hand, a slight increase in body weight is observed in progesterone treated castrated mice, compared with nontreated orchiectomized mice of the same age. Both groups are, however, lighter than control mice. Female rats consumed more food, gaining more weight when given progesterone (Silberberg and Silberberg, 1965b). Male rats do not exhibit these effects (Hervey and Hervey, 1964).

In orchiectomized mice, progesterone administration augments the aging of articular cartilage and promotes the development of senile osteoarthrosis, counteracting the effect of castration which is known to retard and attenuate these events (Silberberg and Silberberg, 1965b).

It would appear from this short discourse that much experimental work remains to be done in this area if we are to begin to comprehend the action of progesterone, specifically and progestogens, in general, on the skeletal system. Difficulties which have plagued investigators involved in studying the effects of estrogen may well be involved here. Knowledge as to dosage effects, age of animals at time of treatment, duration of treatment, synergistic and antagonistic effects, sex and species differences are germane to the problem. Of equal importance is the effect progesterone deficiency or administration exerts on other endocrine organs. All of these parameters must be controlled before statements regarding the skeletal specificity or nonspecificity of progestogens can be put forward with some degree of confidence. Tissue culture studies which allow freedom from other circulating hormones and ultrastructural studies are in order.

VII. ADRENOCORTICOTROPIC HORMONE (ACTH) AND ADRENOCORTICAL HORMONES

A. Adrenocorticotropic Hormone (ACTH)

Highly purified adrenocorticotropic hormone (ACTH) has been isolated from the adenohypophysis of pig, sheep, beef, and man. In all of these species, ACTH consists of a straight chain polypeptide having

39 amino acid residues and a molecular weight of approximately 4500. Although species differences are recognized in the structure, all are able to stimulate the adrenal cortex, promoting the output of adrenocortical steroids. The metabolic changes resulting from ACTH are, by and large, equivalent to those induced by specific adrenocortical steroids.

Skeletal changes induced by administration or deficiency of ACTH and adrenal cortical hormones have been recognized for many decades (Moon, 1937; Ingle *et al.*, 1938; Wells and Kendall, 1940). The antagonistic action of ACTH to growth hormone in hypophysectomized rats is also known (Marx *et al.*, 1943). Using pure sheep ACTH, the body growth of normal and gonadectomized rats is inhibited. Inhibition does not occur, however, if the rats are adrenalectomized (Evans *et al.*, 1943). It was further shown (Becks *et al.*, 1944) that ACTH treatment of normal rats results in retarded chondrogenesis, as well as, osteogenesis. Following hypophysectomy, osteogenesis ceases, whereas, if the animals were treated with ACTH, osteogenesis continues at a reduced rate. The effects are clearly beyond nutritional factors and could not be explained on the basis of reduced food consumption. Since growth inhibition and skeletal changes are not induced by ACTH in the absence of the adrenals, further discussion will focus on the skeletal effects of adrenal glucocorticoids which are released from the adrenal cortex in response to endogenous or exogenous ACTH.

B. Adrenocortical Hormones

Numerous steroids have been isolated from the mammalian adrenal cortex numbering nearly 50. These include androgens, estrogens, progestogens, mineralocorticoids, glucocorticoids, and a large number of steroid precursors or metabolites of the active steroids. The most important adrenal corticoids are derivatives of pregnane and consist of 21 carbon atoms. These may be divided into two groups, those which exhibit a profound effect on mineral metabolism called mineralocorticoids, and those affecting carbohydrate metabolism called glucocorticoids. Corticoids which lack oxygen at C-11 exhibit major effects on water and electrolyte metabolism, e.g., aldosterone, 11-deoxycorticosterone and 11-deoxycortisol. Those which possess oxygen at C-11 produce significant effects on carbohydrate and protein metabolism. The most important natural steroids in this category include cortisone, cortisol (17-hydroxy-corticosterone, also called hydrocortisone), corticosterone, and 11-dehydrocorticosterone.

Of the adrenocortical hormones, only cortisone and its analogs significantly affect the rate of bone growth and resorption with skeletal changes exhibiting osteoporosis to metaphyseal sclerosis. Both crude adrenal cortical extracts and adrenal transplants have been shown to affect skeletal structures (Landauer, 1947a,b; Silberberg et al., 1954). Distinction, however, has to be made between glucocorticoids and mineralocorticoids. The role of the latter hormones is associated with the homeostatic regulation of electrolytes which constitute the internal environment. The nature of this subject is obviously germane to skeletal biology, but it is beyond the intended scope of the present chapter. It must be pointed out, however, that mineralocorticoids, e.g., deoxycorticosterone, have been reported to slightly inhibit the body growth of chick embryos (Stock et al., 1951), retard the growth and fracture repair of dog bones (Fontaine et al., 1952), and counteract the growth spurt observed in rat tail following growth hormone administration (Maassen, 1952). The changes are believed to be indirect and of little significance to skeletal growth and differentiation, save for their role in mineral metabolism. Much remains to be done experimentally with these substances.

With regard to the effects of glucocorticoids on the skeletal system, the growing chick is reported to be less sensitive to cortisone administration than mammals. Large amounts of hormone, however, inhibit embryonic development and produces significant stunting in growth (Karnofsky et al., 1951; Montgomery, 1955; Siegel et al., 1957). Sayeed et al. (1962) have shown that during early embryogenesis, cell proliferation of the mesoderm is depressed by 60 hours and ectoderm by 72 hours. No effect is noted on endoderm. Chondrogenesis is also delayed and the epiphyseal plate becomes narrower (Huble, 1957). In addition, failure of Haversian remodeling sequences results in osteoporosis using pharmacological doses of cortisone (Urist and Deutsch, 1960). Also, the temporary medullary spongiosa induced by estrogen in birds is less affected by cortisone than the lamellar cortical bone (Urist and Deutsch, 1960). Sobel (1958) reported that cortisone and ACTH are antagonistic with cortisone inhibiting the action of growth hormone on the developing chick.

The skeleton of the mouse and guinea pig are claimed to be insensitive to cortisone (Follis, 1951; Storey, 1963). Maturation of bone in infant mice (Howard, 1962) and the retardation of the growth processes in articular cartilage (Silberberg et al., 1954), however, have been observed. Experimental studies in which large doses of cortisone are given to immature rats show inhibition of growth of the facial skeleton (Moss, 1955; Johannessen et al., 1966). Low cortisone dose treatment retards skeletogenesis of the rat tibia, the cranial base, and the temporomandibular joint (Che-Kuo and Johannessen, 1970). In this regard, it is interesting to note the varied dose effects elicited by cortical hormones. Roberts

(1969) reported that in rat periodontal tissues, low doses of cortisol increase osteogenic cell proliferation, osteoclast numbers, and resorption of alveolar bone. At a high dose level, osteogenic cell proliferation and bone formation are depressed without evidence of increased resorption of alveolar bone. It is suggested from the results, that the cellular effect of cortisol on bone arises from the interaction of two antagonistic mechanisms, i.e., direct inhibition of osteogenic cell proliferation and induced secondary hyperparathyroidism, resulting in increased bone resorption. Jee *et al.* (1970) stated that in parathyroidectomized rats, low dose levels of cortisol depress precursor cell proliferation. Follis (1951) reported that the rat does not develop bone rarefaction following cortisone administration, unlike the bird, rabbit, and man; although, young rats exhibit bone growth inhibition, dense metaphyseal bone forms instead. This osteosclerotic bone results from a failure of metaphyseal resorption and remodeling and occurs only following large doses of cortisone (Sissons and Hadfield, 1955; Storey, 1960a). Maturation of endochondral cartilage is delayed. Chondrocytes appear smaller with proportionately greater amounts of intercellular matrix which stains more weakly for mucopolysaccharides. Osteoclasts are reduced in number, in addition to reduced vascularity. Consequently, unresorbed mineralized tissue is retained and subsequently enclosed by new lamellar bone formation (Follis, 1951). It is of interest to note that Storey (1960b) succeeded to induce osteoporosis in the rat, similar to that seen in other animals following cortisone administration, when the rate of bone resorption was increased by maintaining the rat on a calcium deficient diet.

Gross skeletal variations have been reported in rabbits following treatment with adrenal hormones (Blunt *et al.*, 1950; Sissons and Hadfield, 1951; Jee *et al.*, 1966). The striking changes taken from the elegant work of Jee *et al.*, (1966) are shown in Figs. 21–23. These show significant osteoporosis in compact bone, reduction in growth plate thickness, but persistence of some metaphyseal trabeculae. The latter observation differs from that seen in the rabbit and rat as reported by Sissons and Hadfield (1955) and Follis (1951). Such differences not only point to species variations, but experimental variables as well. Histologically, one sees a strikingly wide basophilic matrix at the reduced hypertrophic zone of the epiphyseal plate. At the primary spongiosa, cell modulation in favor of osteoclasts over osteoblasts is observed. In the femoral shaft, cell modulation in favor of osteoclast formation is markedly increased, accounting for the observed rarefaction.

In general, cortisone administration to growing rabbits, as in other animals, results in both inhibition of bone formation and increased bone resorption. The rate of osteoporosis is increased with the cortisone dose level (Storey, 1963), calcium deficiency (Storey, 1961) and increased

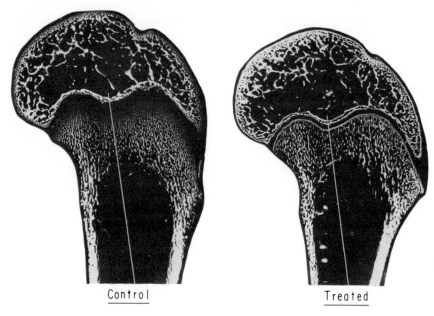

Control   Treated

Fig. 21. The photomicrograph shows microradiographic sections of rabbit distal femora taken from control and 5 mg cortisol/kg/day treated animals. Note the significant thinning of the epiphyseal plate of the treated animal. ×4.5. (Reproduced with the courtesy of the authors and publisher from Jee *et al.*, 1966.)

mechanical stress (Storey, 1958b). On the other hand, the development of osteoporosis is slower when older animals are treated (Storey, 1958a).

It has long been recognized that adrenocortical hormones affect the skeletal system of man. Albright (1942–43), postulated that Cushing's syndrome, opposite to the adrenogenital syndrome, resulted from pan-hypercorticalism. Children with Cushing's syndrome are somewhat retarded in growth (Wilkins, 1955). The syndrome can be induced by the presence of a basophilic adenoma of the pituitary, hyperactive adrenal cortex from a variety of causes, and primary adrenal hyperplasia of adrenocortical tumor. This condition is characterized by excessive gluconeogenesis due to excess secretion of glucocorticoids and is associated with antianabolic connective tissue changes leading to, among other things, the rarefaction of the skeleton. The osteoporosis associated with Cushing's syndrome differs from other types of osteoporoses of diverse etiology. A common feature, however, exists in that bone resorption is increased while the rate of bone formation is variable. The etiology of human osteoporosis to date, however, remains an enigma, with the realization that the antianabolic concept is apparently an oversimplification. Histological study of bone reveals that cortisone can inhibit

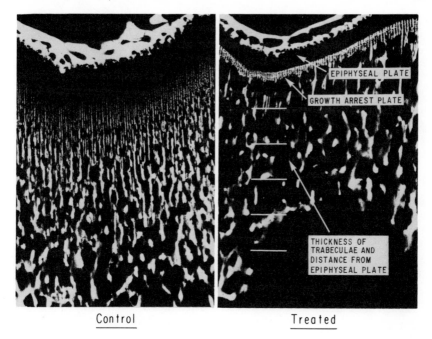

Control   Treated

Fig. 22. The photomicrograph shows a higher magnification of the femoral epiphyseal plates and metaphyseal regions presented in Fig. 21. Thinned plate, shorter trabeculae and reduced total metaphyseal-trabecular volume are demonstrated in the treated animal. ×21. (Reproduced with the courtesy of the authors and publisher from Jee *et al.*, 1966.)

or convert the characteristic combinations of calcium and vitamin D deficiency or low dietary calcium and hyperactivity to that of typical osteoporosis by inhibition of bone matrix formation. Initially, resorption continues rapidly and subsequently, more slowly and extensively (Storey, 1963). Klein *et al.* (1965) reported in a histo-quantitative study of human ribs from patients previously treated with adrenocorticoids, that the number of foci of new bone formation decreases dramatically. Apparently, steroids affect primarily the mesenchymal cell population resulting in a decrease in the number of available osteoblasts (Sissons, 1960). Klein *et al.* (1965) derived information which, contrary to the general opinion, showed no evidence of increased bone resorption and suggested that resorption was possibly decreased.

In congenital adrenal hyperplasia, abnormal amounts of androgenic steroids are secreted. The administration of corticoids in physiological quantity can suppress the glandular hyperactivity. If treatment is initiated early enough, the problem of accelerated skeletal maturation

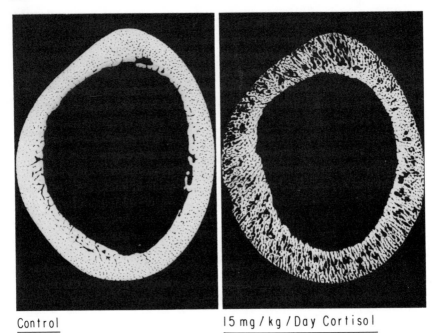

Control   15 mg / kg / Day Cortisol

Fig. 23. The photomicrograph shows a comparison of the microradiographic cross sections taken from midfemoral shafts of a control and cortisol treated rabbits. Note the extensive hormone induced osteoporosis. ×5. (Reproduced with the courtesy of the authors and publisher from Jee *et al.*, 1966).

and eventual growth retardation can be prevented (Wilkins, 1955).

Key *et al.* (1952) reported the failure of cortisone to retard fracture repair in rats. However, inhibition was demonstrated when higher doses were used (Duthie and Barker, 1955; Storey, 1960a). Blunt *et al.* (1950) found that in the rabbit, experimental fractures were very sensitive to cortisone treatment. In a study of the fate of transplants of autogenous tibial grafts in dogs under cortisone treatment, significant retardation was observed in graft incorporation, trabecular formation and alkaline phosphatase activity. The trabeculae lacked the usual osteoblastic lining. These changes were still present 4 weeks after surgery, but observation of a number of fractures in patients undergoing cortisone therapy in dosages comparable to those used in dogs healed uneventfully (Tonna and Nicholas, 1959).

The antianabolic nature of the adrenocortical hormones is expressed through inhibition of protein synthesis (Wool and Weinshelbaum, 1960), as well as, incorporation of sulfate into mucopolysaccharides (Lash and Whitehouse, 1961; Whitehouse and Boström, 1962), including the chondroitin sulfate of costal cartilage (Boström and Odelblad, 1953). Daughaday and Mariz (1962a,b) reported the inhibition of collagen synthesis by cortisone and hydrocortisone. In a biochemical study using ^{3}H-amino acids and ^{35}S as radiotracers, Ebert and Prockop (1963) concluded that although sulfated mucopolysaccharides, collagen, and protein synthesis are inhibited by hydrocortisone, the effects do not appear to be specific for any one of these synthetic pathways. All three were inhibited concomitantly. Rat bone collagen, once synthesized, is apparently stable in the presence of a great loss of body nitrogen induced either by starvation, cortisone, or prednisolone (Sobel and Marmorston, 1954; Sobel and Feinberg, 1970). In an autoradiographic study of collagen synthesis by osteoblasts in rats receiving daily doses of cortisone, it was reported that [^{3}H]proline uptake and turnover in the form of tropocollagen was diminished (Rohr and Wolff, 1967). Essentially similar results were reported by Mankin and Conger (1966) for the utilization of [^{3}H]glycine by rabbit articular cartilage cells following intra-articular injections of hydrocortisone.

Adrenocortical hormone treatment is known to alter epiphyseal cartilage and alveolar bone histochemistry. In cortisone and hydrocortisone treated rats, it was reported that the PAS-stainability and metachromasia of epiphyseal cartilage and bone were increased, implicating a lower state of polymerization of protein-mucopolysaccharides in relation to an interference in the calcification of cartilage and adjacent bone (Bernick and Ershoff, 1963; Bernick and Zipkin, 1967). Meyer and Kunin (1969) reported that cortisone treatment resulted in decreased glycolytic enzyme activity of rat epiphyseal cartilage. Oxidative enzyme activity appeared not to be effected by low doses of cortisone. When the daily dose was raised, however, from 3 to 15 mg/100 gm body weight for 7 days, succinic dehydrogenase activity was depressed.

At the electron microscope level (Silberberg *et al.*, 1966b), it was shown that in cortisone treated mice, chondrocytes of the articular cartilage decrease in size and organellar development is disturbed. Within 24 hours the granular endoplasmic reticulum is disrupted. This change is consistent with interference in protein synthesis. The Golgi apparatus is initially prominent; 'however, after 24 hours atrophic changes occur. Since the Golgi apparatus is considered the site for synthesis and sulfation of mucopolysaccharides, the late atrophic changes indicate interference in mucopolysaccharide synthesis. In addition, the observed accumulation of glycogen is believed to be related to the failure of sulfation, synthesis and subsequent elaboration of mucopolysaccharides. Beyond 24 hours degenerative features become more prominent (Figs. 24–26). These included cytoplasmic vacuolization, increased osmiophilic

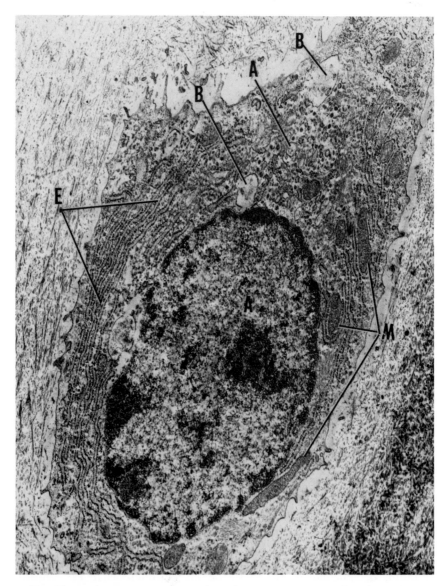

Fig. 24. An electron microradiograph is shown of an upper midzonal chondrocyte of the articular cartilage from the femoral head of an untreated 4-week-old mouse. Note the highly developed granular endoplasmic reticulum (E), Golgi system (A), numerous mitochondria (M) and incompletely membrane-bound multivesicular bodies (B). ×21,000. (Reproduced with the courtesy of the authors and publisher from Silberberg *et al.*, 1966b.)

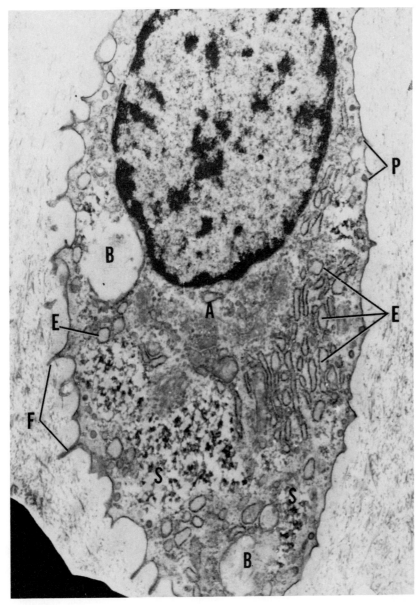

Fig. 25. An electron micrograph is shown of a deep midzonal chondrocyte of the articular cartilage from the femoral head of a 24-hour cortisone acetate-treated mouse. Note the rough endoplasmic reticulum (E), large incompletely membrane-bound multivesicular bodies (B), numerous pinocytotic vesicles (P), the Golgi system (A), the appearance of several glycogen lakes (S) and abundance of footlets (F). Compare with Fig. 24. ×21,000. (Reproduced with the courtesy of the authors and publisher from Silberberg *et al.*, 1966b.)

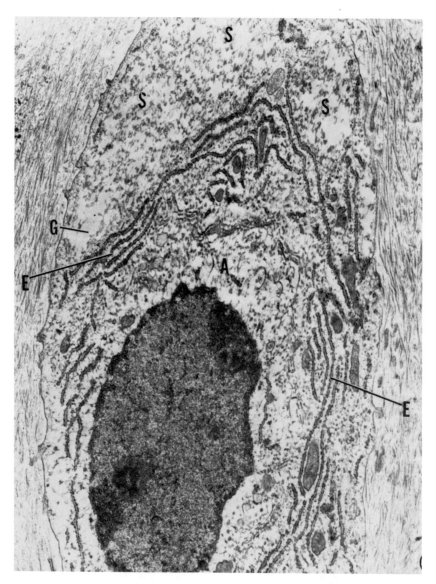

Fig. 26. An electron micrograph is shown of an upper midzonal chondrocyte of the articular cartilage from the femoral head of a 6 days cortisone acetate-treated mouse. Note the diminished, narrow endoplasmic reticulum (E), atrophic Golgi system (A), large accumulations of glycogen (S), electron-lucent ground substance (G) and relative absence of footlets. Compare with Figs. 24 and 25. ×21,000. (Reproduced with the courtesy of the authors and publisher from Silberberg *et al.*, 1966b.)

inclusions and cell death with the formation of fibrillar microscars within the cartilage matrix.

In summary, it would appear that the effects of adrenocortical hormones on the skeletal system are significant and result from complex interactions which are only presently being realized. Although unquestionable species differences exist in sensitivity to the hormones, as can be demonstrated histologically, these differences also appear to be dose and time dependent. This dependency is observed more clearly when studied in one given species. Low doses generally stimulate bone resorption while high doses suppress bone resorption. Recent studies, as noted, suggest direct inhibition of bone precursor cell proliferation after high dose administration and indirect stimulation of parathyroid hormone secretion following low dose administration, as salient features behind the action of cortisol and probably other corticosteroids. One, however, must not lose sight of the fact that these hormones elicit degenerative ultrastructural changes in protein and mucopolysaccharide producing organelles which are essential to cartilage and bone matrix synthesis. As a result of the accumulating research knowledge concerning the action of adrenocortical hormones at different levels of skeletal tissue organization, one cannot help but reflect on the naiveté of our present understanding of the action of such hormones.

VIII. INSULIN

Insulin is a protein hormone elaborated by the β-cells of the islets of Langerhans. In the human, about 2 million islets are scattered widely throughout the pancreatic acinar tissue. Insulin was the first protein hormone whose complete amino acid sequence and structure was determined (Sanger, 1959). The molecular structure of ox insulin consists of 2 polypeptide chains, one, A, having 21 amino acid residues, the other, B, having 30 amino acid residues. Two disulfide bridges link the chains. An additional disulfide bond exists on the A-chain between position 6 and 11. Insulin obtained from all vertebrate species consists of the double chain and comparable disulfide bonds, but species differences in the number and sequence of amino acid residues exist. The molecular weight varies from 6000 in basic media to 12,000 in acid media.

Both insulin deficiency and insulin administration exhibit a recognizable effect on the skeletal system. Some experimental work was reported using chick embryos and chick limb-bone rudiments cultivated *in vitro*. Too little work has been reported on mammals; it is unfortunate that the information is at best exiguous, especially since the appearance of diabetes in humans and the use of insulin are not uncommon. Fell (1953) showed that the study of organized tissue fragments in organ culture is a scientifically useful method which allows the assessment of the direct effects of one or a number of hormones and their interactions. Using embryonic chick limb-bone rudiments cultivated in vitro, Hay (1958) reported an increase in weight and nitrogen content of insulin treated rudiments. The hormone stimulated growth in bone length during the early part of the cultivation period, but the final length of the treated rudiments were shorter than those of the controls. Earlier investigators reported essentially similar findings (Landauer and Bliss, 1946; Landauer, 1947a, b, 1953; Duraiswami, 1950; Chen, 1954). These findings support the view that insulin can stimulate bone growth and induce nitrogen retention (Salter and Best, 1953; Lawrence et al., 1954) in hypophysectomized rats. The histological findings suggest that insulin promotes hypertrophy of cartilage cells and glycogen deposition in hypertrophic cells, as well as, increases the water content per unit weight of bone (Hay, 1958). Unlike hypertrophic mammalian cartilage, glycogen is normally not present in any significant quantity in hypertrophic avian cartilage. The production of dwarfism and skeletal abnormalities in chick embryos (Fig. 27) are well known (Landauer and Bliss, 1946; Landauer, 1947a, b; Duraiswami, 1950, 1955; Zwilling, 1952; Chen, 1954; Anderson et al., 1959). Duraiswami (1955) claims that glycogen and acid mucopolysaccharides are involved in the insulin-induced disturbances in the chick embryo. A single injection of insulin into the chick yolk sac produces a hypoglycemia lasting as long as 72 hours. Animals with the most severe hypoglycemias exhibit significant dwarfism and skeletal deformities (Zwilling, 1952). In order to test this hypothesis Anderson et al. (1959) performed studies in which insulin was injected into 369 chick yolk sacs at 96 hours. The eggs were opened on the eighteenth day of incubation. Seventy-nine chicks showed inhibition of skeletal growth or serious skeletal deformities; 94 died before the eighteenth day of incubation. It was concluded that in insulin dwarfism, no marked change was detected in the glycogen content of preosseous cartilage, ^{35}S uptake, metachromasia, Schiff reaction, alkaline phosphatase, and phosphorylase. No change was detected in the ratio of total cell volume to matrix volume in hypertrophic cartilage. Here cartilage cells were, however, smaller and surrounded by less matrix the greater the degree of dwarfism.

Evaluation of the effects of insulin on the skeletal system based on a method where the hormone is injected into the yolk sac is most difficult. Conclusions must be drawn with great caution, since the endocrine system of the developing chick is intact. Questions naturally arise as to

Fig. 27. The photomicrographs show a comparison of a normal 18-day-old and similarly aged dwarfed chick following yolk sac injection of 6 units of insulin at 96 hours. (Reproduced with the courtesy of the authors and publisher from Anderson *et al.*, 1959.)

the specificity of results and to the possible synergistic or antagonistic effects of insulin on other endocrine secretions.

In mammalian studies, it was reported that insulin-deficient mice consumed less food than untreated controls, necessitating the use of pair-fed controls (Silberberg *et al.*, 1968). The effects of insulin on the mammalian skeleton are far less remarkable than on the developing avian skeleton; therefore, few investigators studied mammals. Earlier observations indicate that administered insulin stimulates the articular chondrocytes of dwarf mice (Silberberg *et al.*, 1966b). In another study (Silberberg *et al.*, 1968), 3-week-old mice were made acutely diabetic by a single injection of guinea pig anti-insulin serum. These animals were observed for 48 hours after the development of glycosuria. Histologically, cartilage from the knee joints and femoral heads exhibited hypertrophy of the

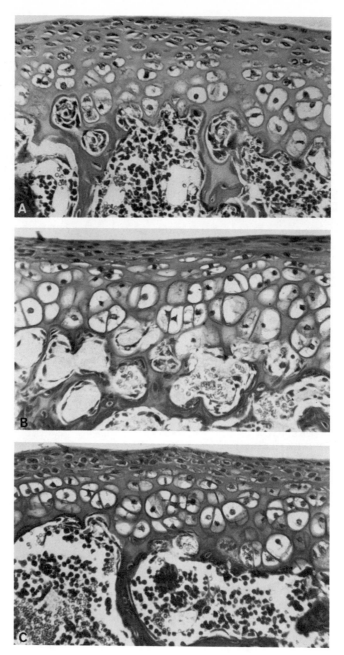

Fig. 28. The photomicrographs show sections of distal femoral articular cartilage of 3-week-old mice. A, Untreated control. B, From a mouse treated with anti-insulin serum 1 day before sacrifice. C, From a pair-fed control to mice receiving anti-insulin serum. Note that the anti-insulin treated (B) animal exhibits flattening of chondrocytes and retained hypertrophic cells resulting from decreased breakdown. ×275. (Reproduced with the courtesy of the authors and publisher from Silberberg *et al.*, 1968.)

chondrocytes (Fig. 28A, B, C). Damage to chondrocytes was manifest in electron microscopic observations. Chondrocytes were smaller in size than those of pair-fed controls. The cell organelles were poorly developed. Some intracellular membrane disruption, derangement of ribosomes and endoplasmic reticulum, and underdevelopment of the Golgi complex were characteristic cell changes (Figs. 29 and 30). Glycogen deposition was delayed. Increased degenerative changes, death of chondrocytes and matrical microscar formation were also recorded. The results point to possible depressive effects on carbohydrate and fatty acid synthesis, as well as, disturbed protein synthesis. Such cell metabolic abnormalities are associated with the absence of insulin. These findings are the opposite to those obtained after administration of insulin to dwarf mice (Silberberg *et al.*, 1966b).

In rats made alloxan diabetic and observed for 2 weeks, it was reported that knee joint articular cartilage matrix revealed increased acid mucopolysaccharides and decreased acid-Schiff positive materials, indicating increased depolymerization of the ground substance (Cicala *et al.*, 1962).

The experimental data are presently too sparse to allow us to form any kind of concrete picture of the effects of insulin deficiency or administration on the skeletal system of both birds and mammals. None of the experimental designs used allow for definitive information, since by and large, the results are obtained in the presence of intact endocrine systems and under states of physiological difficulty or deficiency. At the present time, it is not clear whether the handful of results point to specific insulin effects on the cartilage structures of the skeletal system or reflect the effects of secondary mechanisms. Silberberg *et al.* (1968) point out that actually many of the ultrastructural changes observed following inducement of diabetes resembled those seen in the underfed controls. In the diabetic mice the changes were more marked. In 6-week-old mice, fasting for an additional 48 hours produced similar ultrastructural changes (Silberberg *et al.*, 1967).

The role of insulin in combination with and free from other endocrine relationships is wanting. Experiments of longer duration are necessary, as well as, investigations in which animals of different age are used. The dose effects or the effects of extended insulin deficiency are also germane to the problem. Observations have been generally directed to cartilage, but the possible effects on bone cell maturation, transition and function have not been investigated. Some evidence exists to show that insulin counteracts the effect of growth hormone on *in vitro* epidermal mitotic activity (Bullough, 1954). The effects of insulin on bone cell proliferation, however, is not known. The role of insulin on cartilage

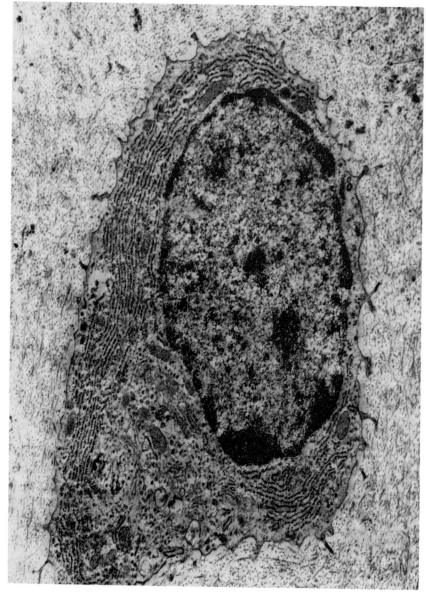

Fig. 29. An electron micrograph is shown of an upper midzonal chondrocyte of the articular cartilage from the femoral head of a 3-week-old C57BL untreated mouse. Note the abundant, well-stacked, rough endoplasmic reticulum, small mitochondria, and well-defined Golgi system. ×20,000. (Reproduced with the courtesy of the authors and publisher from Silberberg *et al.*, 1968.)

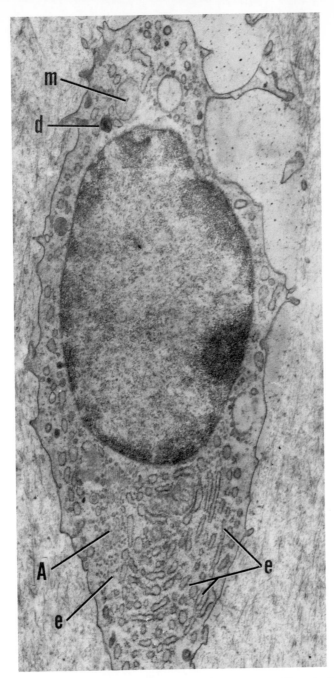

Fig. 30. An electron micrograph is shown of an upper midzonal chondrocyte of the articular cartilage from the femoral head of a 3-week-old C57BL mouse, treated with a single dose of anti-insulin serum 27 hours prior to sacrificing. Note changes in the contours of both the nucleus and plasmalemma. The endoplasmic reticulum (e) is vesiculated, the Golgi system (A) is less conspicuous and few mitochondria (m) are encountered. Compare with Fig. 29. ×20,000. (Reproduced with the courtesy of the authors and publisher from Silberberg *et al.*, 1968.)

or bone mineralization is also largely unknown. Before any reasonable scientific judgment can be made about the effects of insulin deficiency or administration on the skeletal system, additional investigative work is needed.

IX. PARATHYROID HORMONE (PTH)

PTH has been isolated from bovine and porcine parathyroid glands and consists of a straight chain polypeptide of about 75 amino acid residues (Rasmussen *et al.*, 1964). Pure parathyroid hormone exerts a profound affect on both calcium-mobilizing and phosphaturic activities resulting in increased serum calcium and decreased serum phosphate levels by a direct action on bone, kidney and the intestinal mucosa. Hormone secretion is mediated by fluctuating levels of plasma calcium and indirectly by plasma phosphate levels. It is probable that the indirect effect of phosphate on PTH secretion occurs via action on the dynamics involving the concentration gradient which moves calcium from blood to bone rather than on the blood calcium level per se. Elevation of the plasma phosphate following nephrectomy or feeding stimulates PTH production, but the levels of serum calcium remain within normal limits. On the other hand, experimental interruption of movement of calcium from blood to bone results in a marked depression of plasma phosphate accompanied by diminished PTH secretion. Bone is a primary target of parathyroid hormone, in that it serves as the mineral bank for the role of PTH in blood calcium homeostasis. This is not to imply that calcium homeostasis is solely dependent on the action of PTH on bone, since the interrelated role of the action of PTH on the kidneys and gut mucosa, their involvement with phosphate, the action of vitamin D and thyrocalcitonin are germane to this subject. Numerous detailed discussions on the subject exist and the reader is referred to Munson (1955), McLean (1956), Pak (1971) and the many reports found in publications edited by Greep and Talmage (1961) and Talmage and Bélanger (1968).

Parathyroidectomy results in diminished serum calcium levels, involving largely, if not exclusively, the diffusible bone fraction, reduced blood citrate, increased levels of serum inorganic phosphate and, conversely, reduced urinary excretion of calcium and diminished urinary elimination of inorganic phosphate. The deficiency of calcium in the extracellular fluid generally leads to tetany and often death (Turner and Bagnara, 1971). In the rat parathyroid hormone deficiency can be seen to lead to decalcification of bones and disorganization of tooth enamel. Chemical analyses of these bones reveal low ash values with calcium and phosphorus deficiencies. Administration of parathyroid hormone to parathyroidectomized mammals results in a physiological adjustment, reversing the mineral imbalances (Foulks and Perry, 1959; Pullman *et al.*, 1960). Calcium infusion or regulation of the calcium to phosphorus ratio of the diet can also alleviate the tetanic state. Excessive hormone administration induces hypercalcemia, hypophosphatemia, increased urinary excretion of phosphate and calcium, ectopic deposits of calcium in organs and tissues, hyposensitivity of the neuromuscular system and skeletal demineralization accompanied by increased serum alkaline phosphatase. A significant increase in gastrointestinal absorption of calcium (Wills *et al.*, 1970) is also experienced. The stimulation of bone resorption by parathyroid hormone is a prominent feature (Raisz, 1963; Pechet *et al.*, 1967). It must be pointed out, however, that the magnitude of the hypercalcemia and phosphaturia is proportional to the administered dose of PTH. The principal organs responding to the hormone are the kidneys and bones. Experimental evidence is in favor of the direct action of the hormone on both organs. The renal response is rapid, whereas, the skeletal response is slow. The difference in speed of response may be due, in part, to the significant differences in the rates of blood flow through these organs or to the time required for differentiation of bone cellular elements following PTH-cell stimulation.

Experimental hyperparathyroidism has been produced in a wide variety of animals (Thomson and Collip, 1932). The resulting damage to bone and its repair have been extensively studied, resulting in a series of characteristic events closely related to the pathological changes observed in man. The advanced changes of chronic hyperparathyroidism, however, may be seen only in man, since a rapid resistance to exogenous PTH prevents the experimental production and maintenance of the true chronic condition (McLean, 1956). Prolonged administration of sublethal amounts of PTH induce profound effects on the osseous system (Urist, 1967). An osteodystrophy is produced revealing resorption of cortical bone and trabecular bone and resulting in subsequent limb deformation and fractures. The previously mineralized regions stripped of mineral are replaced by fibrous tissue and giant cells resulting in a condition resembling in man osteitis fibrosa cystica. In the rat, even a single toxic dose of PTH leads to rapid bone resorption and associated histological changes, followed by a slower regenerative phase (Heller *et al.*, 1950). The formation of dense fibrous tissue at the metaphyseal regions is especially evident in the rat.

Generalized osteitis fibrosa or von Recklinghausen's disease is a clinical syndrome associated with parathyroid hyperplasia and excessive release of hormone in man. Bone softening resulting from removal of minerals is so severe in cases, that fractures and disabling deformities are experienced. In addition, roentgenographic findings reveal generalized decreased bone density with accompanying bone cysts and distorting bone tumors. The vertebral column may exhibit kyphosis or scoliosis, while long bones commonly reveal bowing. Certain patients, however, can be found with hypercalcemia resulting from hyperparathyroidism in whom skeletal changes are minimal or absent.

Classical microscopic changes are revealed in all animals studied in response to toxic doses of PTH. Resorption of trabecular bone is very prominent within hours after PTH administration. PTH also appears to exert some action on osteocytes and their lacunae via induced osteocytic osteoclasis (Bélanger et al., 1963). Extensive modulation of osteoblasts into osteoclasts is observed resulting in a significant population increase (Young, 1964). Osteoblastic activity is increased in an attempt to repair damaged trabeculae, resulting in excessive deposition of fibrous connective tissue. The cellular modulations which occur are claimed to be reversible (Heller et al., 1950). Recovery from large doses of PTH is accompanied by overgrowths of new bone in areas of trabecular resorption and fibrous tissue formation. These overgrowths or hyperostoses are especially prominent in the rat and mouse (Burrows, 1938; Barnicot, 1945).

Tissue culture studies amplify the influence of PTH on osteoclasts. Gaillard (1959) reported that parathyroid gland tissue of man, mouse and chick embryos, cultivated in direct contact with fragments of mouse embryo parietal bone anlage, induced bone resorption accompanied by osteoclasts. Addition of PTH to cultured anlage resulted in an increased number of osteoclasts suggesting that PTH is indispensable for creating the conditions favoring the survival, formation, and functioning of osteoclasts. Where osteoclasts were absent, bone resorption did not occur.

It must be pointed out that the histological changes described in experimental animals result from PTH treatment in toxic quantity. Often, these changes are accompanied by extensive necrosis of cellular elements, especially osteocytes and those of bone marrow (Heller et al., 1950). The degree to which these histological changes can be attributed to physiological events or to clinical hyperparathyroidism is questionable, since for the most part, they may reflect a toxic response to PTH. Albright and Reifenstein (1948) believe this response to be quite different from that produced by smaller doses of PTH. One must not dismiss the fact, however, that the physiological and toxic effects of PTH are both expressions of the direct effects of PTH on bone. Evidence for the direct effects of PTH on bone are derived from the reports of Barnicot (1948), Carnes (1950), Chang (1951), and Engel (1952). Barnicot (1948) implanted bone and parathyroid tissue intracerebrally in mice. The bone in contact with the parathyroid tissue was resorbed. Chang (1951) confirmed Barnicot's experiment and also showed that transplanting of parathyroid tissue to the cranial cavity in mice caused resorption of the calvarium where contact existed. Carnes (1950) reported that bone matrix may be destroyed by PTH, irrespective of its mineral content, while Engel (1952) found that parathyroid extract induced depolymerization of the ground substance of bone and cartilage. Furthermore, it was shown that PTH stimulated the release of calcium from bone to the extracellular fluid after kidney removal in a number of species. Talmage *et al.* (1957) using peritoneal lavage after the influence of kidney or gut absorption of calcium had been ruled out, showed this to be the case.

Histochemical studies reveal that in surviving mouse embryo bone explants subjected to parathyroid extract, a release of specific lysosomal hydrolases (including β-glucuronidase, β-galactosidase, and β-N-acetyl-aminodeoxyglucosidase) was induced presumably from osteoclasts. The activity is increased with time and is preceded by lowering of the pH (Vaes, 1965). It is believed that the release of hydrogen ions via stimulation of osteoblastic activity results primarily in the formation of lactic acid and secondarily citric acid, shifting the local pH to low acidic levels (Borle *et al.*, 1960; Vaes and Nichols, 1961, 1962). Optimal hydrolase activity occurs at an acid pH. The increase in hydrogen ions is favorable for mineral solubilization, allowing for the attack of the organic matrix by the lysosomal hydrolases. Menczel *et al.* (1970) reported that alkaline phosphatase levels increase at rat tibial ends, but not at the shafts, reaching a peak 24 hours after PTE treatment. Increase in alkaline phosphatase activity is commensurate with the fact that this enzyme appears in elevated quantity in regions where formation of matrical fibrous connective tissue is to take place. It is indicative of the functional stimulation of osteoblasts by PTE which leads them to lay down a meshwork of argyrophilic collagenous fibers (Hancox, 1961). Hekkelmann (1965) revealed by biochemical methods a decrease in isocitric dehydrogenase activity following PTH treatment, whereas, no significant change in oxidative enzymes was noted histochemically by Deguchi and Mori (1969).

Using radiotracer techniques, it has been reported that PTH increases DNA and RNA synthesis of bone cells (Park and Talmage, 1967; Owen and Bingham, 1968). DNA synthesis occurs in response to stimulation of cell proliferation, whereas, the RNA increase is compatible with the accelerated synthesis of new proteins, possibly including enzymes, as well as matrical proteins which are contingent on the production of messenger RNA. PTH also stimulates adenyl cyclase activity in bone and kidney cortex cells (Chase and Aurbach, 1967, 1968). Consequently, there occurs an increase in intracellular 3′,5′-adenosine monophosphate (cyclic AMP). There is increasing evidence that cyclic AMP functions as a "second messenger" in bone and renal target cells mediating some of the known effects of PTH (Chase *et al.*, 1969). Injection in rats of cyclic AMP and its derivatives or addition of the nucleotides to bone fragments *in vitro* results in effects which resemble those of PTH. There exists a need for an electron microscopic—autoradiographic or cyto-chemical method which can assess in specific cell types the distributions and levels of adenyl cyclase and cyclic AMP in response to parathyroid hormone. The mediation of PTH hormone action and those of other protein and polypeptide hormones of diverse function by cyclic AMP, indicates that the intracellular responses to the nucleotide are conditioned by particular target cells. The nucleotide may activate events which stimulate enzyme systems, alter membrane permeability and movement of ions. These events may differ in different target cells as appears to be the case with the action of PTH on bone versus kidney cells (Turner and Bagnara, 1971).

Parathyroid hormone administration results in profound ultrastructural changes, especially in osteoblasts. Cameron *et al.* (1967) reported that in rats, metaphyseal osteoblasts reveal swollen mitochondria, contain dense granules, a distended Golgi system, ribosomes separated from a distended endoplasmic reticulum and the appearance of dense cyto-plasmic bodies within 6–26 hours. Osteocytes were little effected while osteoclasts remained normal. Repeated administration of hormone disrupts the fine structure of many cell types. A recent electron microscopic study by Laird and Cameron (1972) in which rat parietal bone was cultured for 10 hours in 0.1 units PTH per ml of culture medium showed reduction in the amount and complexity of the rough endoplasmic reticulum, but mitochondrial enlargement was rarely observed, raising the question of the possibility of processing artifacts (Fig. 31). These observations appear to reflect toxic effects of parathyroid hormone on cell ultrastructure. Hormone effects on the osteocytes which influence the mineral phase of bone can also be observed at the ultrastructural level. This activity is revealed through osteocytic osteolysis and osteoplasis. Electron microscopic evidence of both these processes (Fig. 32A, B), nevertheless, can be seen in nontreated animals (Tonna, 1972).

It seems appropriate at this time for studies involving much smaller doses, in order to determine the nontoxic effects of parathormone on bone cell ultrastructure, with the hope of relating fine structural varia-

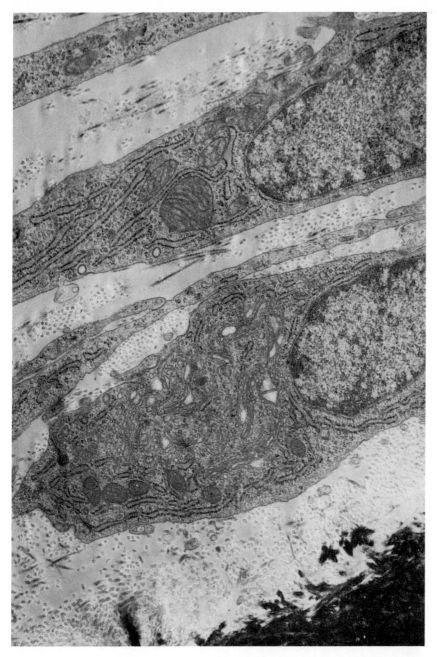

Fig. 31. The electron micrograph shows rat parietal bone cells cultured for 10 hours in 0.1 units of PTH per ml of culture medium. Characteristic elongation of PTH treated cells *in vivo* is noted. There also occurs a reduction in the amount and complexity of the rough endoplasmic reticulum. ×22,000. (The electron micrograph was kindly supplied by Drs. P. P. Laird and D. A. Cameron (1973), Department of Pathology, University of Sydney, Australia.)

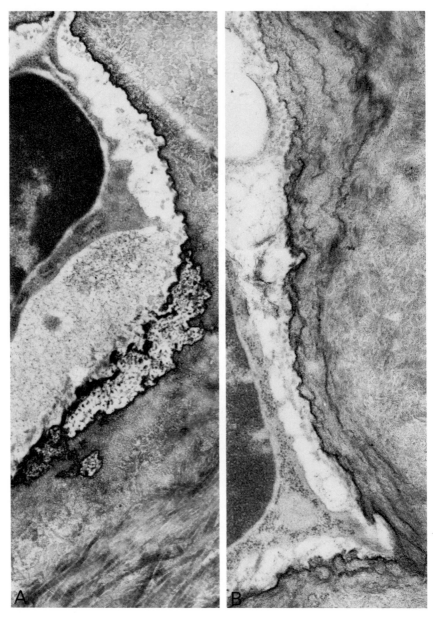

Fig. 32. The electron micrographs show femoral cortical osteocytes of 1-year-old short-lived BNL mice residing in lacunae. The samples were decalcified in preparation for the visualization of perilacunar surfaces. Note in A, that the lacunar surface is limited by the dark osmiophilic lamina which is extensively concave-scalloped. This feature is indicative of bone resorption by the process of osteocytic osteoclasis.

tions to known hormone action, as has been achieved with several other hormones. In this regard, Baud (1968) found that perilacunar wall resorption was greatly increased by parathyroid extract and constitutes the earliest most evident morphological manifestation of hormone activity.

X. THYROCALCITONIN (CALCITONIN, TCT)

An endocrine substance exists which is capable of rapidly lowering the plasma calcium in response to hypercalcemia, an action opposite to that exerted by parathyroid hormone. Copp and his co-workers (1962) identified a calcium-lowering factor which exerted this effect and was present as an impurity in commercial parathyroid extracts. This hormone was called calcitonin, and was believed to originate from the parathyroids. In 1964, Foster et al. reported that perfusion experiments in the goat revealed a fall in plasma calcium only when the thyroid gland and not the parathyroids were perfused. Thus, the thyroid gland was shown to be the origin of calcitonin. The provisional name "thyrocalcitonin" was given by Hirsch et al. (1963) for the identical hormone which he and his colleagues were first to observe as the principle responsible for the calcium-lowering effect in crude rat thyroid extract.

In the mammalian thyroid, thyrocalcitonin is secreted by C cells (Pearse, 1966) which are parafollicular in the dog, but may be epifollicular or follicular in other species. The hormone appears to be a polypeptide with an estimated molecular weight of about 3600 isolated from pig thyroid extracts (MacIntyre, 1967). In birds, reptiles, and fish, the ultimobranchial body which forms two or three discrete glands produces a rich supply of calcitonin. Reynolds et al. (1970) have shown that calcitonins of ultimobranchial origin are more effective by at least 10 to 100 times, at lower doses for longer time periods, than mammalian thyroid preparations.

Administration of thyrocalcitonin to the rat results within minutes in the rapid decline of plasma calcium and plasma phosphate. These ions become minimal by about 1 hour (Hirsch et al., 1964). The loss of plasma phosphate is mediated by the effect of thyrocalcitonin on the renal excretion of phosphate (Robinson et al., 1966). In parathy-

---

In B, the perilacunar surface is extensively convex-scalloped indicative of bone formation by the process of osteocytic osteoplasis. Beneath the limiting osmiophilic lamina, evidence of previous phases of lacunar bone resorption and alternate bone deposition is indelibly outlined by weaker osmiophilic markings. A, ×22,750; B, ×26,000.

roidectomized rats exhibiting a low rate of bone resorption, the renal effect of thyrocalcitonin by which the plasma phosphate is lowered predominates, whereas, no further significant lowering of plasma calcium occurs. Prior nephrectomy prevents the loss of plasma phosphate (Robinson *et al.*, 1967). When bone resorption is active, as in the normal situation, the hormone produces a fall in both plasma calcium and phosphate which is not prevented by prior nephrectomy (Hirsch *et al.*, 1964). From studies using thyroidectomized dogs whose superior parathyroid glands were left intact, Jowsey (1969) reported that there is evidence that the effects of thyrocalcitonin may be more pronounced in the control of serum phosphate than serum calcium.

MacIntyre and co-workers (1967) have shown conclusively in the isolated perfused cat tibia, that the action of thyrocalcitonin on the skeleton is direct. It has been concluded from isotopic studies that the hormone inhibits the resorption of bone (Milhaud *et al.*, 1965). A similar conclusion was derived by a number of investigators from bone tissue culture studies (Friedman and Raisz, 1967; Gaillard, 1967a). This view was confirmed by *in vivo* isotopic hydroxyproline excretion studies (Martin *et al.*, 1966). Hirsch (1967) also showed that in rats, although thyrocalcitonin inhibits release of calcium from both "stable" and "labile" bone, the inhibition is most marked on the stable compartment which is responsive to parathyroid hormone. Foster *et al.* (1966a) studied the effects of thyrocalcitonin on the bones of parathyroidectomized male rats. After 1 week of daily treatment, striking differences in the quantity of trabecular bone was noted in the X-rays of caudal vertebrae. No radiological differences were observed in other bones (Fig. 33A, B). Histologically, the vertebrae of treated animals revealed an increase in trabecular bone of both the primary and secondary spongiosa (Figs. 34A, B, 35). The number of osteoclasts in these regions were significantly reduced. It was concluded that thyrocalcitonin can exert an effect on bone independent of parathyroid hormone. Chronic administration of thyrocalcitonin has been shown to induce cortical bone development in the long bones of rats and rabbits (Wase *et al.*, 1967). Implantation of thyrocalcitonin containing collagen sponges into rat calvaria also resulted in the formation of a thick layer of new bone (Fig. 36A, B) in the immediate vicinity of the implant (Mantzavinos and Listgarten, 1970). The results point to the possibility that the hormone not only acts directly on bone to inhibit resorption, but in certain cases to promote bone formation. Sturtridge *et al.* (1968) are of the opinion that their microradiographic and fluorescent microscopy data indicate that thyrocalcitonin promotes new bone formation, but contrary to other reports does not influence resorption. A numerical increase in osteoblasts and

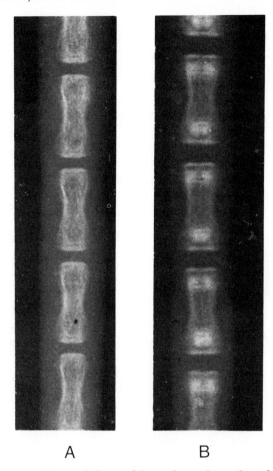

A   B

Fig. 33. X-rays are shown of the caudal vertebrae of parathyroidectomized male rats. A, Nontreated animal. B, Thyrocalcitonin treated animal. Note the striking increase in metaphyseal trabecular bone at both ends of each vertebrae in B. (Reproduced with the courtesy of the authors and publisher from Foster *et al.*, 1966a.)

stimulated bone formation attributable to thyrocalcitonin was also demonstrated in tissue culture studies (Gaillard, 1967b; 1970). Klein and Talmage (1968), however, show that repeated thyrocalcitonin administration inhibits all phases of bone resorption. In thyroparathyroidectomized rats, thyrocalcitonin inhibits calciphylaxis (Selye *et al.*, 1967) and experimentally induced calcergy or cutaneous calcification (Gabbiani *et al.*, 1968). Thus, the hormone also acts to prevent soft-tissue calcification.

In man, clinical trials using thyrocalcitonin indicate that it may be of therapeutic value in the acute treatment of hypercalcemia (Foster

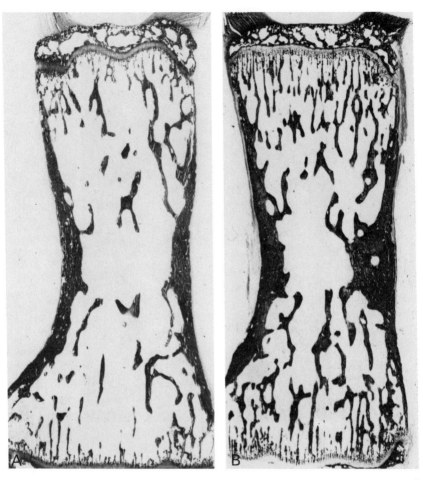

Fig. 34. The photomicrographs show sections of rat caudal vertebrae. A, Nontreated control animal. B, Thyrocalcitonin-treated animal. Note the significant increase in trabecular volume of both the primary and secondary spongiosa. ×24. (Reproduced with the courtesy of the authors and publisher from Foster *et al.*, 1966a.)

et al., 1966b) and may also prove of value in the chronic management of bone disease. The possibility that thyrocalcitonin might be useful in the treatment of osteoporosis has aroused considerable interest in view of the fact that the hormone can act on the skeleton of man. Although the first trials have been somewhat disappointing (Caniggia *et al.*, 1970), it is too early to deny this possibility in view of the fact that one is not sure of the physiological levels of hormone. Furthermore, there exists suggestive evidence that the target organ response to thyro-

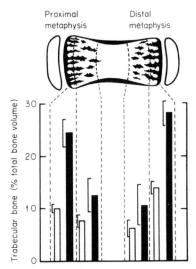

Fig. 35. The volume of rat trabecular bone (10th caudal vertebrae) is expressed in percent (mean ± S.E.) for nontreated control and thyrocalcitonin treated animals. (Reproduced with the courtesy of the authors and publisher from Foster *et al.*, 1966a.)

calcitonin may decrease with advancing age, which if confirmed, would reduce the therapeutic value of thyrocalcitonin in treatment of skeletal problems associated with the aged (Care and Duncan, 1967).

A current hypothesis has been formulated by Talmage (1969, 1970; Talmage *et al.*, 1970) to account for the effects of parathyroid hormome and thyrocalcitonin on the movement of calcium ions necessary to plasma level maintenance and bone remodeling (Fig. 37). The hypothesis is directed to the osteogenic cell compartment consisting of osteoblastic cell precursors, differentiating osteoblasts and osteoblasts capable of bone matrical precursor production. All bone surfaces, i.e., periosteal and endosteal, are lined by members of the osteogenic cell compartment. In this hypothesis the surface cells form a "membrane" dividing the bone into two extracellular fluid compartments. Sufficient space between cells allows for movement of Ca^{2+} from blood to bone via the normal ion concentration gradient. The concentration of calcium and phosphate ions within the bony inner compartment depends upon the solubility products of the ions which are below those in the outer compartment found on the opposite side of the layer of cells. The osteoblastic cells are polarized and possess a calcium pump located near the outer compartment (Talmage, 1969) enabling calcium ions to move out of bone against the concentration gradient. In addition, the pump eliminates

the excessive accumulation of calcium within the cell which is significantly lower than that of the environmental tissue fluids. The cell is
more permeable to calcium ions on the inner compartment side than
on the side of the outer compartment. Therefore, calcium ions enter
into the cell in lieu of the concentration gradient existing between the
cell and its environment and are pumped out through the other side

For legend see opposite page.

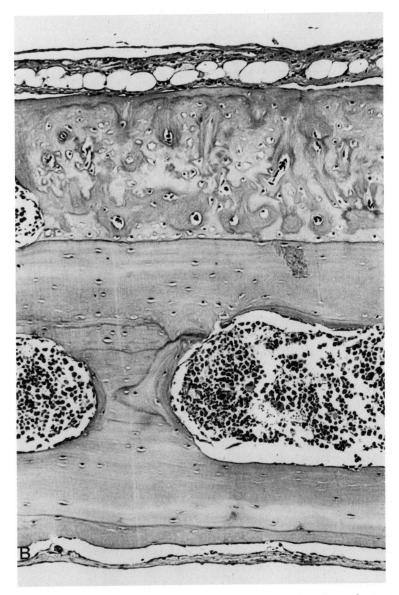

Fig. 36. Sections through rat calvaria are shown 2 months after subcutaneous implantation of collagen sponge (A) without thyrocalcitonin and (B) containing thyrocalcitonin. Note in B the deposition of newly formed bone which is highly cellular and irregular in structure. The layer of new bone formation is separated from the older bone by the presence of a basophilic subperiosteal reversal line. ×200. (Reproduced with the courtesy of the authors and publisher from Mantzavinos and Listgarten, 1970.)

Calcium Transport In Osteoblast

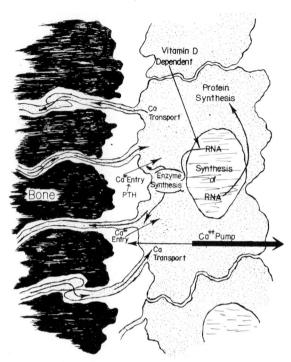

Fig. 37. The diagram illustrates a number of basic relationships believed to exist in the mechanism of PTH action on bone cells. Entry of calcium ions (Ca^{++}) into cells of the osteoblast "membrane" from the mineralized compartment is promoted by PTH. A calcium pump liberates ions into the extracellular fluid compartment on the opposite side of the "membrane." Intracellular spaces permit the reentry of fluid and calcium ions into the mineralized compartment. (Reproduced with the courtesy of the author and publisher from Talmage, 1969.)

of the cell against the normal gradient prevailing between bone and blood. It is held that calcium ions move into the mineralized compartment from the extracellular fluid via the open intercellular spaces between osteoblasts lining the surfaces of bone. In the light of recent information, Talmage believes that following adenyl cyclase activation and increased intracellular cyclic AMP production, parathyroid hormone mediates the rapid transport of calcium ions from bone across the osteoblast into the extracellular environment by activation of the calcium pump in response to the increased concentration of intracellular calcium. At the same time, PTH suppresses collagen synthesis and stimulates DNA synthesis of osteoblast precursors, accelerating the division and differentiation of their progeny to osteoclasts. This latter state, however,

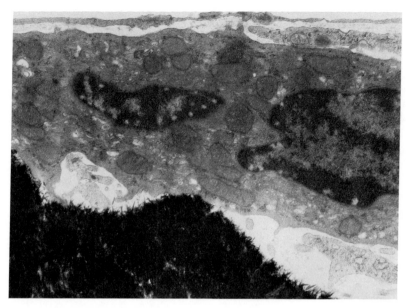

Fig. 38. The electron micrograph shows an osteoclast in an undecalcified sample of tibial metaphysis from a rat treated for 2 weeks with porcine calcitonin. It is noted that the usual intimate contact encountered between the folded cell surface and bone surface is absent. Instead, the cell surface is smooth while the bone surface shows no evidence of resorption. Compare with Fig. 8. ×21,000. (Reproduced with the courtesy of the author and publisher from Zichner, 1971).

must be achieved via cell fusion, since no nuclear division has been encountered to account for the multinuclear state of osteoclasts (Tonna, 1960). Thyrocalcitonin, on the other hand, may act via an enzyme system by promoting the release of organic phosphate which combines with calcium, reducing the concentration of cell transportable calcium. This increases the availability of calcium and phosphate ions for bone accretion, while producing a hypocalcemic effect on circulating body fluids. In this regard, it is interesting to consider the fact that thyrocalcitonin and parathyroid hormone exhibit antagonistic activities. Repeated administration of thyrocalcitonin blocks the action of parathyroid hormone (Klein and Talmage, 1968). This phenomenon was also observed in cultivated explants of mouse calvaria (Heersche, 1969). Wells and Lloyd (1968a,b) suggested a possible mechanism to explain the antagonistic action of thyrocalcitonin and parathyroid hormone. Accepting that parathyroid hormone increases adenyl cyclase which results in elevated cyclic AMP and which in turn mobilized calcium from bone to blood, thyrocalcitonin may well increase phosphodiesterase which catalyzes the degrada-

Fig. 39. An electron micrograph taken from a young rat treated for 2 weeks with porcine calcitonin shows the complete separation and smoothness of the surface of an osteoclast and the smooth collagenous, decalcified bone surface. Compare with Fig. 8. ×10,000. (The electron micrograph was kindly supplied by Dr. L. Zichner, Department of Pathology, University of Zurich, Switzerland.)

tion of cyclic AMP producing a hypocalcemic effect in a manner experienced through the use of imidazole. It is suggested by Gaillard (1970) that both thyrocalcitonin and imidazole are able to induce the formation of osteoblasts and bone.

The hypothesis submitted by Talmage is extremely entertaining. It appears reasonable on morphological grounds and plausible on a chemical basis. Although some electron microscopic work has been reported on the effects of PTH on bone cells, the effects of thyrocalcitonin have yet to be evaluated at this level save for the recent report by Zichner (1971) in which, following porcine calcitonin treatment of rats, osteoclasts appeared to lack the presence of the brush-border characteristic of the active osteoclast (Fig. 38). Instead, these cells were retracted from the surface of bone and exhibited smooth surfaces (Fig. 39). Howship's lacunae, indicative of bone resorption, were also absent. Osteoblasts resembled precursor cells in morphology. No doubt, with the growing body of cytochemical and microscopic information, Talmage's hypothesis will be refined and may well become the most important singular theory on the integrated behavior of the actions of parathyroid hormone and thyrocalcitonin on bone cell-plasma homeostasis and bone cell function.

XI. PAROTIN

Experimental reports spanning over 80 years exist in support of the hypothesis that the salivary glands are, at least in part, endocrine organs. These organs secrete a hormone-like substance called parotin. Ogata (1944a, b) and other investigators (Maruyama, 1950; Ito and Aonuma, 1952; Sato, 1953) noted that salivary extracts from young cattle consistently lowered the serum calcium levels and enhanced the calcification of teeth and bones when administered to rabbits. Ito *et al.* (1954) reported increased uptake of ^{32}P inorganic phosphate by incisors and femurs of rats treated with parotin.

In a number of studies involving the administration of parotin, Fleming (1959, 1960) reported interference with the secretory activity of ameloblasts of mouse mandibular incisors affecting enamel protein matrix formation. In the same animals, a marked increase in the vascularity of the femoral epiphyseal plates was observed. The cartilage plates were thinner than controls and fibrillar in some regions. Osteoclastic activity was significantly enhanced along plate surfaces and femoral shafts. Increased vascularity was not limited to the skeletal structures, but was also observed in the testes and ovaries of parotin treated animals. In a small but interesting study (Chaunecy *et al.*, 1963), it was reported that ten golden hamsters, 1 month of age, were partially desalivated by bilateral removal of the submandibular and sublingual glands. Five of these animals were given no further treatment. The remainder were given twice weekly injections of parotin. An additional ten hamsters were used; five served as untreated controls, the remainder received parotin. Parotin treatment was continued for 9 weeks. Examination of femoral epiphyseal cartilage revealed no demonstrable effects of parotin in intact animals. In the partially desalivated group, the epiphyseal plates were significantly narrower. The zone of proliferation was predominant and indicative of stimulation. These alterations were prevented in partially desalivated animals treated with parotin.

In view of the scarcity of studies in this area and the yet unsettled status of parotin in the realm of endocrine physiology, it may be considered inappropriate to include this topic in the present chapter. The hormone-like nature of parotin and its effects on the skeletal system are, however, interesting and worthy of further investigation. It is hoped that the brief presentation may engender some interest in this largely neglected area leading to the pursuit of laboratory studies on the relationship of parotin to skeletal biology. It may well be determined that the skeletal response is mediated through the action of other endocrine secretions affected by the presence or absence of parotin and not by parotin directly. Experimental evidence of parotin activation of adrenocortical function of the pituitary is known (Ito *et al.*, 1952). Further studies are required to unravel the existing basic relationships.

XII. SUMMARY

Hormonal effects on the growth, development, and response of the skeletal system have been described briefly in the present chapter under the headings of individual hormones for pedagogical reasons. This artificial organization does not imply segregated effects and responses of individual hormones. In fact, ample experimental evidence shows that the presence or absence of a hormone may exert significant influence on the tissue susceptability or response to other hormones; both synergism and antagonism are known. Copp and Kuczerpa (1967) reported that somatotropic hormone alters the response to thyrocalcitonin. Whereas, thyrotropic hormone is synergistic to STH (Marx *et al.*, 1942). Thyrocalcitonin is also antagonistic to parathyroid hormone. Gabbiani *et al.* (1968) showed that thyroxine inhibits changes induced by parathyroid extract overdosage. It must, therefore, be emphasized that the effects of hormones exerted on the skeletal system are not exclusively directed to the skeletal system. As a consequence of affecting the action of other hormones, they also affect other vital tissue systems which, in turn, may alter the demands placed on the skeletal system. Even somatotropic hormone, which above all hormones is claimed to exert its primary affect directly on the skeletal system, is known to produce significant changes in protein, fat, and carbohydrate metabolism of many tissues. With the accumulation of experimental information, there is a growing appreciation and awareness of the complexity which exists in the interrelationships between hormones and tissue systems. No denial of this statement is intended, however, by the fact that the descriptions alluded to in this chapter are indeed simplistic and entertain at best a parochial view of the more complex nature and extensive panorama of the skeletal–hormonal interrelationships which exist.

Some efforts have been made to study the ultrastructural effects of hormones on various cellular compartments of the skeletal system using the electron microscope. This knowledge is, however, exiguous at best. A significant effort in this direction is needed to explore the detailed response of the cell machinery to hormonal influence in cells which are known to be responsive. The cellular response, on the other hand, need not involve the intracellular penetration of the hormone to affect the organelles. Instead, it may act on the cell wall to release an enzyme which is needed for the production of a "second messenger" called cyclic AMP. In view of the more recent findings involving adenyl cyclase and cyclic AMP, the field of electron microscopic cytochemistry is indeed inviting. These techniques hold the promise of significant discovery regarding the target cell effects of hormones. Existing information is at the histological-cytological level. Optimistically, the next decade will raise our level of understanding to that of the ultrastructural order.

Presently the most exciting area, and one which undoubtedly will receive increased attention, involves the prostaglandins. Classified as hormones, prostaglandins make up a group of 14 compounds thus far uncovered and are found in humans and all other mammals investigated. They are believed to be produced by all tissues of the body and act locally. Prostaglandins are involved in the normal body function as well as in pathology. The 14 compounds are intracellular substances, all of which resemble one another in chemical structure and consist of 20 carbon chain, acidic lipids. Small chemical differences which exist between them, however, account for significantly different biological effects. The prostaglandins have been divided into four main categories namely, E, F, A, and B, with E and F consisting of E_1, E_2, E_3 and F_1, F_2, F_3 subgroups. These categories are considered to be the primary prostaglandins in that all the others can be derived from their modification.

Such substances apparently exert a wide variety of biological effects also involving the skeletal system. It is known that prostaglandin (PGE_1) increases cyclic AMP concentration in fetal rat bone. The effects on bone resorption are found to be similar to those of PTH, and like PTH the effects are inhibited by thyrocalcitonin and cortisol (Klein and Raisz, 1970). These authors reported from tissue culture studies that previously incorporated ^{45}Ca is released from embryonic bone in response to prostaglandin. Although the site of hormone action is not known, it was concluded that the similarity in effect between PTH and PGE is consistent with the hypothesis that PGE acts to elevate cyclic AMP concentrations in bone cells.

Within the last decade, beginning investigations into the ultrastructural nature of hormonal effects, including both morphological and biochemical, have already borne fruit, pointing future research into several intriguing directions. Such studies will no doubt intensify, leading subsequently to a better understanding of the direct cellular effects of hormones on bone growth, development, maintenance, repair, and aging. With this information at hand, we will be better prepared to undertake studies involving skeletal pathology.

ACKNOWLEDGMENTS

This presentation was supported under NIH grants HD03677 and DE03014. The author wishes to thank the many investigators who have contributed photomicrographs for this chapter, especially Dr. Ruth Silberberg. Special thanks is extended to Dr. P. J. Collipp for kindly reviewing the manuscript.

REFERENCES

Albright, F. (1942–43). *Harvey Lect.* 38, 123.
Albright, F., and Reifenstein, E. C., Jr. (1948). "The Parathyroid Glands and Metabolic Bone Disease." Williams and Wilkins, Baltimore, Maryland.
Anderson, I. A. (1950). *Quart. J. Med.* 19, 67.
Anderson, C. E., Crane, J. T., and Harper, H. A. (1959). *J. Bone Joint Surg.* 41A, 1094.
Armstrong, W. D. (1942). *Trans. Josiah Macy, Jr., Conf. Bone Wound Healing, 3rd* p. 74.
Asboe-Hansen, G. (1966). *In* "Hormones and Connective Tissue" (G. Asboe-Hansen, ed.), pp. 29–61. Williams and Wilkins, Baltimore, Maryland.
Ascenzi, A., Francois, C., and Bocciarelli, D. S. (1963). *J. Ultrastruct. Res.* 8, 491.
Asling, C. W., and Evans, H. M. (1956). *In* "The Biochemistry and Physiology of Bone" (G. H. Bourne, ed.), pp. 671–703. Academic Press, New York.
Asling, C. W., Simpson, M. E., Li, C. H., and Evans, H. M. (1954). *Anat. Rec.* 119, 101.
Asling, C. W., Simpson, M. E., and Evans, H. M. (1965). *Rev. Suisse Zool.* 72, 1.
Barnicot, N. A. (1945). *J. Anat.* 79, 83.
Barnicot, N. A. (1948). *J. Anat.* 82, 233.
Baud, C. A. (1968). *Schweiz. Med. Worschr.* 98, 717.
Becks, H., Kibrick, E. A., Marx, W., and Evans, H. M. (1941). *Growth* 4, 437.
Becks, H., Simpson, M. E., Evans, H. M., Ray, R. D., Li, C. H., and Asling, C. W. (1946). *Anat. Rec.* 94, 631.
Becks, H., Simpson, M. E., Li, C. H., and Evans, H. M. (1944). *Endocrinology* 34, 305.
Becks, H., Asling, C. W., Simpson, M. E., Li, C. H., and Evans, H. M. (1949). *Growth* 13, 175.
Bélanger, L. F., Robichon, J., and Copp, D. H. (1963). *Proc. Amer. Assoc. Anat.* 145, 206.
Bellieni, G., Bigliani, R., and Forte, D. (1956). *Atti. Accad. Fisiocr. Siena Sez. Med. Fis.* 3, 238.
Bernick, S. (1970), *Calc. Tiss. Res.* 5, 170.
Bernick, S., and Ershoff, B. H. (1963). *Endocrinology* 72, 231.

Bernick, S., and Zipkin, I. (1967). *J. Dental Res.* **46**, 1404.

Blunt, J. W., Plotz, C. M., Lattes, R., Howes, E. L., Meyer, K., and Ragan, C. (1950). *Proc. Soc. Exp. Biol. Med.* **73**, 673.

Borle, A. M., Nichols, N., and Nichols, G., Jr. (1960). *J. Biol. Chem.* **235**, 3323.

Boström, H., and Odelblad, E. (1953). *Ark. Kemi.* **6**, 39.

Breibart, S., Bongiovanni, A. M., and Eberlein, W. R. (1963). *New England J. Med.* **268**, 255.

Brookes, M., and Lloyd, E. G. (1961). *J. Anat.* **95**, 220.

Budy, A. M. (1960). *Clin. Orthop.* **17**, 176.

Budy, A. M., Urist, M. R., and McLean, F. C. (1952). *Amer. J. Pathol.* **28**, 1143.

Bullough, W. S. (1954). *Exp. Cell Res.* **7**, 186.

Burrows, R. B. (1938). *Amer. J. Anat.* **62**, 237.

Canjggia, A., Gennari, C., Bencini, M., Cesari, L., and Borrello, G. (1970). *Clin. Sci.* **38**, 397.

Cameron, D. A., Paschall, H. A., and Robinson, R. A. (1967). *J. Cell Biol.* **33**, 1.

Care, A. D., and Duncan, T. (1967). *J. Endocrin. (Brit.)* **37**, 107.

Carnes, W. H. (1950). *Amer. J. Pathol.* **26**, 736.

Chang, H. Y. (1951). *Anat. Rec.* **111**, 23.

Chase, L. R., and Aurbach, G. D. (1967). *Proc. Nat. Acad. Sci.* **58**, 518.

Chase, L. R., and Aurbach, G. D. (1968). *Science* **159**, 545.

Chase, L. R., Fedak, S. A., and Aurbach, G. D. (1969). *Endocrinology* **84**, 761.

Chaunecy, H. H., Kronman, J. H., Spinale, J. J., and Shklar, F. (1963). *J. Dental Res.* **42**, 894.

Che-Kuo, H., and Johannessen, L. B. (1970). *J. Dental Res.* **49**, 34.

Chen, J. M. (1954). *J. Physiol.* **125**, 148.

Cicala, V., Vergine, A., and Tesauro, P. (1962). *Rev. Anat. Pat. Oncol.* **21**, 819.

Collipp, P. J., Patrick, J. R., Goodheart, C., and Kaplan, S. A. (1966). *Proc. Soc. Exp. Biol. Med.* **121**, 173.

Copp, D. H., Cameron, E. C., Cheney, B. A., Davidson, A. G. F., and Henze, K. G. (1962). *Endocrinology* **70**, 638.

Copp, D. H., and Kuczerpa, A. V. (1967). *Fed. Proc.* **26**, 369.

Daughaday, W. H., and Mariz, I. K. (1962a). *J. Biol. Chem.* **237**, 2831.

Daughaday, W. H., and Mariz, I. K. (1962b). *J. Lab. Clin. Med.* **59**, 741.

Deguchi, T., and Mori, M. (1969). *Histochemie* **20**, 234.

Duraiswami, P. K. (1950). *Brit. Med. J.* **ii**, 384.

Duraiswami, P. K. (1955). *J. Bone Joint Surg.* **37A**, 277.

Duthie, R. B., and Barker, A. M. (1955). *J. Bone Joint Surg.* **37B**, 691.

Ebert, P. S., and Prockop, D. J. (1963). *Biochem. Biophys. Acta* **78**, 390.

Engel, M. B. (1952). *A.M.A. Arch. Pathol.* **53**, 339.

Evans, H. M., Simpson, M. E., and Li, C. H. (1943). *Endocrinol.* **33**, 237.

Fahmy, A. (1968). *Anat. Rec.* **160**, 346.

Fahmy, A., Hillman, J. W., Talley, P., and Long, V. (1969). *J. Bone Joint Surg.* **51A**, 802.

Fahmy, A., Lee, S., and Johnson, P. (1971a). *Calc. Tiss. Res.* **7**, 11.

Fahmy, A., Talley, P., Frazier, H. M., and Hillman, J. W. (1971b). *Calc. Tiss Res.* **7**, 139.

Fell, H. B. (1953). *Sci. Progr. Twent. Cent.* **162**, 212.

Fleming, H. S. (1959). *J. Dent Res.* **38**, 374.

Fleming, H. S. (1960). *Amer. N.Y. Acad. Sci.* **85**, 313.

Follis, R. H., Jr. (1951). *Proc. Soc. Exp. Biol. Med.* **78**, 723.

354   *Edgar A. Tonna*

Fontaine, R., Mandel, P., and Wiest, E. (1952). *Mem. Acad. Chir. Paris* **78**, 351.
Foster, G. V., Baghdiantz, A., Kumar, M. A., Slack, E., Soliman, H. A., and Mac-Intyre, I. (1964). *Nature (London)* **202**, 1303.
Foster, G. V., Doyle, F. H., Bordier, P., and Matrajt, H. (1966a). *Lancet* **2**, 1428.
Foster, G. V., Joplin, G. F., MacIntyre, I., Melvin, K. E. W., and Slack, E. (1966b). *Lancet* **i**, 107.
Foulks, J. G., and Perry, F. A. (1959). *Amer. J. Physiol.* **196**, 567.
Friedman, J., and Raisz, L. G. (1967). *Science* **150**, 1465.
Gabbiani, G., Tuchweber, B., Côté, G., and Lefort, P. (1968). *Calc. Tiss. Res.* **2**, 30.
Gaillard, P. J. (1959). *Develop. Biol.* **1**, 152.
Gaillard, P. J. (1967a). *Proc. Kon. Med. Akad. Wet. Ser. C.* **70**, 309.
Gaillard, P. J. (1967b). *Acta Physiol. Pharmacol. Neerl.* **14**, 501.
Gaillard, P. J. (1970). *Calc. Tiss. Res.* **4**, 86.
Gardner, W. U. (1936). *Amer. J. Anat.* **59**, 459.
Gardner, W. U. (1940). *Anat. Rec.* **76**, 22.
Gardner, W. U. (1943). *Endocrinology* **32**, 149.
Geschwind, I. I., and Li, C. H. (1955). "The Hypophyseal Growth Hormone, Nature and Actions," Chapter 3. Blakiston, New York.
Greep, R. O., and Talmage, R. V. (1961). "The Parathyroids," Thomas, Springfield, Illinois.
Ham, A. W. (1930). *J. Bone Joint Surg.* **12**, 827.
Hamilton, T. H. (1968). *Science* **161**, 649.
Hamolsky, M. W., Stein, M., Fischer, D. B., and Freedberg, A. S. (1961). *Endocrinology* **68**, 662.
Hancox, N. M. (1961). *J. Anat.* **95**, 411.
Harris, W. H., and Heany, R. P. (1969). *New England J. Med.* **280**, 253.
Hay, M. F. (1958). *J. Physiol.* **144**, 490.
Hechter, O., and Halkerston, I. D. K. (1965). *Ann. Rev. Physiol.* **27**, 133.
Heersche, J. N. M. (1969). Thesis, Univ. of Leiden.
Hekkelmann, J. W. (1965). "The Parathyroids," p. 211. Chicago Univ. Press, Chicago, Illinois.
Heller, M., McLean, F. C., and Bloom, W. (1950). *Amer. J. Anat.* **87**, 315.
Henneman, P. H., Forbes, A. P., Moldawer, M., Dempsey, E. F., and Carroll, E. L. (1960). *J. Clin. Invest.* **39**, 1223.
Hervey, G. R., and Hervey, E. (1964). *J. Endocr.* **30**, Proc. VII–VIII.
Hirsch, P. F. (1967). *Endocrinology* **80**, 539.
Hirsch, P. F., Gauthier, G. F., and Munson, P. L. (1963). *Endocrinology* **73**, 244.
Hirsch, P. F., Voelkel, E. F., and Munson, P. L. (1964). *Science* **146**, 412.
Horstmann, P. (1949). *Acta Endocrinol. Suppl.* **5**, 3, 1.
Howard, E. (1962). *Endocrinology* **70**, 131.
Howard, E. (1963). *Endocrinology* **72**, 11.
Huble, J. (1957). *Acta Endocrinol.* **25**, 59.
Illig, R., and Prader, A. (1970). *J. Clin. Endocrinol. Metab.* **30**, 615.
Ingle, D. J. S., Higgins, S. M., and Kendall, E. C. (1938). *Anat. Rec.* **71**, 363.
Ito, Y., and Aonuma, S. (1952). *J. Pharm. Soc. Japan* **72**, 1517.
Ito, Y., Aonuma, S., and Higashi, K. (1952). *J. Pharm. Soc. Japan* **72**, 1465.
Ito, Y., Tsurufuji, S., and Kubota, Y. (1954). *J. Pharm. Soc. Japan* **74**, 350.
Jee, W. S. S., Blackwood, E. L., Dockum, N. L., Haslam, R. K., and Kincl, F. A. (1966). *Clin. Orthop.* **49**, 39.

Jee, W. S. S., Park, H. Z., Roberts, W. E., and Kenner, G. H. (1970). *Amer. J. Anat.* **129**, 477.

Johannessen, L. B., Davidovitch, Z., and Blackey, W. E. (1966). *Arch. Oral Biol.* **11**, 31.

Joss, E. E., Zuppinger, K. A., Sobel, E. H. (1963). *Endocrinology* **72**, 123.

Jowsey, J. (1969). *Calc. Tiss. Res.* **4**, 188.

Kao, K-Y, T., Hitt, W. E., and McGavack, T. H. (1967). *Proc. Soc. Exp. Biol. Med.* **125**, 734.

Karnofsky, D. A., Ridgeway, L. P., and Patterson, P. A. (1951). *Endocrinology* **48**, 596.

Kennedy, B. J., Tibbetts, D. M., Nathanson, I. T., and Aub, J. C. (1953). *Cancer Res.* **13**, 445.

Key, J. A., Odell, R. T., and Taylor, L. W. (1952). *J. Bone Joint Surg.* **34B**, 665.

Keys, P., and Potter, T. S. (1934). *Anat. Rec.* **60**, 377.

Klein, D. C., and Raisz, L. G. (1970). *Endocrinology* **86**, 1436.

Klein, D. C., and Talmage, R. V. (1968). *Proc. Soc. Exp. Biol. Med.* **127**, 95.

Klein, M., Villanueva, A. R., and Frost, H. M. (1965). *Acta Orthop. Scand.* **35**, 171.

Knobil, E. (1961). *In* "Growth in Living Systems" (M. X. Zarrow, ed.), p. 353. Basic Books, New York.

Laird, P. P., and Cameron, D. A. (1972). Personal communication.

Landauer, W. (1947a). *Endocrinology* **41**, 489.

Landauer, W. (1947b). *J. Exp. Zool.* **105**, 317.

Landauer, W. (1953). *Growth* **17**, 87.

Landauer, W., and Bliss, C. I. (1946). *J. Exp. Zool.* **102**, 1.

Lardy, H., Tomita, K., Larson, F. C., and Albright, E. C. (1957). *Colloq. Endocrinol.* **10**, 156.

Larson, F. C., and Albright, E. C. (1958). *Endocrinology* **63**, 183.

Lash, J. W., and Whitehouse, M. W. (1961). *Lab Invest.* **10**, 388.

Lawrence, R. T. B., Salter, J., and Best, C. H. (1954). *Brit. Med. J.* **ii**, 437.

Li, C. H. (1969a). *In* "La Spécificité Zoologique des Hormones Hypophysaires et de Leurs Activitiés" (M. Fontaine, ed.), p. 93. Centre Nationale de la Recherche Scientifique, Paris.

Li, C. H. (1969b). *In* "La Spécificité Zoologique des Hormones Hypophysaires et de Leurs Activités" (M. Fontaine, ed.), p. 175. Centre Nationale de la Recherche Scientifique, Paris.

Lindquist, B., Budy, A. M., McLean, F. C., and Howard, J. L. (1960). *Endocrinology* **66**, 100.

Maassen, A. P. (1952). *Acta Endocrinol.* **9**, 135.

MacIntyre, I. (1967). *Calc. Tiss. Res.* **1**, 173.

Maddaiah, V. T., Rezvani, I., Chen, S. Y., and Collipp, P. J. (1970). *Biochem. Med.* **4**, 492.

Mankin, H. J., and Conger, K. A. (1966). *J. Bone Joint Surg.* **48A**, 1383.

Mantzavinos, Z., and Listgarten, M. A. (1970). *J. Periodont.* **41**, 663.

Martin, T. J., Robinson, C. H., and MacIntyre, I. (1966). *Lancet* **I**, 900.

Maruyama, M. (1950). *J. Biochem. (Tokyo)* **37**, 1.

Marx, W., Simpson, M. E., and Evans, H. M. (1942). *Proc. Soc. Exp. Biol. Med.* **49**, 594.

Marx, W., Simpson, M. E., Li, C. H., and Evans, H. M. (1943). *Endocrinology* **33**, 102.

McLean, F. C. (1956). In "The Biochemistry and Physiology of Bone" (G. H. Bourne, ed.), pp. 705–727. Academic Press, New York.

McLean, F. C., and Urist, M. R. (1961). In "Bone, an Introduction to the Physiology of Skeletal Tissue," 2nd ed., p. 101. Univ. of Chicago Press, Chicago, Illinois.

Menczel, J., Eilon, G., Klein, T., and Tishbee, A. (1970). Calc. Tiss. Res. 4, 51.

Meyer, K. (1956). In "Bone Structure and Metabolism" (G. E. Wolstenholme and C. M. O'Connor, eds.), p. 65. Little, Brown, Boston, Massachusetts.

Meyer, K., Hoffman, P., and Linker, A. (1958). Science 128, 896.

Meyer, W. L., and Kunin, A. S. (1969). Arch. Biochem. 129, 431.

Milhaud, G., Perault, A. M., and Moukhtar, M. S. (1965). C. R. Acad. Sci. Paris 261, 4513.

Miner, R. W., and Henegan, B. J. (1951). Ann. N.Y. Acad. Sci. 54, 531.

Minot, A. S., and Hillman, J. W. (1967). Proc. Soc. Exp. Biol. Med. 126, 60.

Montgomery P. O'B. (1955). Proc. Soc. Exp. Biol. Med. 90, 410.

Moon, H. D. (1937). Proc. Soc. Exp. Biol. Med. 37, 34.

Moon, H. D., Simpson, M. E., Li, C. H., and Evans, H. M. (1951). Cancer Res. 11, 535.

Moss, M. L. (1955). Proc. Soc. Exp. Biol. Med. 89, 648.

Munson, P. L. (1955). Ann. N.Y. Acad. Sci. 60, 776.

Nichols, G., Jr. (1966). In "Calcified Tissues" (H. Fleisch, H. J. J. Blackwood, and M. Owen, eds.), pp. 215–226. Springer, Berlin.

Ogata, T. (1944a). Folia Endocrinol. Japan 20, 9.

Ogata, T. (1944b). Folia Endocrinol. Japan 20, 434.

Owen, M., and Bingham, P. J. (1968). In "Parathyroid Hormone and Thyrocalcitonin (Calcitonin)" (R. V. Talmage and L. F. Bélanger, eds.), p. 203. Excerpta Medica Foundation, Amsterdam.

Pak, C. Y. C. (1971). Ann. N.Y. Acad. Sci. 179, 450.

Park, H. Z., and Talmage, R. V. (1967). Endocrinology 80, 552.

Pearse, A. G. E. (1966). Proc. Roy. Soc. 164B, 478.

Pechet, M. M., Bobadilla, E., Carroll, E. L., and Hesse, R. H. (1967). Amer. J. Med. 43, 696.

Priest, R. E., Koplitx, R., and Benditt, E. (1960). J. Exp. Med. 112, 225.

Puche, R. C., and Romano, M. C. (1968). Calc. Tiss. Res. 2, 133.

Puche, R. C., and Romano, M. C. (1970). Calc. Tiss. Res. 6, 133.

Puche, R. C., and Romano, M. C. (1971). Calc. Tiss. Res. 7, 103.

Pullman, T. N., Lavender, A. R., Aho, I., Rasmussen, H. (1960). Endocrinology 67, 570.

Raisz, L. G. (1963). Nature (London) 197, 1015.

Rasmussen, H., Sze, Y-L., and Young, R. (1964). J. Biol. Chem. 239, 2852.

Ravault, P., Vignon, G., and Fraisse, H. (1950). Rev. Rheum. 17, 247.

Ray, R. D., Asling, C. W., Simpson, M. E., and Evans, H. M. (1950a). Anat. Rec. 107, 253.

Ray, R. D., Simpson, M. E., Li, C. H., Asling, C. W., and Evans, H. M. (1950b). Amer. J. Anat. 86, 479.

Ray, R. D., Asling, C. W., Walker, D. G., Simpson, M. E., Li, C. H., and Evans, H. M. (1954). J. Bone Joint Surg. 36A, 94.

Reifenstein, Jr., E. C., and Albright, F. (1947). J. Clin. Invest. 26, 24.

Reiss, M., Fernandes, J. E., and Golla, Y. M. L. (1946). Endocrinology 38, 65.

Reynolds, J. J., Minkin, C., and Parsons, J. A. (1970). Calc. Tiss. Res. 4, 350.

Riddle, O., Rauch, V. M., and Smith, G. C. (1944). Anat. Rec. 90, 295.

Riddle, O., Rauch, V. M., and Smith, G. C. (1945). *Endocrinology* 36, 41.
Riekstniece, E., and Asling, C. W. (1966). *Proc. Soc. Exp. Biol. Med.* 123, 258.
Rigal, W. M. (1964). *Proc. Soc. Exp. Biol. Med.* 117, 794.
Roberts, W. E. (1969). Ph.D. Thesis, Dept. of Anat., Univ. of Utah, Salt Lake City.
Robinson, C. J., Martin, T. J., and MacIntyre, I. (1966). *Lancet* II, 83.
Robinson, C. J., Matthews, E. W., MacIntyre, I. (1967). *J. Endocrinol. (Brit.)* 39, 71.
Rodin, A. E., and Kowaleski, K. (1963). *Can. J. Surg.* 6, 229.
Rohr, H., and Wolff, H. (1967). *Z. Ges. Exp. Med.* 142, 155.
Rubenstein, H. S., and Solomon, M. L. (1941a). *Endocrinology* 28, 112.
Rubenstein, H. S., and Solomon, M. L. (1941b). *Endocrinology* 28, 229.
Rubenstein, H. S., Kurland, A. A., and Goodwin, M. (1939). *Endocrinology* 25, 724.
Salter, J., and Best, C. H. (1953). *Brit. Med. J.* ii, 353.
Sanger, F. (1959). *Science* 129, 1340.
Sato, T. (1953). *Gunma. J. Med. Sci. (Japan)* 2, 183.
Saunders, F. J., and Elton, B. L. (1959). *In* "Recent Progress in the Endocrinology of Reproduction" (C. W. Lloyd, ed.), pp. 227–254. Academic Press, New York.
Sayeed, M. M., Blumenthal, D. S., and Blumenthal, H. T. (1962). *Proc. Soc. Exp. Biol. Med.* 109, 261.
Schiff, M. (1966). *In* "Hormones and Connective Tissue" (G. Asboe-Hansen, ed.), pp. 282–341. Williams and Wilkins, Baltimore, Maryland.
Scow, R. O., Simpson, M. E., Asling, C. W., Li, C. H., and Evans, H. M. (1949). *Anat. Rec.* 104, 445.
Selye, H., Pahk, U. S., and Tuchweber, B. (1967). *Fed. Proc.* 26, 430.
Siegel, B. V., Smith, M. J., and Gerstl, B. (1957). *A.M.A. Arch. Pathol.* 63, 562.
Siffert, R. S. (1951). *J. Exp. Med.* 93, 415.
Silberberg, M., and Silberberg, R. (1941). *Arch. Pathol.* 31, 85.
Silberberg, M., and Silberberg, R. (1946). *Anat. Rec.* 95, 97.
Silberberg, M., and Silberberg, R. (1956). *In* "The Biochemistry and Physiology of Bone" (G. H. Bourne, ed.), p. 653. Academic Press, New York.
Silberberg, M., and Silberberg, R. (1965). *Growth* 29, 311.
Silberberg, M., Silberberg, R., and Opdyke, M. (1954). *Endocrinology* 54, 26.
Silberberg, M., Silberberg, R., and Hasler, M. (1965). *Anat. Rec.* 151, 297.
Silberberg, M., Hasler, M., and Silberberg, R. (1966a). *Pathol. Microbiol.* 29, 137.
Silberberg, M., Silberberg, R., and Hasler, M. (1966b). *Arch. Pathol.* 82, 569.
Silberberg, M., Silberberg, R., and Hasler, M. (1967). *Pathol. Microbiol.* 30, 283.
Silberberg, R., and Hasler, M. (1968). *Pathol. Microbiol.* 31, 25.
Silberberg, R., and Silberberg, M. (1954a). *Amer. J. Anat.* 95, 263.
Silberberg, R., and Silberberg, M. (1954b). *Lab. Invest.* 3, 228.
Silberberg, R., and Silberberg, M. (1957). *Endocrinology* 60, 67.
Silberberg, R., and Silberberg, M. (1965a). *Exp. Med. Surg.* 23, 157.
Silberberg, R., and Silberberg, M. (1965b). *J. Gerontol.* 20, 228.
Silberberg, R., and Silberberg, M. (1970). *Gerontology* 16, 201.
Silberberg, R., Goto, G., and Silberberg, M. (1958a). *A.M.A. Arch. Pathol.* 65, 438.
Silberberg, R., Thomasson, R., and Silberberg, M. (1958b). *A.M.A. Arch. Pathol.* 65, 442.

Silberberg, R., Hasler, M., and Silberberg, M. (1965). *Amer. J. Pathol.* **46**, 289.

Silberberg, R., Hasler, M., and Silberberg, M. (1966a). *Acta Anat.* **65**, 275.

Silberberg, R., Hasler, M., Silberberg, M. (1966b). *Anat. Rec.* **155**, 577.

Silberberg, R., Hasler, M., and Young, D. A. (1968). *Lab. Invest.* **19**, 7.

Simmons, D. J. (1963). *Clin. Orthop.* **26**, 176.

Simpson, M. E., Marx, W., Becks, H., and Evans, H. M. (1944). *Endocrinology* **35**, 309.

Simpson, M. E., Asling, C. W., and Evans, H. M. (1950). *Yale J. Biol. Med.* **23**, 1.

Sinex, F. M. (1968). *In* "Treatise on Collagen" (G. N. Ramachandran and B. S. Gould, eds.), Vol. 2, p. 410. Academic Press, New York.

Sissons, H. A. (1960). *In* "Bone as a Tissue" (K. Rodahl, J. T. Nicholson, and E. M. Brown, Jr., eds.). McGraw-Hill, New York.

Sissons, H. A., and Hadfield, G. J. (1951). *Brit. J. Surg.* **39**, 172.

Sissons, H. A., and Hadfield, G. J. (1955). *J. Anat.* **89**, 69.

Sobel, H. (1958). *Proc. Soc. Exp. Biol. Med.* **97**, 495.

Sobel, H., and Feinberg, S. (1970). *Calc. Tiss. Res.* **5**, 39.

Sobel, H., and Marmorston, J. (1954). *Endocrinology* **55**, 21.

Sobel, H., Raymond, C. S., Quinne, K. V., and Talbot, N. B. (1956). *J. Clin. Endocrinol.* **16**, 241.

Stidworthy, G., Masters, Y. F., and Shetlar, M. R. (1958). *J. Gerontol.* **13**, 10.

Stock, C. C., Karnfosky, D. A., and Sugiura, K. (1951). *In* "Symposium on Steroids in Experimental and Clinical Practice" (A. White, ed.). Blakiston, New York.

Storey, E. (1958a). *J. Bone Joint Surg.* **40B**, 103.

Storey, E. (1958b). *J. Bone Joint Surg.* **40B**, 558.

Storey, E. (1960a). *Aust. New Zeal. J. Surg.* **30**, 36.

Storey, E. (1960b). *Brit. J. Exp. Pathol.* **41**, 207.

Storey, E. (1961). *Endocrinology* **68**, 533.

Storey, E. (1963). *Clin. Orthop.* **30**, 197.

Sturtridge, W. C., Jowsey, J., and Kumar, M. A. (1968). *Calc. Tiss. Res.* **2**, 99.

Suzuki, H. K. (1958). *J. Bone Joint Surg.* **40A**, 435.

Swislocki, N. I., and Szego, C. M. (1965). *Endocrinology* **76**, 665.

Talmage, R. V. (1969). *Clin. Orthop.* **67**, 210.

Talmage, R. V. (1970). *Amer. J. Anat.* **129**, 467.

Talmage, R. V., and Bélanger, L. F. (1968). *In* "Parathyroid Hormone and Thyrocalcitonin (Calcitonin)." Excerpta Medica Foundation, New York.

Talmage, R. V., Elliott, J. R., and Enders, C. A. (1957). *Endocrinology* **61**, 256.

Talmage, R. V., Cooper, C. W., and Park, H. Z. (1970). *In* "Vitamins and Hormones," Vol. 28, p. 103. Academic Press, New York.

Tapp, E. (1966). *J. Bone Joint Surg.* **48A**, 526.

Tarsoly, E., Hajer, Gy., and Urbán, I. (1965). *Acta Chir. Hung.* **6**, 435.

Tata, J. R. (1958). *Biochem. Biophys. Acta* **28**, 91.

Thomson, D. L., and Collip, J. B. (1932). *Physiol. Rev.* **12**, 309.

Tinacci, F., Tosetti, D., and Cioni, P. (1962). *Coll. Currents* **3**, 3.

Tonna, E. A. (1960). *Nature (London)* **185**, 405.

Tonna, E. A. (1961). *J. Biophys. Biochem. Cytol.* **9**, 813.

Tonna, E. A. (1965). *In* "Use of Radioautography in Investigating Protein Synthesis" (C. P. Leblond and K. B. Warren, eds.). *Symp. Int. Soc. Cell Biol.* Vol. 4, p. 215. Academic Press, New York.

Tonna, E. A. (1971). *Gerontologia* **17**, 273.

14. Hormonal Influence on the Skeleton

Tonna, E. A. (1972). Connective Tissue Research, 1, 221.
Tonna, E. A., and Nicholas, J. A. (1959). J. Bone Joint Surg. 41A, 1149.
Tonna, E. A., and Pavelec, M. (1971). J. Gerontol. 26, 310.
Tonna, E. A., and Pentel, L. (1972). Lab. Invest. 27, 418.
Trueta, J. (1968). "Studies of Development and Decay of the Human Frame," pp. 45–56. Saunders, Philadelphia, Pennsylvania.
Turner, C. D., and Bagnara, J. T. (1971). "General Endocrinology," pp. 74–181. Saunders, Philadelphia, Pennsylvania.
Udupa, K. N., and Gupta, L. P. (1965). Indian J. Med. Res. 53, 623.
Urist, M. R. (1967). Amer. Zool. 7, 883.
Urist, M. R., Budy, A. M., and McLean, F. C. (1948). Proc. Soc. Exp. Biol. Med. 68, 324.
Urist, M. R., and Deutsch, N. M. (1960). Endocrinology 66, 805.
Urist, M. R., Budy, A. M., and McLean, F. C. (1950). J. Bone Joint Surg. 32A, 143.
Vaes, G. (1965). Exp. Cell Res. 39, 470.
Vaes, G., and Nichols, G., Jr. (1961). J. Biol. Chem. 236, 3323.
Vaes, G., and Nichols, G., Jr. (1962). Endocrinology 70, 546.
Van den Hooff, A. (1964). Acta Anat. (Basel) 57, 16.
Verzár, F., and Spichtin, H. (1966). Gerontologia 12, 48.
Walker, D. G., Asling, C. W., Simpson, M. E., Li, C. H., and Evans, H. M. (1952). Anat. Rec. 114, 19.
Wase, A. W., Solewski, J., Ricker, E., and Seidenberg, J. (1967). Nature (London) 214, 388.
Wells, H. B., and Kendall, E. C. (1940). Proc. Staff. Meeting Mayo Clin. 15, 324.
Wells, H., and Lloyd, W. (1968a). Endocrinology 83, 521.
Wells, H., and Lloyd, W. (1968b). In "Parathyroid Hormone and Thyrocalcitonin (Calcitonin) (R. V. Talmage and L. F. Bélanger, eds.), p. 332. Excerpta Medica Foundation, Amsterdam.
Whitehouse, M. W., and Boström, H. (1962). Biochem. Pharmacol. 11, 1175.
Whitson, S. W., Yee, J., Park, H. Z., and Jee, W. S. S. (1971). I.A.D.R. Progr. 106.
Wilhelmi, A. E., and Mills, J. B. (1969). In "La Spécificité Zoologique des Hormones Hypophysaires et de Leurs Activitiés" (M. Fontaine, ed.), p. 165. Centre Nationale de la Recherche Scientifique, Paris.
Wilkins, L. (1955). Ann. N.Y. Acad. Sci. 60, 763.
Wills, M. R., Wortsman, J., Pak, C. Y. C., and Bartter, F. C. (1970). Clin. Sci. 39, 89.
Wool, I. G., and Weinshelbaum, E. I. (1960). Amer. J. Physiol. 198, 360.
Young, R. W. (1964). In "Bone Biodynamics" (H. M. Frost, ed.), p. 117. Little, Brown, Boston, Massachusetts.
Zichner, L. (1971). In "Calcified Tissue Structural, Functional and Metabolic Aspects" (J. Menczel and A. Harell, eds.). Proc. Eur. Symp. Calcified Tissues, 8th pp. 27–34. Academic Press, New York.
Zwilling, E. (1952). Ann. N.Y. Acad. Sci. 55, 196.

15

TRAUMA AND TUMOR GROWTH WITH SPECIAL EMPHASIS ON WOUND STRESS AND "WOUND HORMONES"

Philip Ferris, Norman Molomut, and Joseph LoBue

I. INTRODUCTION

The main theme of this chapter is intended to be the effect of surgical and wound trauma on tumor growth. However, other forms of stress will be examined, since all stress may produce physiological states affecting tumor cell growth.

II. LONG-TERM EFFECTS OF TRAUMA ON TUMOR INDUCTION

Before discussing the immediate effects of surgery or wound on tumor cell proliferation, we shall briefly discuss the evidence for possible long-term effects. Elucidation of causal effects of such trauma is important from a basic scientific, as well as a medicolegal point of view.

There have been many reported cases of patients manifesting cancers at the site of previous surgery or wounds, many months or years after the original trauma (Shapiro, 1965). Arons et al. (1965) cite a study based on 22 cases of scar carcinoma in patients sustaining previous burns. They report many cases in which patients were shown to demonstrate similar latent periods and anatomic sites and all suggest trauma as a causative agent of the subsequent tumors. Stoll and Crissey (1962) report two cases in which epitheliomas appeared as a result of a single trauma and currently accepted criteria of trauma and cancer causation are discussed by these authors. The first case involved a male in whom a basal cell carcinoma appeared at the site of a former blackhead several months after it was struck by a steel clip. The second case was that of a steel worker burned by hot slag. Three months later a basal cell epithelioma developed at the site of injury.

Ghadially et al. (1963) present a review of the traumatic etiology of keratoacanthoma observed in both man and in experimental animals. Based on their work as well as that of others cited in their review, they concluded that injury may act as a cocarcinogenic agent, since it stimulates mitotic activity. This was based on the observation that in carcinogenically treated rabbits, tumor cell growth readily followed injury whereas mechanical trauma alone in normal rabbits was not tumorgenic.

Turner and Laird (1966) in their review, reported studies indicating a high correlation between trauma and development of menengiomas; trauma accounting for two-thirds of the cases reported. Recently, Depaoli and Baldi (1969) cited the case of a 37-year-old man who developed a schlerodermiform undifferentiated basal cell carcinoma 4 months after receiving an injury to the shoulder incurred by a fall. This tumor corresponded precisely to the injured area and was surrounded by normal skin.

In summary, although there seems to be some difficulty in accepting the direct effect of local trauma on subsequent tumor induction, many reports tend to support such a supposition (Pack, 1950; Bowers and Young, 1960; Byrd et al., 1961; Ridgon, 1962; Naji et al., 1969; Shearman, et al., 1970). Experimental studies in laboratory animals showing short-term effects are more convincing and examples of these will now be considered in more detail.

III. IMMEDIATE EFFECTS OF TRAUMA ON TUMOR GROWTH IN EXPERIMENTAL ANIMALS

Noonan et al. (1970) studied the effect of bone fracture on preimplanted lymphosarcoma cells, and the effect of biopsy postimplantation of this tumor in C₃H mice. They found that both fracture following preimplantation, and biopsy post implantation, significantly shortening the survival time in these mice.

The influence of surgical trauma on tumor growth in C_3H mice was demonstrated by Gottfried *et al.* (1960). Fragments of mammary tumor H2712 were implanted 15 mm to the left edge of a wound in the dorsal axillary region either by trocar implant or in the form of gel foam saturated with tumor cells; the latter were used to prolong the latent period of growth. Wounds were maintained by daily opening, using sterile surgical techniques. Tumors in wounded animals appeared earlier and grew larger than in controls. When introduced by trocar, the latent period was from 4 to 5 days, whereas the gel foam implant increased this to 12 to 15 days. Thus, it would appear that wounding shortened the latent period and increased the growth rate of the transplanted tumor.

Additional studies by Gottfried *et al.* (1961a,b) also support the concept that wound trauma directly effects the growth rate of a grafted tumor. Mammary adenocarcinoma H2712 when implanted into mice, adjacent to surgically prepared and maintained wounds, displays a shorter latent period and an accelerated growth rate compared to nonwounded controls bearing the same tumor. Wounding appears not only to affect the rate and the latent period of tumor growth, but seems to be quantitatively proportional to the degree of trauma (Gottfried and Molomut, 1964). Effects were noted even when the trauma was distant from the area of tumor development. Most interesting is the fact that trauma effects were cumulative.

Sizikov (1965) studied the temporal effects of wounds on tumor growth using sarcoma-45. Rats subjected to laparotomies and round skin wounds showed no difference in tumor growth whether rats were wounded 8 days after tumor implantation or if wounding was done 15 to 22 days after; in fact tumors grew faster in control rats than wounded rats. However, when wounding occurred 1 day after implant, the tumors were almost twice as large as controls when measured 25 days after wounding. Early wounding seems to have a temporary effect which lasts for 2 weeks and coincides with the wound granulation phase. Sizikov's results are similar to those of Vaitkevicius *et al.* (1962) who investigated the DC-5 tumor in Ma/my inbred mice. In this study, the tumor was implanted during the period of acute inflammation, following chemical, thermal, or mechanical injury. Results obtained suggested that the effect of injury upon tumor growth parallels the extent of neutrophil exudate at the injured site. It was therefore concluded that tumor stimulation was related to either neutrophil infiltration or to the proliferation of fibroblasts.

Buinauskas *et al.* (1958) studied the effect of surgical trauma on the "takes" of Walker 256 carcinoma cells in rats following celiotomy or visceral trauma. Although the latent period in traumatized rats was longer than in controls, twice as many takes were found in traumatized rats as in controls and such surgically stressed rats died earlier. These investigations suggested that increased tumor growth resulted from lower resistance due to operative stress.

IV. EFFECT OF CELL AND WOUND TISSUE FACTORS

It appears that injury to a variety of tissues releases factors which in some way stimulates the proliferative activity of dormant cells, both normal and neoplastic. Rosen *et al.* (1962) studied nonprotein portions of canine wound fluids and observed that ultrafiltrates of wound fluid markedly increased the growth of *Lactobacillus caesii* and mouse fibroblasts *in vitro*. *In vivo* studies using stainless steel cylinders suggested the existence of small molecular weight compounds which initiated fibroplasia. Wound fluids were obtained at different times after implantation of these cylinders and their components deproteinized by pressure filtration through cellophane at 4°C. Menkin (1960, 1961) also reported a diffusable substance, obtained from inflammatory exudates, which accelerated epithelial hyperplasia in rabbits.

A purified protein derived from mouse submaxillary glands (Turkington, 1969; Turkington *et al.* 1971) seems to induce growth of both normal mammary and mammary tumor cells and has been designated as epithelial growth factor (EGF). EGF has also been found in rats and rabbits and both are biologically active in mouse mammary glands. Turkington in fact (1969) has demonstrated that EGF is a potent stimulus for cell proliferation of C_3H mouse mammary carcinoma. It seems to initiate DNA synthesis but does not alter the rate of DNA synthesis per cell. The effects appear to be similar to the stimulatory effect of insulin, both in magnitude and time course (Turkington and Hilf, 1968).

Regenerating liver as well as certain liver extracts may release or contain substances which enhance tumor growth (Paschkis, 1958; Paschkis *et al.*, 1958). Fisher and Fisher (1963) demonstrated that liver, kidney, brain, and spleen homogenates all accelerated the onset and growth of Walker carcinoma cells growing subcutaneously in Sprague-Dawley rats. Tumor cells suspended in these tissue homogenates also increased markedly the growth of the tumor. Even when homogenates were inoculated at sites distant from the tumor cells, tumors appeared earlier and grew more rapidly than in control rats. These results are similar to those reported by Paschkis (1958) in which he describes similar tumor stimulation by aqueous rat liver extracts and beef liver fractions. Paschkis also describes the effects of tumor homogenates, embryo extracts, and those from other tissues, especially from skin, lung, and thymus as effectively increasing tumor growth. Fisher and Fisher (1963), using parabiotic rats, found that the injection of liver homogenate into one parabiont had a growth promoting effect on the other. These experiments suggest that a specific factor was present which directly stimulated tumor cell proliferation and that this factor is released from damaged liver cells. Such results support the observations of Menkin (1941) who demonstrated that factors liberated by severely damaged cells display growth promoting properties.

Schneyer (1955) found a 6- to 10-fold increase in the growth of spontaneous murine mammary adenocarcinoma when grown in extracts of embryonic tissues. *In vitro* studies by Rubin (1970) have indicated that addition of sonically disrupted embryonic cells resulted in a 6-fold increase in DNA synthesis of chick embryo cells infected with Rous sarcoma virus in cell cultures inhibited by a high population density but had little effect on sparce cell cultures. The rapid increase in growth of group-inhibited cells appears to be due to the release of some material from the sonicated tissue.

Joseph and Dyson (1970) studied the effect of abdominal wounding on the regeneration of ear wound tissue. In both male and female rabbits, tissues from the ear were removed immediately after inflicting a longitudinal incision 3 cm long. In other groups the ear tissue was removed 14 days after wounding. In all groups studied, there was a significant increase in ear tissue regeneration 14 and 21 days after the injury. These observations suggested that the release of a wound hormone stimulated ear tissue regeneration. These investigators caution, however, that abdominal wounding could also have reduced tissue growth inhibitor levels or that the general stress of the wounding may have set into motion complex hormonal changes which enhanced tissue repair processes.

V. WOUND HORMONES

Evidence for the existence of systemic factors released by injured tissues is presented in a review by Cook and Fardon (1942). They describe experiments that demonstrate the release of proliferation-promoting factors into the intercellular fluids by injured cells. Such a substance or substances have been termed "wound hormones."

The existence of wound hormones or any systemic growth promoting factor has subsequently had a controversial history. A number of studies discredit this concept (Savlov and Dunphy, 1954; Calnan et al., 1964; and Schilling, 1968), yet more recent reports suggest once again, the possible existence of such factors (Hell, 1970; Alexander et al., 1971).

Antagonists of the "wound hormone" concept base their arguments on the lack of enhanced wound healing in secondary wounds produced at time intervals after the primary wound. Douglas (1959) demonstrated that the healing of a primary wound occurred more rapidly than a secondary wound in the same host. Calnan et al. (1964) studied the tensile strength of healing wounds in rats under various conditions and, according to them, no evidence of a wound hormone could be found. Bruckner and Longmire (1969) using guinea pigs, inflicted two wounds 7 days apart on the skin of the back. They photographed both sets of wounds during healing, and by use of these photographs, estimated the healing rate. No difference between early and late wounds were found. Based on the concept of the wound hormone, the 7-day wound was expected to heal earlier than the 1-day wound. Schilling (1968), in a recent review of the present knowledge of the events in wound healing, discredits the concept of systemic factors released by wounds enhancing the healing process.

Despite the antiwound hormone school, there is data suggesting the possibility that such an agent may exist. For example, 4 hours after injury to the ear epidermis of guinea pigs, Hell (1970) found increased DNA synthesis, and an apparent diversion of stratum spinosum cells from keratinization and G_0 phase, back into DNA synthesis. These changes were attributed to the release of substances from the damaged epidermis. To test this, fluids were extracted from ear wound sites and individual skin slices incubated in it. Tritiated thymidine was added and both stimulated and nonstimulated slices were autoradiogramed. Slices in crude wound extracts showed significant increases in DNA synthesis which were proportional to the extent of the injury. Such observations suggest that a wound hormone might be involved. It is also possible that wounding may have removed an inhibitor, but the author ruled this out on the basis of the numbers of cells recruited into DNA synthesis.

There appears to be no reports in which the serum from a wounded animal was tested for its systemic effects in tumor-bearing, nonwounded hosts. In a recent series of experiments (Ferris, unpublished), the effect of laparotomy and back skin wounds on the growth of Ehrlich's ascites carcinoma (EAC) was studied. Swiss mice were injected subcutaneously with 5×10^6 EAC cells and divided into the following groups: (1) mice injected with EAC cells alone; (2) mice wounded on the same day as tumor implant by laparotomy; (3) mice given back incisions both prior to, and on the same day of tumor implant; (4) mice injected intraperitoneally (IP) with 0.4 ml of serum removed from donor mice 4 hours after back incisions and injected as above; and (5) mice injected with normal serum from nonwounded donors. Tumor growth was measured volumetrically and these measurements converted into a growth index by dividing the final size of a tumor after 25 days growth by the size of the same tumor after 6 days growth and before the wound serum injections. Mice wounded by laparotomy and skin incisions showed a 5- to 10-fold increase in tumor growth over nonwounded controls confirming the findings of Gottfried *et al.* (1960) using the H2712 mammary carcinoma in C_3H mice. Skin wounds seemed to have a greater effect on EAC growth than mice traumatized by laparotomy. The groups of mice receiving wound serum from laparotomy and skin wound donors, revealed a markedly greater increase in growth than the wounded tumor-bearing mice, with the skin wound serum appearing more potent. Mice receiving normal mouse serum showed growth rates similar to nonwounded controls. At present it is not possible to explain the differences observed between skin wounded and laparotomy stressed mice with respect to the effect on tumor growth. However, these experiments suggest a systemic influence of trauma on the growth of the EAC cells.

VI. THE EFFECT OF STRESS ON METASTASES

The effect of trauma on metastases of tumor cells was studied by Fisher and Fisher (1959). After intraportal injection of 250 Walker carcinoma cells into Sprague-Dawley rats, only 19% were found to have metastasized after 2 weeks. Surgery increased the rate of metastases. Partial hepatectomy caused a 3-fold increase over controls as did hepatic manipulation. From these experiments it was concluded that hepatic trauma stimulated metastases.

Further investigations by these investigators (Fisher and Fisher, 1962) indicated that thrombosis was most important in the development of hepatic metastases. Most important, of 230 rats injected with 250 tumor cells, only 20% showed metastases 2 to 3 weeks later, however, when many of these rats were inspected by laparotomy and resutured, 83% exhibited metastatic lesions. This increase was believed to be due more to surgical manipulation than any other procedures used throughout the experiment and was also found to occur in adrenalectomized animals.

Injection of ^{51}Cr-labeled Walker tumor cells via intraportal injection into normal rats or rats whose hind limbs had been traumatized unilaterally resulted in greater lodgement of cells within stressed limbs. In uninjured rats, on the other hand, the distribution of tumor cells was equal in both limbs (Fisher et al., 1967).

The damaging effects of ischemia and nitrogen mustard injury to rabbit spleens was demonstrated by Alexander and Altemeier (1964). These stresses, augmented by a concommitant increase in endothelial adhesiveness and capillary permeability in the injured tissues, in addition to surgical trauma, were considered important factors in the metastases of VX-2 carcinoma. Metastases were 20 times greater in rabbits 1 week after injury than in normal controls and diminished with time thereafter.

Previous work by Agostino et al. (1961) showed that heparin greatly decreased the growth of intravenously injected Walker 256 carcinoma cells. This seems to have been related to alterations in the coagulation mechanism which prevented tumor cells from lodging themselves. This hypothesis was further tested by Agostino and Cliffton (1965) using fewer Walker carcinoma cells than in ealier investigations. The following groups of rats were studied: (1) controls injected with tumor cells alone; (2) rats injected in the right thigh with 0.5 ml turpentine followed by tumor cells 48 hours later; (3) rats injected with turpentine as above, but given 5000 units of fibrolysin, and 15 minutes later, tumor cells; and (4) rats injected with turpentine as above but given 50 units of heparin 15 minutes later, followed by tumor cells. Sixty percent of the controls developed liver metasteses, 10% unilateral leg tumors, and 28% bilateral leg tumors. In those rats given turpentine in the right hind leg, 51% showed metasteses in the liver, 64% unilateral leg tumors and 23% bilateral leg tumors. The group receiving turpentine and fibrolysin showed 15% liver metastases, 17% unilateral tumors, 6% bilateral tumors and 15% no evidence of metastatic tumors. Rats receiving turpentine and heparin showed 19% liver metastases, 44% unilateral tumors, 4% bilateral tumors, and 16% no tumors. These data suggest that inflammation increased metastases and tumor incidence probably due to an increased influx of blood borne cells into the inflammed site. The significantly lower incidence of unilateral tumors in fibrolysin- and heparin-treated groups suggests that stickiness of the damaged vessels may have been a factor in trapping cells.

On the other hand, recent work by Hagmar and Boeryd (1969) indicates that formation of thromboses inhibit metastases. They suggest that heparin affects formation of lung metastases by mechanisms other than a decrease in blood coagulation.

It has also been reported that the surgical stress of amputation of a primary tumor does not increase metastases (Ketcham *et al.*, 1961), nor does nonsurgical stress, such as electrical shock and confinement, increase tumor growth (Marsh *et al.*, 1959). In fact, these investigators report inhibition of Ehrlich's ascites carcinoma growth by these stresses. Crowding, audiogenic stress and electric.shock were also shown to be ineffective by Kaliss and Fowler (1968).

VII. ENDOCRINE EFFECTS OF TRAUMA

A complete discussion of this subject (reviewed by Cole *et al.*, 1961) is beyond the intended scope of this chapter; however, because of the important role of hormones in stress, a brief discussion of pertinent findings will now be presented.

With regard to surgical stress, Slawikowski (1960) studied the growth of a subcutaneous implant of Walker 256 carcinoma cells in both adrenalectomized and normal rats, and found that growth was retarded in adrenalectomized rats, although its incidence appeared increased. Moolten *et al.* (1970) have reported an increase in DNA synthesis in the regenerating liver of rats subjected to surgery and bovine growth hormone a few hours to 3 days prior to partial hepatectomy, yet cortisone, hydrocortisone, and ACTH were ineffective. Laparotomies, sham adrenalectomies, and adrenalectomies were performed at the same time intervals before hepatectomy and produced similar responses to bovine growth hormone. Surgical procedures alone, however, did not accelerate DNA synthesis in intact livers. Based on these observations, it was hypothesized that both the prior hormone treatment and the surgical stress together, "primed" the liver in such a way as to induce an increased rate of DNA-synthesis once hepatectomy was performed. Comparable findings have been reported by Charters *et al.* (1969), and Soroff *et al.* (1967). The conflicting data regarding the role of the endocrine system and its effect on accelerated cell growth and stress may be closer to resolution as a consequence of the recent work of MacManus and his associates (MacManus and Whitfield, 1969a, b; Rixon *et al.*, 1970; MacManus *et al.*, 1971). Their studies show that exogenous cyclic AMP stimulates rat lymphocytes into DNA synthesis, mitosis, and cell proliferation, and that adrenal, pituitary, and parathyroid hormones may be involved in initiating adenyl cyclase activity. However, Brown *et al.* (1970), studying the cyclic AMP system in slow-growing hepatic tumors, reported that hepatic tumor cells do not respond to epinephrine by increased adenyl cyclase activity

VIII. OTHER EFFECTS OF STRESS ON CELL PROLIFERATION

It may be that many of the systemic effects discussed earlier are not caused by cell proliferating substances. Perhaps surgical and wound stress, in some way, depress the immune response of the host thereby permitting the tumor to grow more rapidly. There is some presumptive evidence for this. Swiss mice, subjected to audiogenic stress (Ferris, unpublished), showed a small but significantly lower immunological response to sheep red blood cells when compared to nonstressed mice. The recent work of Lappe and Steinmuller (1970) indicates that the carcinogen urethan, depresses the immune response to cancer growth in mice as measured by prolongation of allograft survival. Likewise, Park *et al.* (1970) found a depressed immunocompetence in patients who had undergone cholecyctectomy, cardiac valve replacement, and esophagectomy when compared to nonoperated controls. They suggest that acceleration of metastases in these patients may have occurred as a result of the immunosuppression possibly provoked by surgical trauma. However, it is important to keep in mind that cancer patients as a group seem to possess reduced immune competence when compared to noncancerous populations (Southam, 1968, 1969). The important relationships between stress and immunity, especially with regard to neoplastic disease, have of course, far-reaching implications. For further information on this most relevant subject, which is clearly beyond the scope of this chapter, the reader is referred to Bahnson (1969).

IX. SUMMARY

This brief review has attempted to summarize the consequences that many kinds of stress appear to have upon the growth of normal and tumorous cells. A major unsolved problem facing workers studying wound and surgical stress is whether their effect on metastases and healing are exerted locally or systemically via humoral agents. Evidence for both have been suggested, but it seems that the final resolution of this problem remains to be achieved. Attention must be given to the possibility that the end result, be it enhanced repair of injured tissues or increased tumor growth, may be due to numerous subtle, complex combinations of events, so that what appears to be a simple cause and effect relationship on the surface may in fact be a series of interrelated events culminating in increased growth. New experimental approaches must take into account the potential effect of diurnal cycles, the nutritional status of the animals used, the type of tumors, the site of implantation, the effect of anesthesia, the immunocompetence of the host before, during and after treatment, and the effect of hormones on those biochemical events involved in cell proliferation. A better understanding of the events which trigger a cell to move from one phase of its cycle to the next may help to elucidate what roles if any, hormones and/or growth factors play in the proliferation of cells under stress.

REFERENCES

Agostino, D., and Cliffton, E. E. (1965). *Ann. Surg.* **161**, 97.

Agostino, D., Grossi, C. E., and Cliffton, E. E. (1961). *J. Nat. Cancer Inst.* **27**, 17.

Alexander, J. W., and Altemeir, W. A. (1964). *Ann. Surg.* **159**, 933.

Alexander, J. W., Bossert, J. E., McClellen, M., and Altemeir, W. A. (1971). *Arch. Surg.* **103**, 167.

Arons, M. S., Lynch, J. B., Lewis, S. R., and Blocker, T. G. (1965). *Ann. Surg.* **161**, 170.

Bahnson, C. B. (ed.), (1969), *Conf. Psychophysiolog. Aspects Cancer, 2nd, Ann. N.Y. Acad. Sci.* **164**, 307.

Bowers, R. F., and Young, J. M. (1960). *A.M.A. Arch. Surg.* **80**, 564.

Brown, H. D., Chattopadhyah, S. K., Morris, H. P., and Pennington, S. N. (1970). *Cancer Res.* **30**, 123.

Bruckner, W. L., and Longmire, W. P. (1969). *J. Surg. Res.* **9**, 279.

Buinauskas, P., MacDonald, G. O., and Cole, W. H. (1958). *Ann. Surg.* **148**, 642.

Byrd, B. F., Jr., Munoz, A. S., and Ferguson, H. (1961). *South Med. J.* **54**, 1262.

Calnan, J., Fry, H. J. H., and Saad, N. (1964). *Brit. J. Surg.* **51**, 448.

Charters, A. C., Odell, W. D., and Thompson, J. C. (1969). *J. Clin. Endocrinol. Metab.* **29**, 63.

Cole, W. H., MacDonald, G. O., Roberts, S. S., and Wouthwick, H. W. (1961). *In* "Dissemination of Cancer—Prevention and Therapy," Chapter 8. Appleton, New York.

Cook, E. S., and Fardon, J. C. (1942). *Surg. Gynecol. Obstet.* **75**, 220.

DePaoli, M., and Baldi, A. (1969). *Carcinogenesis Abstr.* **7**, 194.

Douglas, D. M. (1959). *Brit. J. Surg.* **46**, 401.

Fisher, B., and Fisher, E. R. (1959). *Ann. Surg.* **150**, 731.

Fisher, B., and Fisher, E. R. (1962). *In* "Henry Ford Hospital Symposium," Chapter 52. Little, Brown, Boston, Massachusetts.

Fisher, B., and Fisher, E. R. (1963). *Cancer Res.* **23**, 1651.

Fisher, B., Fisher, E. R., and Feduska, N. (1967). *Cancer* **20**, 23.

Ghadially, F. N., Barton, B. W., and Kerridge, D. F. (1963). *Cancer* **16**, 603.

Gottfried, B., and Molomut, N. (1964). *Un. Int. Cancer* **20**, 1617.

Gottfried, B., Smith, L. W., and Molomut, N. (1960). *Proc. Amer. Ass. Cancer Res.* **3**, 114.

Gottfried, B., Molomut, N., and Patti, J. (1961a). *Cancer Res.* **21,** 658.
Gottfried, B., Molomut, N., and Skaredoff, L. (1961b). *Ann. Surg.* **153,** 138.
Hagmar, B., and Boeryd, B. (1969). *Pathol. Eur.* **4,** 103.
Hell, E. (1970). *Brit. J. Derm.* **83,** 632.
Kaliss, N., and Fuller, J. L. (1968). *J. Nat. Cancer Inst.* **41,** 967.
Joseph, J., and Dyson, M. (1970). *J. Anat.* **106,** 181.
Ketcham, A. S., Wexler, H., and Mantel, N. (1961). *J. Nat. Cancer Inst.* **27,** 1311.
Lane, M., Liebelt, A., Calvert, J., and Liebelt, R. A. (1970). *Cancer Res.* **30,** 1812.
Lappé, M. A., and Steinmuller, D. S. (1970). *Cancer Res.* **30,** 674.
MacManus, J. P., and Whitfield, J. F. (1969a). *Proc. Soc. Exp. Med. Biol.* **132,** 409.
MacManus, J. P., and Whitfield, J. F. (1969b). *Exp. Cell Res.* **58,** 188.
MacManus, J. P., Whitfield, J. P., and Youdale, T. (1971). *J. Cell Physiol.* **77,** 103.
Marsh, J. T., Miller, B. T., and Lamson, B. G. (1959). *J. Nat. Cancer Inst.* **22,** 961.
Menkin, V. (1941). *Cancer Res.* **1,** 548.
Menkin, V. (1960). *Progr. Exp. Tumor Res.,* **1,** 279–310.
Menkin, V. (1961). *Pathol. Biol.* **9,** 869.
Moolten, F. L., Oakman, N., and Bucher, N. L. R. (1970). *Cancer Res.* **30,** 2353.
Noonan, C. D., Alexander, A. R., and Stroughton, J. A. (1970). *Radiology* **96,** 661.
Naji, A. F., Bruch, W. E., and Brenner, D. (1969). *Ohio State Med. J.* **65,** 597.
Pack, G. T. (1950). *Compens. Med.* **3,** 5.
Park, S. K., Brody, J. I., Wallace, H. A., and Blackmore, W. S. (1970). *Lancet* **1,** 55.
Paschkis, K. E. (1958). *Cancer Res.* **18,** 981.
Paschkis, K. E., Canarow, A., Goddard, S. S., and Zagerman, J. (1958). *Fed. Proc.* **17,** 121.
Ridgon, R. H. (1962). *South Med. J.* **55,** 441.
Rixon, R. H., Whitfield, J. F., and MacManus, J. P. (1970). *Exp. Cell Res.* **63,** 110.
Rosen, H., Levenson, S. M., Sabatine, P. L., Horowitz, R. E., Shurley, H. M., and Schilling, J. A. (1962). *J. Surg. Res.* **2,** 146.
Rubin, H. (1970). *Proc. Nat. Acad. Sci.* **67,** 125.
Savlov, E. D., and Dunphy, J. E. (1954). *New England J. Med.* **250,** 1062.
Schneyer, C. A. (1955). *Cancer Res.* **15,** 268.
Schilling, J. A. (1968). *Physiol. Rev.* **48,** 374.
Shapiro, H. A. (1965). *J. Med.* **12,** 89.
Shearman, D. J. C., Arnott, S. J., Finlayson, N. D. C., and Pearson, J. G. (1970). *Lancet* **1,** 581.
Sizikov, A. I. (1965). *Tr. Kirg. Navch. Issled Inst. Onkol. Radiol.* **2,** 223.
Slawikowski, G. J. M. (1960). *Cancer Res.* **20,** 316.
Soroff, H. S., Rozin, R. R., Mooty, J., Lister, J., and Rabin, M. S. (1967). *Ann. Surg.* **166,** 739.
Southam, C. M. (1968). *Cancer Res.* **28,** 1433.
Southam, C. M. (1969). *Ann. N.Y. Acad. Sci.* **164,** 473.

Stoll, H. L., Jr., and Crissey, J. T. (1962). *N.Y. J. Med.* **62**, 496.

Turkington, R. W. (1969). *Cancer Res.* **29**, 1457.

Turkington, R. W., and Hilf, R. (1968). *Science* **160**, 1457.

Turkington, R. W., Males, J. L., and Cohen S. (1971). *Cancer Res.* **31**, 252.

Turner, O. A., and Laird, A. T. (1966). *J. Neurosurg.* **24**, 96.

Vaitkevicius, V. K., Sugimoto, M., and Brennen, M. J. (1962). *In* "Henry Ford Hospital Symposium," Chapter 53. Little, Brown, Boston, Massachusetts.

Vasiliew, J. M. (1962). *In* "Henry Ford Hospital Symposium," Chapter 24. Little, Brown, Boston, Massachusetts.

Werden, A. A., Harden, C. A., and Garth, R. S. (1959). *Surgery* **45**, 642.

16

GENERAL SUMMARY

Joseph LoBue and Albert S. Gordon

In the lead chapter of this volume, Dr. Bullough elaborates on the physiological significance of a class of hormones, "chalones," which are intimately associated with mitotic homeostasis in a rather unique way—that is—as *inhibitors* of cell division. As such, these agents would seem to make excellent candidates for operators in the negative feedback control of proliferation in many cell renewal systems. Although first discovered in epidermal tissues, chalones have now also been implicated in regulation of blood cell formation and other tissue growth. Of particular significance is the occurrence of chalones in tumors, suggesting that some remnant of feedback control may be operative even under conditions of abnormal growth. Interestingly, Dr. Ferris and his colleagues, later in the monograph, present quite diagrammatically, an example of specific feedback control of leukemic cell growth that supports this aspect of the "chalone concept" most forcefully. They describe experiments in which growth of chloroleukemic cells in diffusion chambers is specifically retarded when these cells are grown in chloroleukemic hosts.

Chemically, epidermal chalone is a protein (or glycoprotein) of 30,000 MW. Lymphocytic chalone, also proteinaceous, has a weight of about 50,000, whereas the erythrocytic and granulocytic chalones seem to be rather small (4000 MW) glycopolypeptides. Chemical features of other chalones have not yet been established.

Hematopoiesis seems to be regulated not only by chalones, but also by unique differentiating principles. The most completely characterized of these is erythropoietin (Ep). Its biogenesis, assay, and physiology have been discussed in great detail in this volume.

Hypoxia, of course, is the fundamental stimulus of erythropoiesis. This operates indirectly via a renal–hepatic humoral axis: reduced oxygen tension somehow stimulates the kidney to increase Ep production through enhanced formation (or activation) of a renal enzyme "erythrogenin," which, in turn, acts upon a serum substrate ("erythropoietinogen") of hepatic origin converting it to the active hormone. Formation of Ep is subject to negative feedback control, and this, in concert with inhibitor substances and possibly erythrocytic chalones, is responsible for the ultimate regulation of circulating red cell mass.

Chemically, Ep (of anemic sheep plasma origin) is a sialic acid-containing glycoprotein of about 45,000 MW composed of about 30% carbohydrate. The sialic acid moiety may protect against *in vivo* degradation of the molecule and may also be important for attachment to carrier molecules within the circulation and at the cell membrane. The mechanism of action of Ep appears to be manifold, stimulating RNA synthesis and ALA-synthetase activity, among other things. A net result of these events, of course, being the conversion of an undifferentiated precursor cell (the Erythropoietin Responsive Cell) into a recognizable erythroid element.

In addition to erythropoiesis, granulocytopoiesis and platelet formation are also under humoral control. Doctor Metcalf has described a "granulopoietin," the colony-stimulating-factor (CSF), which is active both *in vivo* and *in vitro*, functioning to establish granulocyte differentiation. Once again, remarkably, we find this humoral principle also to be a glycoprotein (40,000–60,000 MW) of intriguingly ubiquitous origin—being extractable from sources as diverse as kidney and salivary gland. The occurrence of high circulating levels of CSF in granulocytic leukemias of mice and man is of considerable interest but its etiological significance remains to be established. Once again, inhibitors (this time of granulocytic differentiation) are prominent. Just how these lipid inhibitors, CSF, and any granulocytic chalones are integrated in the control of granulocytopoiesis is presently unknown.

Not only is differentiation under specific humoral regulation but—as Dr. Schultz and his associates indicate—also neutrophil release through a "leukocytosis-inducing-factor" (LIF). Unfortunately, little is known of the biogenesis or mechanism of action of this principle and its place in the overall scheme of leukocyte production is conjectural.

Regarding eosinophil formation, Dr. Cohen and colleagues have described informative experiments that suggest an "eosinopoietin" may also exist. However, the data is far from conclusive on this point.

Turning our attention to thrombocytopoiesis, Dr. Odell, in preceding pages, has evaluated its humorally-based feedback control features in some detail. The similarity to regulation of erythropoiesis by Ep is indeed striking. Thus, "thrombopoietin" levels are elevated in thrombocytopenic states. The hormone appears to be a glycoprotein, and the kidney has been implicated as one potential site of formation. Unhappily, assay techniques are still relatively primitive (although immunologic assay shows promise); platelet counts are difficult and morphologic evaluation of megakaryocytopoiesis requires great courage and skill. These problems have slowed progress in this important area of investigation.

Doctors Fredrickson and Goetinck took on the heroic task of reporting on the humoral changes occurring in diseases of the blood and other pathologic states. It is clear that Ep production is decreased in uremic anemia and chronic inflammation. Uremic anemia may thus be ascribed, in part, to direct effects of renal pathology on Ep production. In addition to this, reduced responsiveness of erythroid marrow to Ep, as well as the occurrence of lipid inhibitors, may also play a role. It is also well established that a fairly large number of neoplastic (e.g., cerebellar hemangioblastoma) and non-neoplastic (e.g., renal cysts) lesions may be associated with Ep overproduction and erythrocytosis. The pathophysiological significance of this is not always readily apparent, particularly when such lesions are not renal or hepatic in origin. Aside from effects of pathologic lesions, there are also inborn metabolic defects in Ep production to contend with. Such states are exemplified by CAF_1 and BALB/c mice which are genetically inefficient Ep producers.

Although investigations into the humoral regulation of erythropoiesis have been conducted primarily in mammalian species, available evidence indicates that red cell formation in lower vertebrates is also under hormonal influence. Moreover, in its broad outlines, control features are remarkably similar to their mammalian counterparts. Thus, as Dr. Zanjani and associates indicate, the primary stimulus for erythropoiesis in lower vertebrates is also hypoxia. Certain fishes, amphibia, reptiles, and birds respond to this kind of stress with enhanced erythropoiesis. Furthermore, hypoxic fish, frog, and bird plasmas have been found to possess erythropoietic activity, albeit generally class specific. Of pronounced theoretical interest is the finding that rat erythrogenin, when incubated with duck sera, generates erythropoietic activity, indicating that avian and mammalian substrate (erythropoietinogen) and the mechanism of Ep formation are similar in these two classes of vertebrates.

Studies of lymphocyte transformation and growth, *in vitro,* have lead to a sound understanding of lymphocyte function and have served as an excellent model for investigations of the immune response. In their chapter, Dr. Havemann and colleagues analyze those humoral factors influencing lymphocyte proliferation in culture. As might be anticipated, these agents fall into two main categories, namely, stimulators and inhibitors (including lymphocytic chalones) of lymphoproliferation. Among the growth stimulators are (1) a "Blastogenic Factor," which is derived from soluble HL-A antigens of mixed leukocyte reactions and which may possess *in vivo* significance and (2) a "Mitogenic Factor," an agent released by lymphocytes in response to specific antigen. An interesting, although nonspecific, growth-promoting substance is "fetuin." This is a component of fetal calf serum consisting of two proteins, both of which possess antiprotease activity. The mechanism of action of these factors is not known but could be associated with deactivation of cytotoxic proteolytic enzymes.

Recent advances in the study of nerve growth factor (NGF) were reviewed by Dr. Angeletti and co-workers. This agent, which is essential, primarily, for growth and development of the sympathetic nervous system, is a protein of about 40,000 MW. It is composed of three subunits (α, β, γ), and only the β subunit appears to possess biological activity. The mouse salivary gland is the best source of NGF, but high activity is also found in snake venom. Although the earliest primary event induced by NGF is unknown, overall metabolic effects include the following: (1) increased glucose utilization (2) stimulation of de novo mRNA synthesis; (3) enhanced polysome formation and protein synthesis; (4) stimulation of de novo production of neurotubular subunit elements; and (5) increased synthesis of adrenergic neurotransmitters.

Two most enigmatic aspects of possible humoral regulation relate to the control of liver regeneration and compensatory renal hypertrophy. Doctors Becker and Malt, respectively, have critically examined these controversial subjects.

Regarding liver regeneration, there is no question that proliferative activity is dramatically increased following partial hepatectomy. This is especially remarkable when one recalls that the unperturbed liver is so extremely quiescent, mitotically. Doctor Becker has summarized the hypotheses developed to explain compensatory hepatic regeneration and these include (1) the concept that "residual" hepatocytes may actively produce a stimulating factor; (2) the idea that a stimulant of hepatocyte mitosis is always present, but the normal complement of cells inactivates this factor to the extent that it is maintained at a level too low to be effective; and (3) the thought that a hepatic chalone produced in effective concentration by nonperturbed livers normally contains the proliferative potential of hepatocytes. Unfortunately, the evidence is so contradictory that it appears impossible to select any one mechanism over the other. Hence, a unifying multiphasic control system is proposed consisting of an initiation phase, a response phase, and a control phase. During initiation, "residual" hepatocytes may be exposed to an absolute or relative alteration in serum factors already present, possibly producing an optimal balance between inhibitors and stimulators. This then initiates proliferative activity. In the response phase, the initiated liver may actively secrete an additional stimulant so that proliferative events may actually be realized. Finally, during control, as liver mass returns to normal, humoral balance would be restored and divisional activity terminated.

Focusing upon the kidney, Dr. Malt interprets the available data as pointing to the existence of a blood-borne regulator of compensatory renal hypertrophy, that is, a "renal growth factor." The most compelling evidence for this comes from heterotopic renal transplantation and vascular parabiosis experiments conducted in anephric animals, both of which result in renal hypertrophy. According to Dr. Malt, invoking "work hypertrophy" (i.e., that the added work load of the remaining kidney following unilateral nephrectomy is the stimulus for increased growth) to explain compensatory renal hypertrophy is inappropriate since hypertrophic changes are known to precede increased metabolic activity.

The question of the existence of "wound hormones" is also controversial. Dr. Bullough in his stimulating discussion of chalones suggests, in passing, that such hormones probably do not exist. Conversely, Dr. Ferris—while indicating the important criticism of the antiwound hormone school, namely, the failure of enhanced healing of secondary wounds—does nonetheless present data to indicate that serum from wounded mice stimulates the growth of intraperitoneal tumors grown in intact hosts. Such a finding would seem to suggest the presence of some growth-promoting principle. Hence, it might be stated (with "tongue-in-cheek," of course) that the "wound hormone" controversy seems to have extended itself into the very pages of this volume.

Doctor Tonna has presented the reader with an extensive review of effects of the more orthodox hormones upon skeletal growth and regeneration. He has especially pointed out the extremely complex nature of the physiological and humoral interrelations that exist to regulate functioning of this organ system. For instance, growth hormone is well established in its role in linear bone growth and the maintenance of the epiphyseal plate through stimulation of cartilage cell proliferation. However, the action of this important humoral agent is dependent upon the age and overall endocrine status of the organism generally. This is true for the other endocrines that exert an effect on skeletal development. Thus, thyrotropin acts synergistically with growth hormone, and, although growth hormone alone can support osteogenesis, thyroid hormone is required to support chondrogenesis. Hence, this latter hormone is necessary for full realization of skeletal development. Androgens also exert an effect on the skeleton, in part via epiphyseal plate chondrocyte proliferation—but also by a dose and age dependent effect on secretion of growth hormone. Estrogens, PTH, thyrocalcitonin, cortical steroids, and possibly even "parotin" (a salivary gland hormone) play a definite part in the development, maintanence, and repair of the skeleton. What, in final analysis, is most astonishing is the multivarious alterations in physicochemical environment, in biosynthetic activity, and in actual functional behavior of the different cells (comprising the living element of skeleton) that must be regulated humorally in order to achieve functional homeostasis in this organ system.

AUTHOR INDEX

Numbers in italics refer to the pages on which the complete references are listed.

SUBJECT INDEX